Advances in Testing and Computation Methods for Disaster Mitigation of Engineering Structures

Guest Editors

Liqiang Jiang
Fengqi Guo
Linli Duan
Yi Hu

MDPI

Basel • Beijing • Wuhan • Barcelona • Belgrade • Novi Sad • Cluj • Manchester

Guest Editors
Liqiang Jiang
School of Civil Engineering
Central South University
Changsha
China

Fengqi Guo
School of Civil Engineering
Central South University
Changsha
China

Linli Duan
School of Civil Engineering
Central South University
Changsha
China

Yi Hu
School of Civil Engineering
Central South University of
Forestry and Technology
Changsha
China

Editorial Office
MDPI AG
Grosspeteranlage 5
4052 Basel, Switzerland

This is a reprint of the Special Issue, published open access by the journal *Buildings* (ISSN 2075-5309), freely accessible at: https://www.mdpi.com/journal/buildings/special_issues/CO9FF76RZI.

For citation purposes, cite each article independently as indicated on the article page online and as indicated below:

Lastname, A.A.; Lastname, B.B. Article Title. *Journal Name* **Year**, *Volume Number*, Page Range.

ISBN 978-3-7258-4995-6 (Hbk)
ISBN 978-3-7258-4996-3 (PDF)
https://doi.org/10.3390/books978-3-7258-4996-3

Advances in Testing and Computation Methods for Disaster Mitigation of Engineering Structures

Contents

buildings

MDPI

Article

Lateral Performance Analysis of Trapezoidal Orthogonal Stiffened Steel Plate Shear Walls

Hong Zheng [1], Lele Cai [1], Jiawei Qin [1], Yuxi Mi [1], Liqiang Jiang [2,*], Xiaoming Ma [1] and Zhiwei Sun [1]

1 School of Civil Engineering, Chang'an University, Xi'an 710061, China; cehzheng@chd.edu.cn (H.Z.); 2021128028@chd.edu.cn (L.C.); 2021228037@chd.edu.cn (J.Q.); 2021228030@chd.edu.cn (Y.M.); 2020228041@chd.edu.cn (X.M.); 2019228021@chd.edu.cn (Z.S.)
2 School of Civil Engineering, Central South University, Changsha 410075, China
* Correspondence: jianglq2019@csu.edu.cn

Abstract: This paper introduces a trapezoidal orthogonal stiffener steel plate shear wall (TSW). The finite element model of the TSW was developed following the validation of low-cycle repeated tests conducted on a single-span double-layer steel plate shear wall. The paper studies the effects of the flat steel plate thickness, stiffener thickness, stiffener height, and stiffener bottom width on the seismic performance of TSW. Building upon these findings, a theoretical formula for the ultimate shear capacity of TSW was developed. The results prove the following: (1) By changing the flat steel plate thickness, the stiffener thickness, and the stiffener height, the seismic behavior of TSW can be enhanced. It is suggested that the flat steel plate thickness is 4~6 mm, the stiffener thickness is 4~6 mm, and the stiffener height is not more than 60 mm, while the effect of the stiffener bottom width on the seismic behavior of TSW can be neglected. (2) The maximum error is 22.16%, compared to the theoretical value of TSW ultimate shear capacity with the finite element simulation value. However, as the finite element results surpass the test results, it indicates that the formula-derived results are unsafe, necessitating a recommendation for correction.

Keywords: orthogonal stiffening; finite element model; seismic performance; ultimate shear capacity

1. Introduction

An edge column, edge beam, fishtail plate, and embedded steel plate compose a steel plate shear wall (SPSW). Compared with other structural systems, it has some merits [1]: (1) compared with the pure bending frame, under the same amount of steel, they have higher lateral stiffness, which can save the amount of steel; (2) compared with the buckling restrained brace, the structure, manufacture, and construction are relatively simple, with higher economy; and (3) compared with reinforced concrete walls, its thickness is very small, making it lightweight and has limited available space, and the production and installation time can be completed in advance in the factory.

SPSW can be divided into three types according to their height–thickness ratio λ: thick plate ($\lambda < 100$), moderate thick plate ($100 \leq \lambda \leq 300$), and thin plate ($\lambda < 300$) [2]. The height–thickness ratio of the shear wall plate has a significant influence on the hysteretic performance of the SPSW. As the height–thickness ratio of the shear wall plate increases, the wall plate easily buckles, the hysteresis loop inclines to pinch, and the energy dissipation capacity decreases [3]. The lateral stiffness of the thick SPSW is large. Although it has a good energy dissipation effect, there is a serious waste of steel during use. Therefore, most of the projects use thin steel plates. Under the serviceability limit state, the thin unstiffened SPSW may experience overall elastic buckling. The plate has a certain out-of-plane deformation and is accompanied by a huge noise of bang, which has a great impact on the normal use of the building [4]. However, setting stiffeners can solve the problem that the hysteretic loop of a thin wall is not fuller than that of a thick plate, enhancing the hysteretic performance of the structure. After setting stiffeners on the steel plate wall, the plate is broken up into multiple panels, the initial lateral stiffness of the structure is improved, the critical elastic buckling load is increased, and the seismic performance is better.

In engineering practice, the height of SPSW is too high. There will be great difficulties in the transportation process, and it is limited in the arrangement. The usual practice is to cut the steel plate wall into two parts and transport it to the construction site. To solve the problem of insufficient strength at the splicing, multichannel transverse slab stiffeners are generally placed at the splicing of two steel plates. In particular, contrasting to SPSW with only vertical stiffeners, the arrangement of transverse stiffeners can also restrict the outside surface deformation of vertical stiffeners.

SPSW has become a hot research topic in recent years due to the attention of researchers at home and abroad. Alinia and Dastfan [5,6] studied the mechanical properties of SPSW with stiffeners and proposed that it can conspicuously enhance energy dissipation capacity and ultimate bearing capacity. Chen et al. [7] and Chen and Guo [8] conducted experimental research on three forms of SPSW and proposed the greater the thickness of the plate, the stronger the constraint effect of the stiffener, which can not only increase the hysteresis loop area but also make the hysteresis curve fuller. Guo et al. [9] studied a semirigid connection frame cross-stiffened SPSW and proposed the structure not only has good plastic deformation ability but also has good ultimate bearing capacity. Haddad et al. [10] analyzed three stiffened and two unstiffened specimens and concluded that stiffened SPSW markedly enhanced shear stiffness and energy dissipation capacity. Mu and Yang [11] analyzed the seismic behavior of two single-span SPSW with obliquely stiffening and proposed that obliquely stiffening improved bearing capacity and stiffness. Alinia and Shirazi [12] analyzed the seismic behavior of a stiffened SPSW by changing the number and arrangement of stiffeners. It is concluded that the greater the number of stiffeners, the stronger the lateral resistance, but it will decrease the ductility of the structure. Compared with a unilateral arrangement, a bilateral arrangement of stiffeners is more effective. Nie [13] proposed that the buckling critical load of asymmetric cross diagonal stiffened SPSW is the largest and the least prone to instability through the characteristic buckling method of ANSYS for six kinds of stiffeners. Alavi [14] analyzed the mechanical properties of SPSW with diagonal stiffeners on both sides through experiments. It is concluded that the hysteretic curve of the SPSW with a diagonal stiffener is fuller, and other seismic performance is also improved. Sigariyazd [15] studied the hysteretic performance of SPSW with diagonal stiffeners and the proposed bearing capacity, energy dissipation, and ductility of the structure were significantly improved. Combined with the outcomes of the test and finite element analysis, the calculation formula of steel plate strength was proposed. Guo [16] studied the critical elastic buckling load and buckling mode of fully stiffened SPSW with slits on both sides and proposed its elastic buckling design formula. Jahanpour [17] introduces the semi-supported steel shear wall (SSSW) system and proposes new means to estimate the ultimate shear capacity of a given SSSW. Nie and Zhu [18] analyzed the seismic performance of two stiffened SPSW with openings and one stiffened SPSW without openings and proposed a simple formula for the shear strength reduction factor and a simplified devise means for computing the lateral bearing capacity of stiffened SPSW. Wang [19] analyzed the mechanic performances of vertical stiffened SPSW, studied the influence of stiffeners on the lateral stiffness of SPSW, and developed a computing formula for the lateral bearing capacity of SPSW. Refs. [20–22] proposed a lightweight design and structural optimization method for node dampers and sliders and an innovative MSB fully prefabricated lifted connection (FPLC), which can significantly reduce energy consumption and is of great significance for achieving green production. The interaction for rib-to-deck welded joints of orthotropic steel deck (OSD) is analyzed using the finite element method in Ref. [23]. The results show that the numerical results indicate that fatigue crack growth in welds is unstable due to inclusions. In Ref. [24], partially encased composite columns (PECCs) with different positions of shear studs are studied and analyzed. Test results illustrate that the position of studs is a considerable parameter for the shear performance. In Ref. [25], the flexural performance of composite beams with different materials was studied, and the SFRC layer thickness and steel rebar diameter had a great influence on the bearing capacity of the beam. The seismic performance of concrete-filled steel tubular (CFST) composite columns was studied and analyzed in Ref. [26]. The results show that the composite columns have good seismic performance.

Based on the above SPSW development background, this paper proposes a new type of SPSW—trapezoidal orthogonal stiffened steel plate shear wall (TSW). The structural form is revealed in Figure 1. The arrangement of TSW is different. One side is vertically arranged and the other side is horizontally arranged. The advantage of this construction method is that there is no cross connection when the stiffeners are arranged, and the structural measures at the joints are simplified, which is convenient for actual construction. The transverse stiffener will constrain the out-of-plane deformation of the vertical stiffener. The vertical stiffener near the column edge can limit the out-of-plane deformation of the steel plate edge under horizontal –, avoid premature buckling, and improve the ductility and energy dissipation capacity of the steel plate wall. The steel plate is divided into several high-thickness comparison grids under the combined action of stiffeners on both sides. The elastic initial stiffness of the steel plate is improved and the seismic performance is better. Variable parameters analysis of the TSW using ABAQUS 2016 and a formula for the shear capacity of TSW is developed. The formula error is within the acceptable range and the accuracy is good.

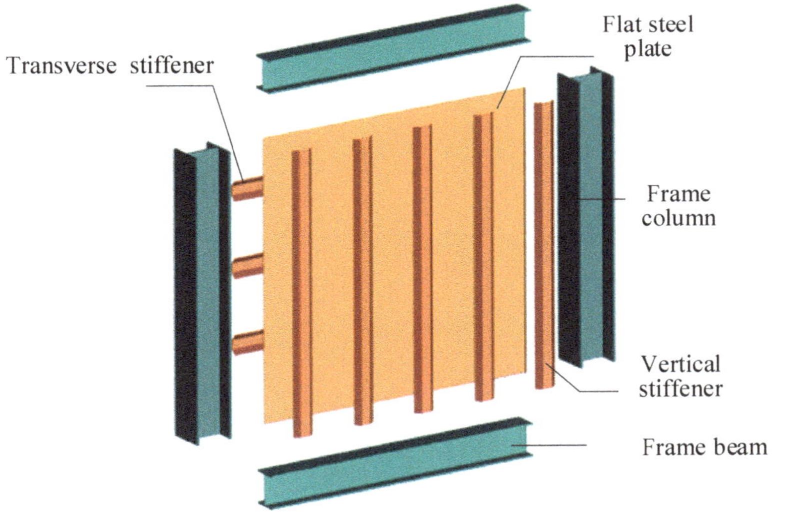

Figure 1. Composition of TSW.

2. Numerical Modelling

2.1. Specimen Design

The TSW is composed of steel frame beams, flat steel plates, and vertical and transverse trapezoidal stiffeners. According to the American codes AISC360-16 [27], the section of the frame column is H350 × 350 × 12 × 19, the section of the upper and lower top beams is H350 × 250 × 10 × 16, and the mid-beam section is H300 × 200 × 9 × 14. The dimensions and materials of H-shaped steel are shown in Table 1, and the size of the trapezoidal stiffener is seen in Table 2. According to the Steel Plate Shear Wall Technical Specification5.4.1 provisions of the grid width–height ratio and the ratio of width to thickness, the design of the TSW division of the steel plate grid width–height ratio is close to 1, and the ratio of width to thickness is in line with regulatory requirements; the specific sizes of the TSW are seen in Figure 2 [28].

Table 1. H-beam section size and material.

	Items	*H* (mm)	*B* (mm)	t_w (mm)	t_f (mm)	Steel Type
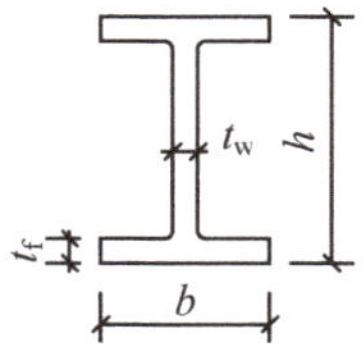	Frame column	350	350	12	19	
	Top and bottom beam	350	250	10	16	Q355B
Section of H shape steel	Middle beam	300	200	9	14	

Table 2. Section size and material of trapezoidal stiffener.

	Items	*H* (mm)	*B* (mm)	*A* (mm)	*T* (mm)	Steel Type
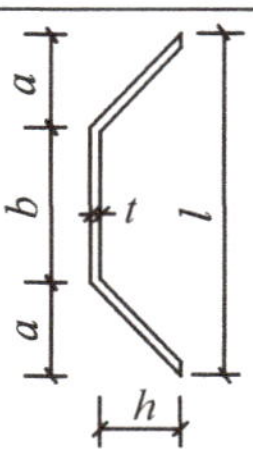	Trapezoidal stiffener	50	100	60	6	Q235B
Trapezoidal stiffener section						

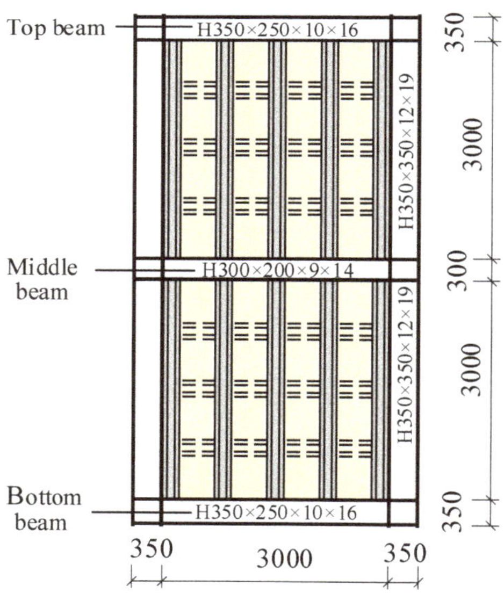

Figure 2. TSW structure and size.

2.2. Finite Element Model (FEM)

This paper selects ABAQUS 2016 to model and analyze TSW. The material constitutive relationship adopts the three broken-line model. The frame beam and column adopt Q355 steel, the flat steel plate adopts Q235 steel, and the trapezoidal stiffener adopts Q235 steel. The mechanical properties of Q235 and Q355 steel are based on the standard design values of Q235 and Q355 steel in GB50017-2003 [29]. The specific indicators of the material mechanical properties are shown in Table 3.

Table 3. Mechanical performances of materials.

Steel Type	Modulus of Elasticity E (MPa)	Yield Strength f_y (MPa)	Poisson Ratio υ	ε_y	ε_{st}	ε_u
Q235	2.06×10^5	235	0.3	0.0011	0.025	0.20
Q355	2.06×10^5	355	0.3	0.0017	0.017	0.16

The finite element model adopts the S4R shell element. The framework beams and columns are connected via bolted–welded joints using the Merge command in ABAQUS 2016. The embedded steel plates and stiffeners are welded together using the Tie command for connection, as seen in Figure 3. The iterative calculation step is 1×10^{-6}. At the same time, all degrees of freedom at the bottom part of the column are limited to simulate the high-intensity bolt friction links at the bottom of the test, coupling the upper part of frame columns on both sides to the center of the top beam, as seen in Figure 4. In order to prevent local stress concentration at the loading position, the top of the frame columns on both sides is coupled to a positive reference point in the z-axis of the top beam center, and two analysis steps are set during the loading process: (1) apply vertical load to the model; (2) apply horizontal load to the model using displacement loading, because the embedded wallboard will produce obvious out-of-plane deformation during the loading process, and in order to ensure the convergence of the model calculation process, both analysis steps open the geometric nonlinearity. After adjusting the grid size according to the calculation results, the grid division of each component is finally determined as follows: the grid size of the embedded steel plate is 40 mm, and the grid size of the outer frame is 60 mm.

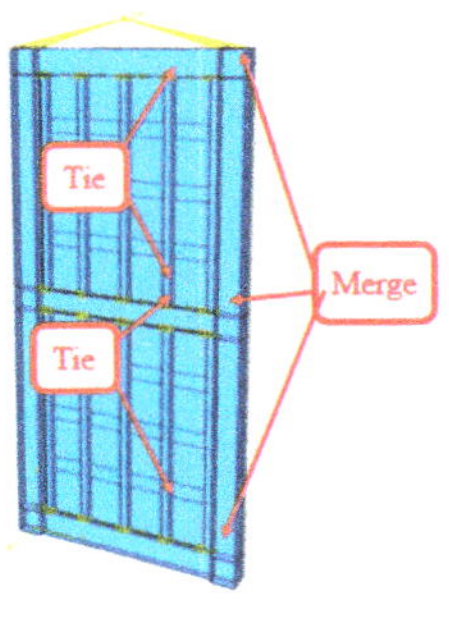

Figure 3. Contact relationship.

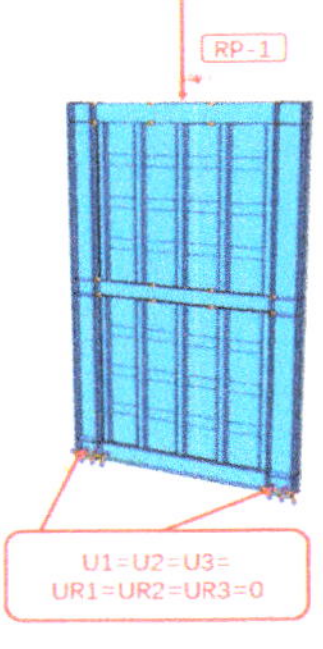

Figure 4. Boundary conditions and loads.

2.3. Validation of the FEM

In this paper, the steel frame-ribbed stiffened low yield point SPSW test in reference is selected for finite element verification [30]. The GLRBS specimen is selected. The specimen size and material attributes are seen in Table 4. The experiment structure is seen in Figure 5, and the loading system used in the experiment is seen in Figure 6.

Table 4. Specimen dimensions and material properties.

Item	Sectional Size (mm)	Length (mm)	Steel Type
Frame column	□ 180 × 180 × 10 × 10	3600	Q235
Upper and lower top beam	HN300 × 150 × 6.5 × 9	1200	Q235
Mid-beam	HN200 × 100 × 5.5 × 8	1200	Q235
Flat steel plate	1200 × 3	1200	BLY160
Ribbed slab	60 × 6	1000	Q235

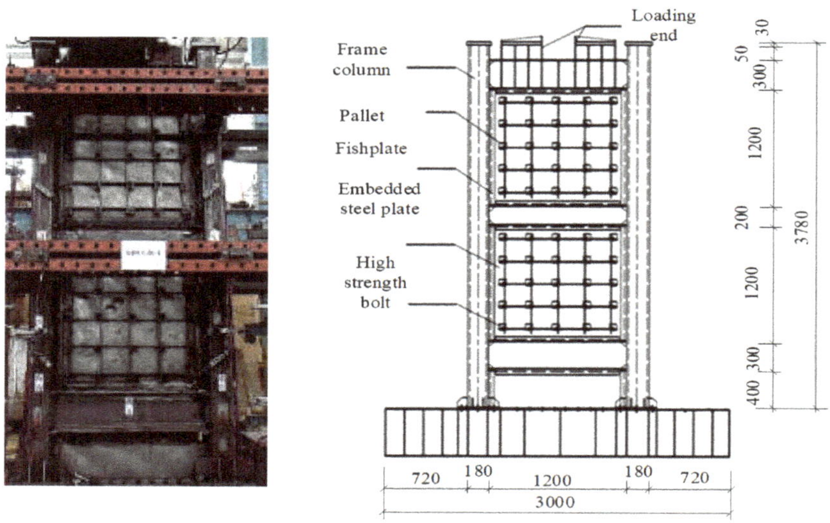

(**a**) Experiment specimen (**b**) Experiment structure and geometric size (mm)

Figure 5. Experiment specimen.

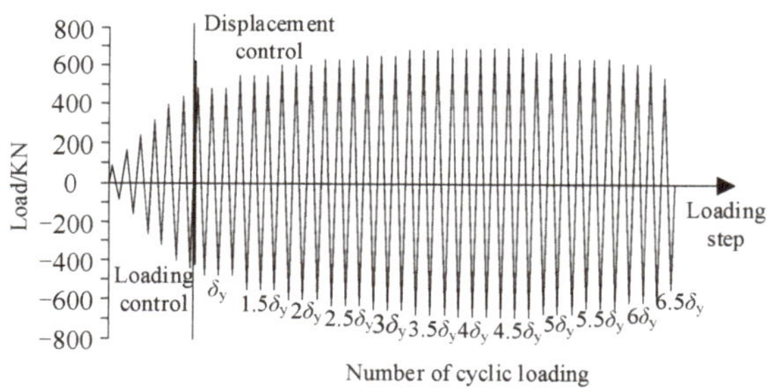

Figure 6. Experiment loading system.

This test was carried out according to the relevant provisions of the JGJ101-96 [31]. The loading process includes two stages: pre-loading and formal loading. In the pre-loading stage, a vertical load of 310 kN was applied synchronously on the top of the columns on both sides, and a horizontal load of 50 kN (about 10% of the yield load) was applied after the axial compression of the side frame column was ensured. The push–pull cycle ensures that the two actuators work synchronously. The specimen was ensured to be in good contact with the measuring device. After the test device worked normally, it was completely unloaded and entered the formal loading stage. The specimen GLRBS was verified with ABAQUS 2016. The shell element components were established according to the size of each part in Table 4. The material attributes of each part of the component were measured by the standard specimen using the standard method of the "metal material room temperature tensile test method" and averaged [32]. The steel three broken-line model was used as the constitutive relationship of the model material, and the FEM established by the above modeling method was used.

The load–displacement curves of the experiment and FEM are contrasted as seen in Figure 7. The hysteresis loops of the two have a similar change trend. The hysteresis curve of FEM is fuller than the pinch effect of the test, which indicates that FEM has a higher energy dissipation capacity. For the skeleton curve, the slope of FEE is larger than that of the test at the initial stage of loading, which indicates that the initial stiffness of FEM is larger than that of the test. Continue to load, the curve of the test specimen appears inflection point, and the specimen enters the buckling stage earlier; however, in the later stage of loading, the downward trend of the two is relatively close, and both show good ductility. The specific results are shown in Table 5. The maximum error is 17.90%. This is mainly because the finite element is too idealistic, resulting in errors in FEM and test.

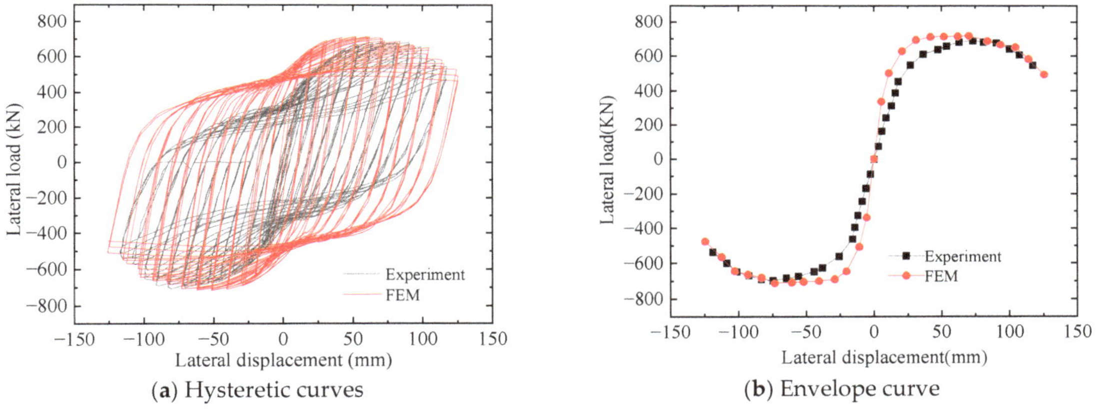

Figure 7. Comparison of load–displacement curves.

Table 5. Comparison of key indexes between test and finite element.

	Yield Load (kN)	Peak Load (kN)	Yield Displacement (mm)	Peak Displacement (mm)	Initial Stiffness (kN/mm)
Experimental	502.00	687.83	20.21	81.47	42.63
Finite element	567.32	715.49	19.08	72.16	50.27
Error	13.02%	4.02%	4.47%	11.43%	17.90%

The comparison between the specimen and the final failure mode of the FEM is seen in Figure 8. The comparison shows that the outer wall of the column bottom is seriously buckled, the weld at the beam–column connection joint is damaged and the plasticity hinge is shaped at the column bottom. The experiment and the FEM obtained failure patterns that are in good agreement. Therefore, the modeling method is acceptable and can be used for subsequent parameter analysis.

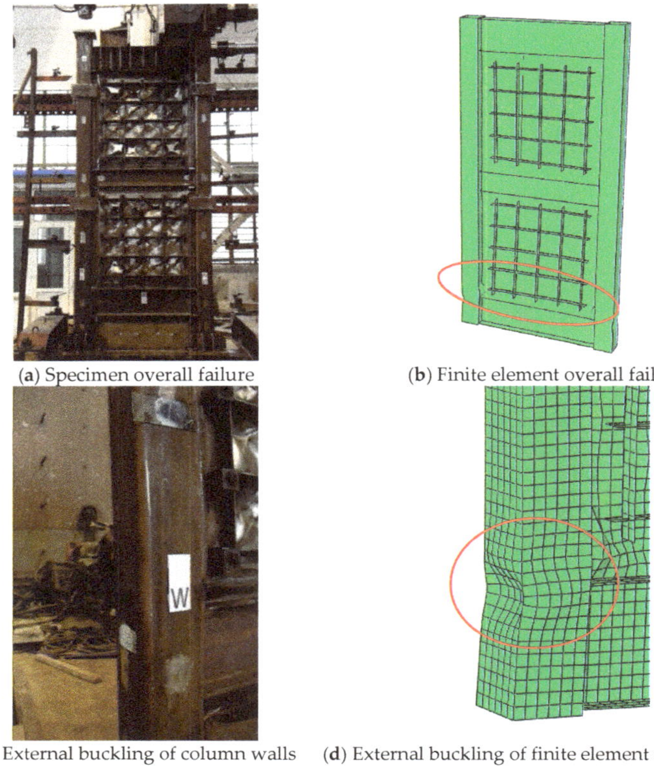

(a) Specimen overall failure (b) Finite element overall failure

(c) External buckling of column walls (d) External buckling of finite element column walls

Figure 8. Comparison of experiment and the FEM failure modes.

3. Parametric Analyses of TSW

To study the effects of the flat steel plate thickness t_b, stiffener thickness t_t, stiffener height h, and stiffener bottom width l on the hysteretic performance of TSW, this paper uses ABAQUS 2016 to establish four groups of finite element models for the TSW. The number of specimens in each group is four. The geometric parameters are shown in Figure 9. The parameter design of the specimen is shown in Table 6. The bottom width l of the trapezoidal stiffener is changed by changing the size of b. Then, by comparing the hysteresis curve, skeleton curve, and strength stiffness degradation curve, the reasonable value range of each design parameter is given.

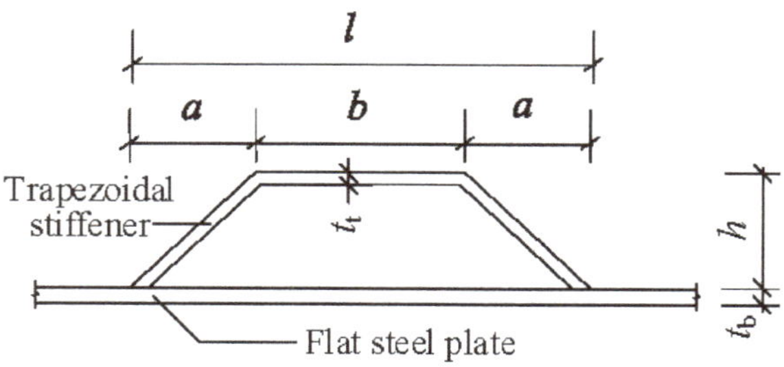

Figure 9. Geometric parameters diagram.

Table 6. Parameter design of four groups of specimens.

Serial Number	t_b (mm)	t_t (mm)	H (mm)	L (mm)
TSW-BT	t_b	6	50	220
TSW-TT	6	t_t	50	220
TSW-TH	6	6	h	220
TSW-L	6	6	50	l

The most important parameter to measure the influence of stiffeners on the out-of-plane restraint of wall panels is the rib stiffness ratio, that is, the ratio of the out-of-plane bending stiffness of the stiffener to the out-of-plane flexural stiffness of the steel plate. The rib stiffness ratio is expressed as η, and the calculation formula is shown in Equation (1) [28].

$$\begin{cases} \eta = \dfrac{E_s I_s}{Dc} \\ D = \dfrac{Et^3}{12(1-\upsilon^2)} \\ I_s = \dfrac{2}{3} t_t h^3 + \dfrac{1}{12} b t_t{}^3 + b t_t h^2 \\ c = (c_x + c_y)/2 \end{cases} \tag{1}$$

In the formula, E_s—the elastic modulus of the stiffener is 2.06×10^5 N/mm^2; I_s—the inertial moment of the stiffener section parallel to the center of gravity axis of the stiffened plate section; D—cylindrical rigidity of stiffened plate; c—average spacing of stiffeners on both sides; E—the elastic modulus of flat steel plate s 2.06×10^5 N/mm^2; υ—Poisson's ratio of flat steel plate is 0.3.

3.1. Effects of Plate Thickness t_b

In this section, four models are established. The FEM is named the TSW-BT series, and the specific size is shown in Table 7.

Table 7. Parameters of TSW-BT specimens.

Serial Number	t_b (mm)	Rib Stiffness Ratio
TSW-BT1	2	5155.58
TSW-BT2	4	644.45
TSW-BT3	6	190.95
TSW-BT4	8	80.56

The load–displacement curves are given in Figure 10. The comparison shows that from TSW-BT2, the hysteresis curves of the subsequent specimens are by degrees transformed into an anti- "S" shape. The hysteresis loop area of TSW-BT3 and subsequent specimens gradually increases, and the shape is fuller. The hysteresis loop area gradually enlarges following the thickness of the flat steel plate. Comparing the skeleton curves, it can be seen that when entering the plasticity, the thicker the flat steel plate, the more the inflection point of the envelope curve occurs first, and the specimens enter the buckling stage earlier; at the later stage of loading, the envelope curves decreased gently, showing good ductility performance. The specific values are shown in Table 8. Compared with TSW-BT1, the yield load of TSW-BT2~TSW-BT4 increased by 24.57%, 21.91%, and 6.44%, respectively. The growth rates of the peak load are 18.29%, 16.87%, and 6.02%; the lateral stiffness increases by 34.11%, 17.97%, and 9.32%. The peak displacement decreased by 1.35%, 1.05%, and 31.7%. The greater the flat steel plate thickness is, the stronger its bearing capability and lateral stiffness.

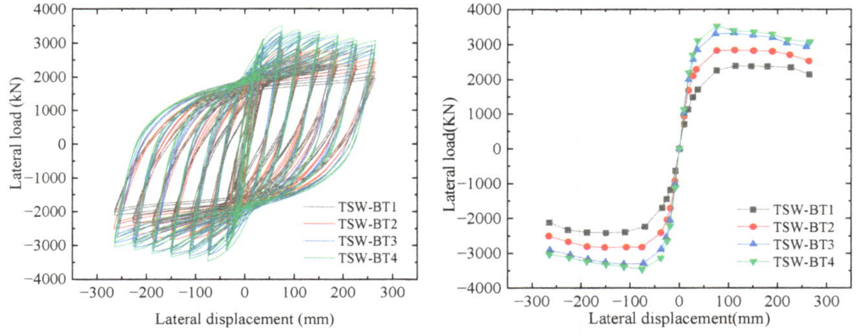

(**a**) Hysteretic curves (**b**) Envelope curve

Figure 10. Comparison of the load–displacement curves of various specimens.

Table 8. Comparison of key indicators of TSW-BT specimens.

Serial Number	Yield Load (kN)	Peak Load (kN)	Yield Displacement (mm)	Peak Displacement (mm)	Lateral Stiffness (kN/mm)
TSW-BT1	1872.84	2395.96	45.47	113.63	74.52
TSW-BT2	2332.98	2834.13	38.04	112.10	99.94
TSW-BT3	2844.06	3312.24	36.23	110.92	117.90
TSW-BT4	3027.31	3511.79	35.56	75.76	128.89

The relative energy dissipation index is the equivalent viscous damping coefficient h_e, which reflects the fullness of the hysteresis loop (see Figure 11) and is calculated by Equation (2).

$$h_{\mathrm{e}} = \frac{S_{(ABC+CDA)}}{2\pi \times S_{(OBE+ODF)}} \tag{2}$$

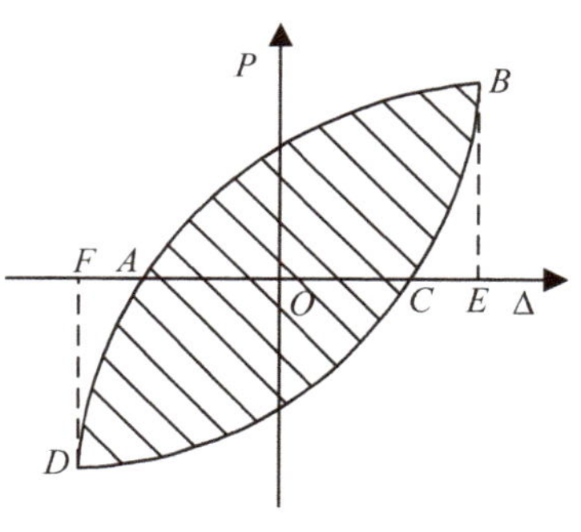

Figure 11. Hysteresis loop area.

In the formula, $S_{(ABC+CDA)}$—the area of hysteresis loop ABCD; $S_{(OBE+ODF)}$—the sum of the area of triangles obe and odf.

The energy dissipation capacity is seen in Figure 12. With the increase in thickness, the absolute energy dissipation value is larger. The energy dissipation values of specimens TSW-BT1, TSW-BT2, and TSW-BT3 at each loading stage are greatly different, while the energy dissipation values of TSW-BT3 and TSW-BT4 at each loading stage are almost close. The equivalent viscous damping coefficient enlargements as increasing thickness of the steel plate before 5.0δy, and the equivalent viscous damping coefficient of the thinner steel plate is higher than rather thick after 5.0δy. A comprehensive comparison of various energy consumption performance indicators shows that the absolute index and relative index of TSW-BT3 (flat steel plate thickness of 6 mm) are better.

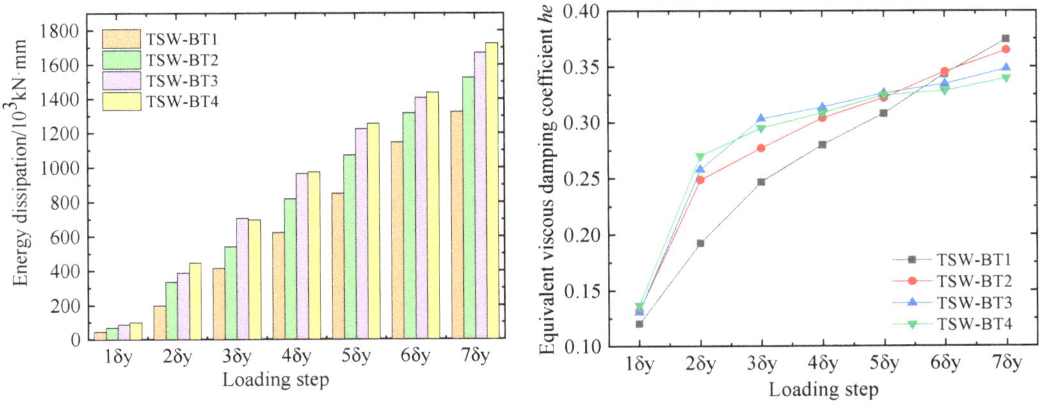

(**a**) Energy dissipation (**b**) Equivalent viscous damping coefficients

Figure 12. Comparison of energy dissipation capacity.

According to the given indicators, the thickness of TSW has a greater impact on its hysteretic capacity. This can effectively enhance the full degree of hysteretic curves and improve its energy dissipation ability. Additionally, the initial stiffness, yield load, and peak load of the specimen are improved to a certain extent. The steel plate thickness for the TSW is suggested to be in the range of 4 to 6 mm.

3.2. Effects of the Stiffener Thickness t_t

In this section, four models are established. The FEM is named the TSW-TT series, and the specific variable parameter size is shown in Table 9.

Table 9. Parameters of TSW-TT specimens.

Serial Number	t_t (mm)	Rib Stiffness Ratio
TSW-TT1	2	63.60
TSW-TT2	4	127.23
TSW-TT3	6	190.95
TSW-TT4	8	254.77

The load–displacement curves are given in Figure 13. The comparison shows that the hysteresis curves of specimens TSW-TT1 and TSW-TT2 have an obvious "pinching" phenomenon. Starting from TSW-TT3, the hysteresis curves gradually become an anti-"S" shape, and the shape is fuller. The comparison of the envelope curves shows that the initial lateral stiffness and peak load increase gradually. In the middle of loading, the greater the thickness of the stiffener is, the greater the peak load. At the end of loading, the skeleton curves decreased gently and showed good ductility. The specific values are shown in Table 10. Compared with TSW-TT1, the yield load of TSW-TT2~TSW-TT4 increased by 13.13%, 15.71%, and 0.33%, respectively. The increase in peak load is 9.93%, 14.72%, and 0.98%. The increments of lateral stiffness are 6.01%, 9.18%, and 0.55%. The peak displacement decreased by 5.89%, 36.97%, and 2.66%. The larger the stiffener thickness is, the stronger its bearing capability and lateral stiffness.

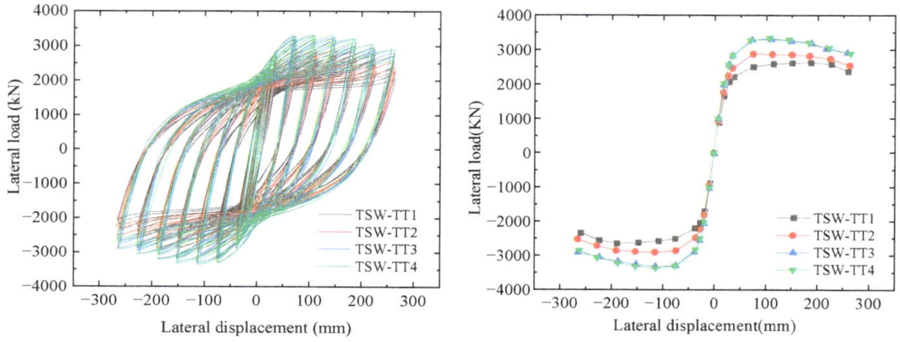

(**a**) Hysteretic curves (**b**) Skeleton curves

Figure 13. Comparison of the load–displacement curves of various specimens.

Table 10. Comparison of key indicators of TSW-TT specimens.

Serial Number	Yield Load (kN)	Peak Load (kN)	Yield Displacement (mm)	Peak Displacement (mm)	Lateral Stiffness (kN/mm)
TSW-BT1	1872.84	2395.96	45.47	113.63	74.52
TSW-BT2	2332.98	2834.13	38.04	112.10	99.94
TSW-BT3	2844.06	3312.24	36.23	110.92	117.90
TSW-BT4	3027.31	3511.79	35.56	75.76	128.89

Figure 14 shows the energy dissipation capacity. As the thickness of the trapezoidal stiffener increases, the absolute energy dissipation value is larger. The energy dissipation values TSW-TT1 and TSW-TT2 at each loading stage are larger, while the energy dissipation values TSW-TT3 and TSW-TT4 at each loading stage are smaller. The equivalent viscous damping coefficient between the 2.0δy and 5.0δy specimens is obviously different, and the numerical difference between the loading stages of TSW-TT3 and TSW-TT4 is not significant. At 6.0δy~7.0δy, the equivalent viscous damping coefficients are similar to each other, and the changing trend is basically the same. Increasing the thickness of the stiffener can significantly improve the energy dissipation performance of the specimens, and the effect is most obvious in the middle of loading, but the thickness of the stiffener is too large, and it will also increase the amount of steel used.

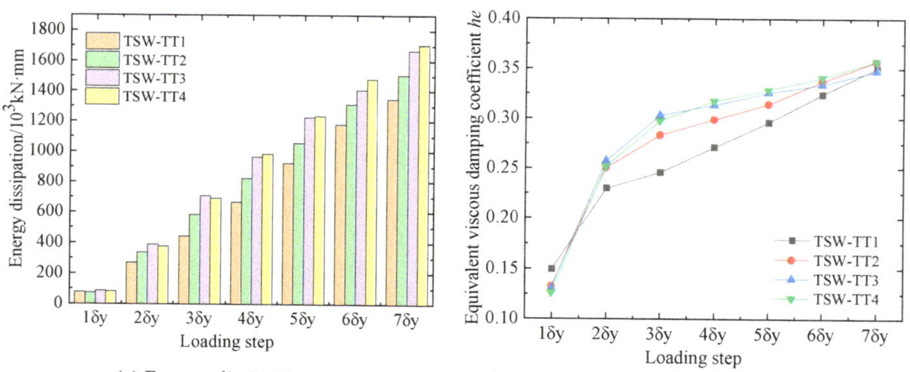

(**a**) Energy dissipation (**b**) Equivalent viscous damping coefficients

Figure 14. Comparison of energy dissipation capacity.

Based on the above indexes, trapezoidal stiffener thickness has an obvious effect on the hysteretic capacity of TSW. The increase in the thickness of the stiffener solves "pinching" phenomenon of the hysteresis curves and improves the energy dissipation ability. The initial stiffness, yield load, and peak load of the specimens are improved to a certain extent, and the stiffness and strength degeneration rate are slowed down. Therefore, this paper suggests a TSW trapezoidal stiffener thickness of 4~6 mm.

3.3. *Effects of the Stiffener Height h*

In this section, the variable parameter analysis is carried out for the height of the trapezoidal stiffener. This series also includes four specimens. The finite element model is named the TSW-TH series, and the specific variable parameter size is shown in Table 11.

Table 11. Parameters of the TSW-TH specimens.

Serial Number	h (mm)	Rib Stiffness Ratio
TSW-TH1	30	61.98
TSW-TH2	40	116.16
TSW-TH3	50	190.95
TSW-TH4	60	288.62

The load–displacement comparison is given in Figure 15. Compared with the hysteresis curves, it shows that after entering the elastic–plastic stage, the hysteresis curves of TSW-TH1 and TSW-TH2 have an obvious "pinch" phenomenon. With the height of the trapezoidal stiffener increase, the shape of the hysteresis curve of the subsequent specimens gradually changes to the anti-"S" shape, and the shape is fuller. Comparing the skeleton curves, it can be seen that with the increases of the trapezoidal stiffener height when comparing the specimens TSW-TH1~TSW-TH4, the initial lateral stiffness gradually decreases, and the maximum stiffness of the wallboard also decreases. The concrete numerical values are shown in Table 12, and the yield load growth rates of TSW-TH1~TSW-TH4 are 6.99%, 5.00%, and 3.53%, respectively. The growth rates of the peak load are 7.19%, 4.96%, and 4.18%, and the peak displacement decreased by 7.61%, 17.86%, and 0.59%, respectively. The lateral stiffness changes little. Therefore, it is recommended that the height of the trapezoidal stiffener is 50 mm.

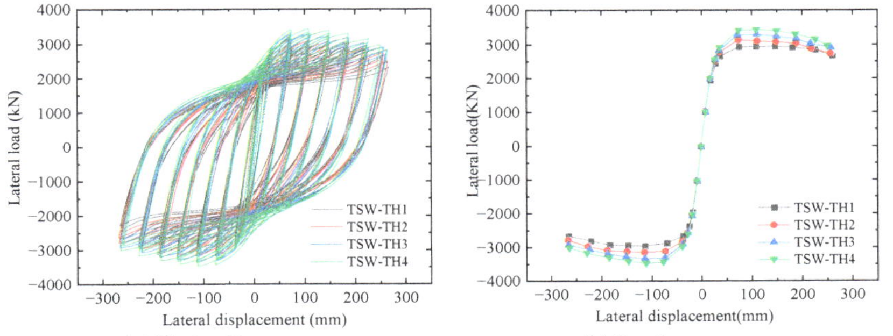

(a) Hysteretic curves (b) Envelope curves

Figure 15. Comparison of the load–displacement curves of various specimens.

The energy dissipation capacity is seen in Figure 16. When the wallboard enters the yield stage, the opening degree of hysteretic curves is low and the energy dissipation is limited. Continuing to loading, as the height of the trapezoidal stiffener increases, the trend of the energy dissipation value of the specimen increases first and then decreases. The equivalent viscous damping coefficient between the 2.0δy and 5.0δy specimens is obviously different. At 6.0δy~7.0δy, the equivalent viscous damping coefficients are close to each other, and the variation trend is basically the same.

Table 12. Comparison of key indicators of TSW-TH specimens.

Serial Number	Yield Load (kN)	Peak Load (kN)	Yield Displacement (mm)	Peak Displacement (mm)	Lateral Stiffness (kN/mm)
TSW-TH1	2531.58	2944.15	32.38	146.17	118.37
TSW-TH2	2708.54	3155.73	34.46	135.04	118.13
TSW-TH3	2844.06	3312.24	36.23	110.92	117.90
TSW-TH4	2944.36	3450.65	38.38	110.27	117.65

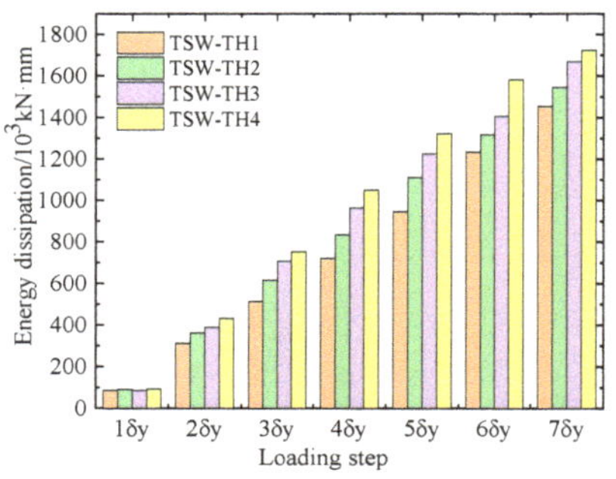

(**a**) Energy dissipation

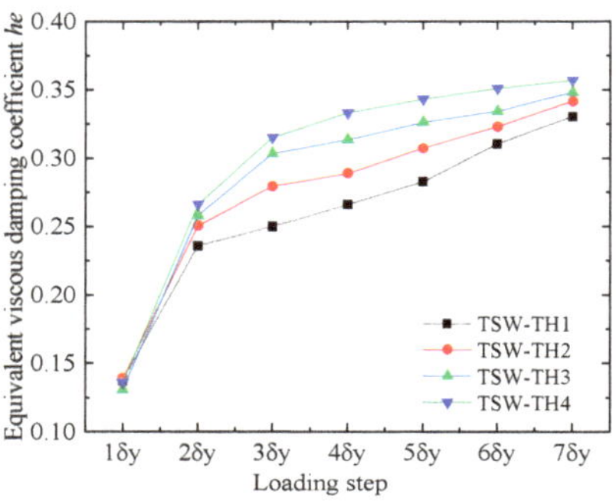

(**b**) Equivalent viscous damping coefficients

Figure 16. Comparison of energy dissipation capacity.

In summary, trapezoidal stiffener height has an obvious influence on the hysteretic capacity of TSW. Increasing the height of the stiffener can obviously solve the "pinching" phenomenon of the hysteretic curve. The trapezoidal stiffener plays an active role in enhancing the hysteretic properties of the specimen in the proper height range, but after exceeding a certain size, the height of the stiffener continues to increase, which will have a reverse effect on the hysteretic performance of the specimen. Therefore, it is suggested that the height of the stiffener should not exceed 60 mm.

3.4. *Effects of the Stiffener Bottom Width l*

In this section, four specimens are designed. The FEM is named the TSW-L series. The specific variable parameter size is shown in Table 13.

Table 13. Parameters of the TSW-L specimens.

Serial Number	*L* (mm)	Rib Stiffness Ratio
TSW-L1	180	177.55
TSW-L2	200	184.00
TSW-L3	220	190.95
TSW-L4	240	198.44

The load–displacement comparison is given in Figure 17. Compared with the hysteresis curves, it shows that the hysteresis curves of the series are close, and the hysteresis loop area is not much different. The influence of changing the bottom width of the trapezoidal stiffener on the hysteresis curves is not obvious. Compared with the previous three parameters, the bottom width of the stiffener has little impact on the hysteretic performance. The comparison of the skeleton curves shows that the envelope curves of this series are very similar during the whole loading process. In the middle stage of loading, the curves of all specimens were basically in a state of coincidence. The descending section of the skeleton curve is different at the end of loading. The larger the trapezoidal stiffener bottom width is, the gentler the downward trend is. This is because the wider the width of the bottom edge of the stiffener is, the better the effect of restraining the inelastic buckling of the flat steel plate is. Comparing the data in Table 14, it shows that with trapezoidal stiffener bottom width increases, the increase and decrease in each key indicator are not obvious, indicating that the bottom width of the trapezoidal stiffener does not affect the bearing capability and initial stiffness.

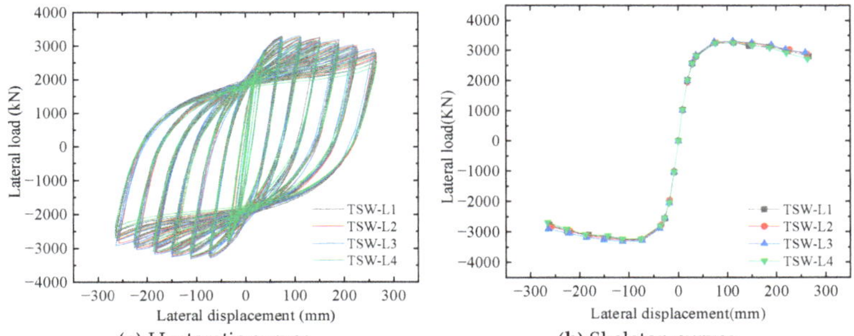

(**a**) Hysteretic curves (**b**) Skeleton curves

Figure 17. Comparison of the load–displacement curves of various specimens.

Table 14. Comparison of key indicators of TSW-L specimens.

Serial Number	Yield Load (kN)	Peak Load (kN)	Yield Displacement (mm)	Peak Displacement (mm)	Lateral Stiffness (kN/mm)
TSW-L1	2801.79	3261.76	35.52	109.17	118.96
TSW-L2	2781.37	3263.19	36.14	110.83	116.96
TSW-L3	2844.06	3312.24	36.23	110.92	117.90
TSW-L4	2763.53	3245.19	35.63	112.13	118.11

The energy dissipation capacity is given in Figure 18. The displacement loading is less than 3.0δy, and the energy dissipation values of this series are scarcely equal. After 3.0δy, the difference gradually appears. As the bottom width of the trapezoidal stiffener increases, the energy dissipation value of the specimen increases first and then decreases. The equivalent viscous damping coefficients of this series are almost the same before 3.0δy, and the difference appears after 3.0δy.

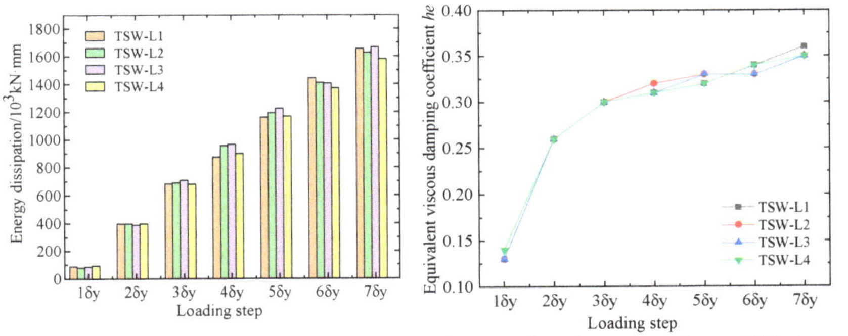

(**a**) Energy dissipation (**b**) Equivalent viscous damping coefficients

Figure 18. Comparison of energy dissipation capacity.

In summary, the TSW-L series specimens have preferable bearing capability stability, but the influence of the bottom width of the trapezoidal stiffener on the initial lateral stiffness, skeleton curve, and energy dissipation capacity is not obvious. Therefore, the influence of the trapezoidal stiffener bottom width on the key indicators of the specimens can be ignored.

4. Theoretical Calculation of the Ultimate Shear Capacity

The outer frame and the flat steel plate form the TSW. They, together, resist the action of lateral forces. When the shear yield member reaches the maximum horizontal bearing capacity, the upper and lower of the outer frame column and the beam end basically form plastic hinges, and the flat steel plate yields, but no out-of-plane buckling occurs, forming a plastic hinge mechanism [33–35].

According to the energy method in Ref. [33], TSW is composed of an embedded steel plate and outer frame, which resist the lateral force together. When the bearing capacity reaches the maximum, the plastic hinge is basically formed at the top and bottom of the outer frame column and the beam end of the frame beam. The embedded steel plate is uniformly yielded but no out-of-plane buckling occurs, forming a plastic hinge mechanism, as shown in Figure 19. The shear-bearing capacity of each layer of the TSW under the plastic hinge mechanism can be deduced. The calculation formula is shown in Equation (2).

$$\sum_{i=1}^{n} F_i H_i = M_P^f + \sum_{i=1}^{n} (V_w^i - V_w^{i+1}) H_i \tag{3}$$

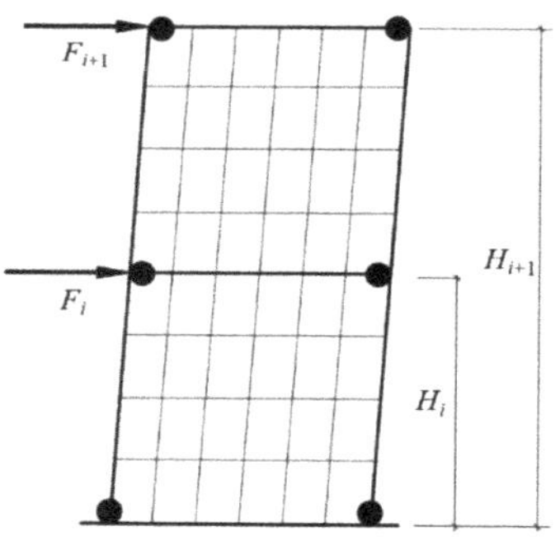

Figure 19. Plastic hinge mechanism of TSW.

In the formula, F_i is the shear bearing capacity of the ith-layer structure, H_i is the distance from the ith-layer structure to the ground, V_w^i is the shear bearing capacity of TSW in the *i*-layer structure, and M_P^f refers to the flexural bearing capacity of the outer frame. It requires a certain reduction. The reduction factor is 0.8.

$$M_P^f = 0.8 \times 4W_P f_y = 3.2W_P f_y \tag{4}$$

In the formula, W_P is the resistance moment of the column plastic section and f_y is the yield strength. In Equation (4), there is only one unknown quantity V_w^i, and the embedded steel plate shear capacity is solved below.

When TSW yields, the steel plate and the stiffeners offer shear-bearing capacity. The calculation formula is shown in Equation (5):

$$V_w = V_C + V_S \tag{5}$$

In the formula, V_C refers to the shear capacity provided passes through steel plates, and V_S refers to the shear capacity provided through stiffeners.

When calculating the maximum shear capacity of SPSW, according to the tension field model proposed, further analysis was carried out on the basis of Refs. [36–38]. Compared with the rectangular section beam, the SPSW on two-side connections has a wider shear section. When the wallboard is sheared and yielded, the shear stress distribution of the section can be boiled down to trapezoid, and the shear stress on both sides is 0, which is evenly distributed after reaching it. The stress distribution diagram is proposed, as given in Figure 20. The shear stress at different positions can be calculated according to Equation (6):

$$\tau(x) = \begin{cases} \frac{x}{B_1}\tau_y & 0 \le x \le B_1 \\ \tau_y & B_1 \le x \le B - B_1 \\ \frac{B-x}{B_1}\tau_y & B - B_1 \le x \le B \end{cases} \quad (6)$$

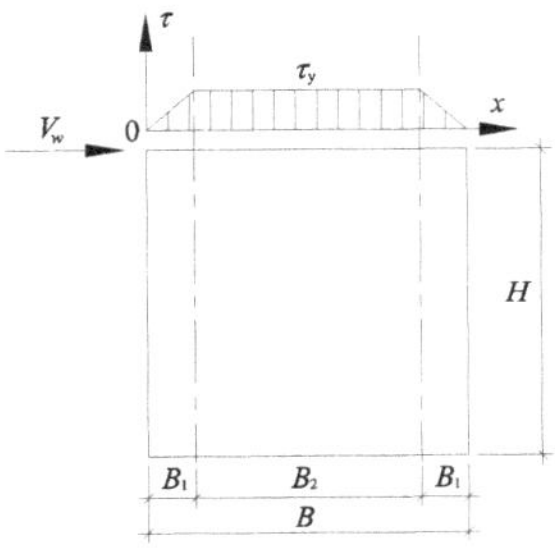

Figure 20. Distribution of shear stress of steel plate with two-side connection.

τ_y refers to shear yield stress, B_1 is the width of shear stress not reaching τ_y, and steel plate width is B.

The integral is the shear capacity, as shown in Equation (7):

$$V_C = t\int_0^B \tau(x)dx = t(B - B_1)\tau_y \quad (7)$$

Steel plate thickness is t. Except for B_1, the rest are known parameters. The following uses the method in reference to calculate B_1 [39].

The distribution of stress is seen in Figure 21:

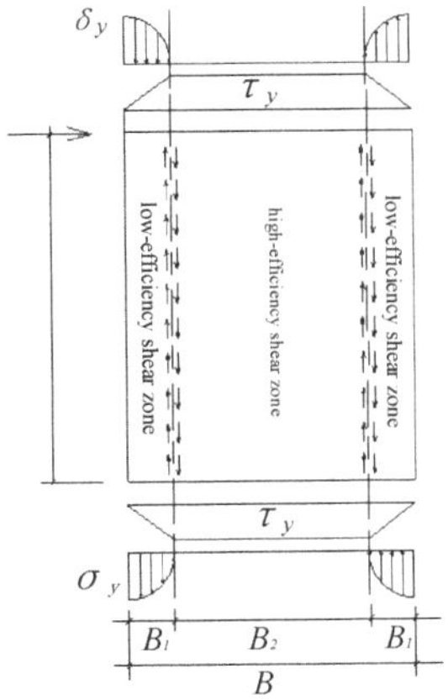

Figure 21. Distribution of stress of steel plate with two-side connection.

According to the von Mises yield equation, the relationship between the normal stress on the $\sigma_y(x)$ steel plate element and the shear stress can be obtained. Equation (8) is shown. In the main shear zone $\tau(x) = \tau_y$, $\sigma_y(x) = 0$ can be obtained by substituting Equation (8), so positive stress is primarily distributed within the secondary shear zone.

$$\sigma_y(x) = \sqrt{f_y^2 - 3\tau^2(x)} \tag{8}$$

The secondary shear zone is isolated, and the equilibrium equation is established, as shown in Equation (9).

$$\tau_y t H = 2\int_0^{B_1} \sigma_y(x) t_e dx \tag{9}$$

In the formula, the plate net height is H.

When the section area of the plate edge restraint member satisfies Equation (10), the "low-efficiency shear zone" will be completely transformed into the "high-efficiency shear zone", that is, $B_1 = 0$.

$$A_s \geq \frac{\tau_y t H}{5 f_{y,s}} = A_{s,\min} \tag{10}$$

A_s denotes the stiffener section area at the plate edge; $A_{s,\min}$ represents the minimum cross-sectional area threshold that the edge restraint member can make the whole wallboard fully exert the shear efficiency; $f_{y,s}$ denotes the material yield stress of the plate-edge restraint member.

When $0 < A_s < A_{s,\min}$, only part of the "low-efficiency shear zone" is transformed into "high-efficiency shear zone". The ratio of the section area of the edge restraint member to the minimum section area threshold of the edge constraint component is defined as ξ.

$$\zeta = \frac{A_s}{A_{s,\min}} \tag{11}$$

Then, Equation (9) is changed to Equation (12):

$$(1-\zeta)\tau_y t H = 2\int_0^{B_1} \delta = \sigma_y(x) t_e dx \tag{12}$$

By substituting Equations (6), (8), (11) and $f_y = \sqrt{3}\tau_y$ into Equation (12), the B_1 expression can be obtained, as shown in Equation (13):

$$B_1 = \frac{2\sqrt{3}H(1-\zeta)}{3\pi} \tag{13}$$

Substituting Equation (13) into Equation (6), embedded steel plate shear-carrying capacity is attained, as shown in Equation (14).

$$V_C = \begin{cases} t(B - \dfrac{2\sqrt{3}H(1-\zeta)}{3\pi})\tau_y & 0 < \zeta < 1 \\ tB\tau_y & \zeta \geq 1 \end{cases} \tag{14}$$

Under the action of lateral load, the stiffeners on both sides of the plate are subjected to tension and compression, respectively. Because the compression stiffeners experience easily inflectional instability, the influence of compression stiffeners is neglected when calculating V_S. Since the tensile stiffener does not reach full yield when the component reaches the peak load, tensile strength is boiled down to $0.8f_y$, and V_S is calculated according to Equation (15):

$$V_S = 5A_s \times 0.8 f_y \cos\theta_s = 4A_s f_y \cos\theta_s \tag{15}$$

In the formula, the angle among the horizontal orientation and stiffeners is θ_s.

In summary, substituting Equations (14) and (15) into Equation (5) can attain the shear bearing capacity of the TSW at shear yield. The expression is shown in Equation (16):

$$V_w = \begin{cases} t(B - \dfrac{2\sqrt{3}H(1-\zeta)}{3\pi})\tau_y + 4A_s f_y \cos\theta_s & 0 < \zeta < 1 \\ tB\tau_y + 4A_s f_y \cos\theta_s & \zeta \geq 1 \end{cases} \quad (16)$$

Due to the FEM proposed in this paper being a double-layer structure, the parameters in Equation (3) are $n = 2$, $F_1 = 0$. Therefore, the ultimate shear capacity of TSW is computed by Equation (17):

$$V = F_2 = \begin{cases} 3.2W_p \dfrac{f_y}{H_2} + t(B - \dfrac{2\sqrt{3}H(1-\zeta)}{3\pi})\tau_y + 4A_s f_y \cos\theta_s & 0 < \zeta < 1 \\ 3.2W_p \dfrac{f_y}{H_2} + tB\tau_y + 4A_s f_y \cos\theta_s & \zeta \geq 1 \end{cases} \quad (17)$$

To verify the accuracy of the above formula, ABAQUS 2016 was used to calculate the finite element of TSW with different sizes under cyclic loading, and the bearing capacity of each component was extracted. The theoretical and FEM bearing capacity comparison is shown in Table 15.

Table 15. Comparison of theoretical bearing capacity and FEM bearing capacity.

Parameters	t_b (mm)	t_t (mm)	h (mm)	L (mm)	Theoretical Results/kN	FEM Results/kN	Error
Thickness of the flat steel plate	2	6	50	220	2710	2396	13.10%
	4				3315	2834	16.96%
	6				3895	3312	17.59%
	8				4154	3512	18.28%
Thickness of the stiffener	6	2	50	220	2489	2626	−5.23%
		4			3191	2887	10.53%
		6			3895	3312	17.60%
		8			3981	3345	19.03%
Height of the stiffener	6	6	30	220	3597	2944	22.16%
			40		3796	3156	20.29%
			50		3895	3312	17.60%
			60		4006	3451	16.10%
Bottom width of the stiffener	6	6	50	180	3664	3262	12.32%
				200	3773	3263	15.64%
				220	3895	3312	17.60%
				240	3871	3245	19.30%

It can be seen from Table 15 that under different size parameters, the error between the theoretical bearing capacity calculated by Equation (17) and the experimental value is between 5.23% and 22.16%; both of them increase with increasing flat steel plate thickness, stiffener thickness, and stiffener height, and increase first and then decrease with increasing stiffener bottom width. This is because with the increase of the thickness of the flat steel plate and the stiffener, the utilization rate of the steel in the FEM gradually decreases and does not reach the maximum, resulting in the error of the ultimate bearing capacity of the theory and FEM gradually increases. With the increase of the stiffener height, the steel utilization of the FEM gradually increases, so that the error of the ultimate bearing capacity of the theory and FEM gradually decreases. However, because the finite element results are greater than the test results, this shows that the results obtained by the formula are not safe, and it is recommended to be appropriately modified for future applications.

5. Conclusions

In this paper, a trapezoidal orthogonal steel plate shear wall (TSW) is put forward. ABAQUS 2016 is applied to establish TSW and analyze its hysteretic performance. The conclusions are as follows:

(1) A single-span double-layer TSW FEM was built, and the FEM was contrasted with the test results of a steel frame-stiffened low yield point SPSW. The results show that the maximum error of peak load and initial stiffness among FEM and experiment result is 17.9%, which is within the acceptable range.

(2) The influence of flat steel plate thickness, stiffener thickness, stiffener height, and stiffener bottom on the hysteretic capability is analyzed. The results show that the flat steel plate thickness, stiffener thickness, and stiffener height can improve the seismic performance of the TSW. It is suggested that flat steel plate thickness should be 4~6 mm, stiffener thickness should be 4~6 mm and stiffener height should not exceed 60 mm. The influence of stiffener bottom width on the key indicators of TSW can be ignored.

(3) The ultimate shear capacity of the TSW is also analytically formulated and then compared with the FEM results. The results indicate maximum error of the ultimate shear capacity of TSW among theoretical calculation value and FEM calculation value is 22.16%, and most of the errors are within 20%. However, because the finite element results are greater than the test results, this shows that the results obtained by the formula are not safe, and it is recommended to correct.

Author Contributions: Methodology, Y.M.; Validation, J.Q.; Formal analysis, X.M.; Data curation, Z.S.; Writing—original draft, L.C.; Writing—review & editing, H.Z. and L.J. All authors have read and agreed to the published version of the manuscript.

Funding: This research was funded by the National Natural Science Foundation of China (52208224, 52008398, 51278061).

Data Availability Statement: Data is contained within the article.

Conflicts of Interest: This article no conflict of interest.

References

1. Guo, Y.L.; Zhou, M. Categorization and Performance of Steel Plate Shear Wall. *J. Archit. Civ. Eng.* **2009**, *26*, 1–13. [CrossRef]
2. Kharrazi, M.H.; Prion, H.G.; Ventura, C.E. Implementation of M-PFI method in design of steel plate walls. *J. Constr. Steel Res.* **2008**, *64*, 465–479. [CrossRef]
3. Sun, G.H.; Gu, Q.; He, R.Q.; Fang, Y.Z. Energy dissipation capacity of steel plate shear walls. *Chin. J. Comput. Mech.* **2013**, *30*, 422–428.
4. Li, N. *Experimental Research on Seismic Behavior of Corrugated Steel Plate Shear Wall System*; Tianjin University: Tianjin, China, 2016.
5. Alinia, M.; Dastfan, M. Cyclic behaviour deformability and rigidity of stiffened steel shear panels. *J. Constr. Steel Res.* **2007**, *63*, 554–563. [CrossRef]
6. Alinia, M.; Dastfan, M. Behaviour of thin steel plate shear walls regarding frame members. *J. Constr. Steel Res.* **2006**, *62*, 730–738. [CrossRef]
7. Chen, G.D.; Guo, Y.L.; Fan, Z.; Han, Y. Cyclic test of steel plate shear walls. *J. Build. Struct.* **2004**, *25*, 19–26. [CrossRef]
8. Chen, G.D.; Guo, Y.L. Ultimate shear-carrying capacity of steel plate shear wall with cross stiffeners. *J. Build. Struct.* **2004**, *25*, 71–78. [CrossRef]
9. Guo, H.C.; Li, Y.L.; Liang, G.; Liu, Y.H. Experimental study of cross stiffened steel plate shear wall with semi-rigid connected frame. *J. Constr. Steel Res.* **2017**, *135*, 69–82. [CrossRef]
10. Haddad, O.; Ramli Sulong, N.H.; Ibrahim, Z. Cyclic performance of stiffened steel plate shear walls with various configurations of stiffeners. *J. Vibroeng.* **2018**, *20*, 459–476. [CrossRef]
11. Mu, Z.; Yang, Y. Experimental and numerical study on seismic behavior of obliquely stiffened steel plate shear walls with openings. *Thin-Walled Struct.* **2020**, *146*, 106457. [CrossRef]
12. Alinia, M.; Shirazi, R.S. On the design of stiffeners in steel plate shear walls. *J. Constr. Steel Res.* **2009**, *65*, 2069–2077. [CrossRef]
13. Nie, J.G.; Li, Y.X. Finite element research about the elastic buckling of six types of stiffening steel plate shear walls. *Steel Constr.* **2013**, *28*, 1–3. [CrossRef]
14. Alavi, E.; Nateghi, F. Experimental study on diagonally stiffened steel plate shear walls with central perforation. *J. Constr. Steel Res.* **2013**, *89*, 9–20. [CrossRef]

15. Sigariyazd, M.A.; Joghataie, A.; Attari, N.K.A. Analysis and design recommendations for diagonally stiffened steel plate shear walls. *Thin-Walled Struct.* **2016**, *103*, 72–80. [CrossRef]
16. Guo, Y.L.; Miao, Y.W.; Dong, Q.L. Elastic buckling behavior of stiffened steel plate shear walls slotted at two edges. *Prog. Steel Build. Struct.* **2007**, *16*, 58–62. [CrossRef]
17. Jahanpour, A.; Moharrami, H.; Aghakoochak, A. Evaluation of ultimate capacity of semi-supported steel shear walls. *J. Constr. Steel Res.* **2011**, *67*, 1022–1030. [CrossRef]
18. Nie, J.G.; Zhu, L.; Fan, J.S.; Mo, Y.L. Lateral resistance capacity of stiffened steel plate shear walls. *Thin-Walled Struct.* **2013**, *67*, 155–167. [CrossRef]
19. Wang, Y.S.; Jia, G.Q.; Wang, X.T.; Xie, C.D.; Liang, X.D. Investigation of lateral-resistant behaviour of thin steel plate shear walls with vertical stiffeners. *J. Xian Univ. Arch. Tech.* **2020**, *52*, 495–504. [CrossRef]
20. Yang, L.; Ye, M.; Huang, Y.; Dong, J. Study on mechanical properties of displacement-amplified mild steel bar joint damper. *Iran. J. Sci. Technol. Trans. Civ. Eng.* **2023**, 1–14. [CrossRef]
21. Liang, F.; Wang, R.; Pang, Q.; Hu, Z. Design and optimization of press slider with steel-aluminum composite bionic sandwich structure for energy saving. *J. Clean. Prod.* **2023**, *428*, 139341. [CrossRef]
22. Deng, E.F.; Wang, Y.H.; Zong, L.; Zhang, Z.; Zhang, J.F. Seismic behavior of a novel liftable connection for modular steel buildings: Experimental and numerical studies. *Thin-Walled Struct.* **2024**, *197*, 111563. [CrossRef]
23. Luo, Y.; Liu, X.; Chen, F.; Zhang, H.; Xiao, X. Numerical simulation on crack–inclusion interaction for rib-to-deck welded joints in orthotropic steel deck. *Metals* **2023**, *13*, 1402. [CrossRef]
24. Huang, H.; Yao, Y.; Zhang, W. A push-out test on partially encased composite column with different positions of shear studs. *Eng. Struct.* **2023**, *289*, 116343. [CrossRef]
25. Zhang, W.; Zhang, S.; Wei, J.; Huang, Y. Flexural behavior of SFRC-NC composite beams: An experimental and numerical analytical study. In *Structures*; Elsevier: Amsterdam, The Netherlands, 2024; Volume 60, p. 105823. [CrossRef]
26. Wei, J.; Ying, H.; Yang, Y.; Zhang, W.; Yuan, H.; Zhou, J. Seismic performance of concrete-filled steel tubular composite columns with ultra high performance concrete plates. *Eng. Struct.* **2023**, *278*, 115500. [CrossRef]
27. Ansi, B. *AISC 360-16*; Specification for Structural Steel Buildings; Chicago AISC: Chicago, IL, USA, 2005.
28. *JGJ/T 380-2015*; Technical Specification for Steel Plate Shear Wall. China Construction Industry Press: Beijing, China, 2015.
29. *GB50017-2003*; Code for Design of Steel Structures. China Planning Press: Beijing, China, 2003.
30. Guo, Z. Experimental Study and Numerical Analysis of Semi-Rigid Frame with Multi-ribbed Stiffening Low Yield Point Steel Plate Shear Walls. *Xian Univ. Arch. Tech.* **2015**. [CrossRef]
31. *JGJ/101-96*; Code for Seismic Test Methods of Buildings. China Planning Press: Beijing, China, 1996.
32. *GB/T 228-2010*; Metallic Materials Tensile Testing at Ambient Temperature. China Standard Press: Beijing, China, 2010.
33. Jiang, L.Q.; Ye, J.H.; Zheng, H. Macroscopic modelling of steel boundary frames equipped with bolt-connected reinforced concrete panel wall. *Eng. Struct.* **2020**, *213*, 110549. [CrossRef]
34. Shi, G.; Gao, Y.; Wang, X.; Zhang, Y. Low Cycle Fatigue Properties of Low Yield Point Steels. *China Civ. Eng. J.* **2019**, *52*, 20–26+52. [CrossRef]
35. Berman, J.; Bruneau, M. Plastic analysis and design of steel plate shear walls. *J. Struct. Eng.* **2003**, *129*, 1448–1456. [CrossRef]
36. Hu, Y.; Zhao, J.H.; Zhang, D.F.; Zhang, H.Q. Cyclic tests of fully prefabricated concrete-filled double-skin steel tube/moment-resisting boundary frames with beam-only connected steel plate shear walls. *Thin-Walled Struct.* **2020**, *146*, 106272. [CrossRef]
37. Guo, L.H.; Rong, Q.; Qu, B.; Liu, J.P. Testing of steel plate shear walls with composite columns and infill plates connected to beams only. *Eng. Struct.* **2017**, *136*, 165–179. [CrossRef]
38. Wang, W. *Research on Lateral Performance and Seismic Behavior of Asymmetric Cross-Stiffened Corrugated Steel Shear Wall*; Chang'an University: Xi'an, China, 2021. [CrossRef]
39. He, B.C. *Cyclic Test and Mechanism of Vertically Corrugated Steel Plate Shear Walls with Various Openings*; Tianjin University: Tianjin, China, 2021. [CrossRef]

buildings

MDPI

Article

Theoretical Analysis of the Plastic Property for Equal Angle Sections Subjected to Axial Force and Biaxial Bending

Yun Sun [1], Da Song [1], Qi Cai [2,*], Yangbing Liu [1], Shuxuan Sun [1] and Yuting Yang [3]

1 School of Civil Engineering, Nanyang Institute of Technology, Nanyang 473004, China; sunyun@nyist.edu.cn (Y.S.); songda@nyist.edu.cn (D.S.); ybliu1123@sohu.com (Y.L.); shuxuansun2024@163.com (S.S.)
2 College of Mechanics and Engineering Science, Hohai University, Nanjing 211189, China
3 College of Civil Engineering, Shenyang Urban Construction University, Shenyang 110170, China; dq_yyt@syucu.edu.cn
* Correspondence: 20230919@hhu.edu.cn

Abstract: To fully harness the design development potential of plastic angle sections, this study employs a theoretical analysis approach to examine the plastic behavior of equal angle sections subjected to axial force and biaxial bending. Based on the simplified angle section results, the full plasticity correlation equations were derived. Subsequently, section shape coefficients were computed. Finally, a methodology for calculating the plastic development coefficients of angle sections was explored. The findings indicate that the full plasticity correlation equations lack the necessary safety margins in designs. Notably, the angle sections possess a greater plastic development capacity along the weak axis compared with the strong axis. It is advisable, for both regular-size and large-size angle sections, to consistently adopt the plastic development coefficients in designs as follows: $\gamma_u = 1.05$ for the strong axis and $\gamma_v = 1.15$ for the weak axis, thereby addressing the shortcomings of the specification in design.

Keywords: angle section; plastic behavior; theoretical analysis; correlation equation; plastic development coefficient

1. Introduction

Plastic section development can improve economic efficiency when designing steel members [1–4]; however, it is important to note that, in theory, further plastic development to the point of forming plastic hinges could result in an unbounded increase in deflection, posing a significant structural security risk. To effectively utilize the plastic properties of steel members, the steel structure design specifications permit plastic design in the case of flexural or beam–column members [5–7]. Nevertheless, the extent of plastic development is typically restricted to prevent excessive radicals and the resulting significant deformations. Taking the Chinese steel structure design standard GB50017 [5] as an example, it is specified that the depth of plastic development should not exceed 1/8 of the section's height. This means that a portion of the section transitions into an elastic–plastic working state, serving as a limit state for plasticity design. This approach not only leverages the section's plastic performance but also ensures the section's comprehensive safety. The design standard GB50017 employs the plastic development coefficient γ to assess the plastic properties of a section. Based on the plate width-to-thickness ratio, five grades, denoted as S1 to S5, are established to characterize the plastic development capacity. These grades also provide width-to-thickness ratio limits for the H-section, cox-section, and circular pipe-section, facilitating the determination of γ values. However, it is worth noting that angle sections are not included in this classification. Consequently, utilizing the plastic properties of angle sections poses challenges, and the design may not always be economically efficient.

Angle sections are widely utilized in various engineering projects, including building structures, bridges, transmission towers, transportation machinery, and industrial settings. They serve as essential load-bearing components or connectors because of their easy fabrication and convenient assembly [8–11]. The mechanical performance of angle section members continues to be a topic of interest, attracting research efforts from multiple groups focused on studying their buckling characteristics. Design methods are constantly evolving, especially concerning high-strength steel angles, to achieve economically efficient design objectives. Shi et al. [12–14] conducted a study on Q420 high-strength steel equal angles, examining various aspects, including residual stress distribution models, local buckling, and overall buckling. Cao [15] assessed the performance of Q460 high-strength steel angle members under axial loads using a combination of numerical simulations and experimental studies. Bezas [16] explored the performance of high-strength steel angle members under both centric and eccentric compression loads while evaluating the resistance of these designed members. Zhang et al. [17–19] conducted a study on the buckling behavior of S690 and S960 high-strength equal-angle members, employing both testing and numerical simulation.

Using large-size (limb width $\geq$ 220 mm, limb thickness $\geq$ 16 mm) and high-strength steel angles has gained popularity in fulfilling engineering needs, including reducing assembly work, construction complexity, and steel consumption [20,21]. Cao et al. [22–24] investigated the buckling behavior and design methods of large-size and high-strength steel angle members subjected to axial loads. They analyzed the influence of various cross-section sizes, strength grades, and slenderness ratios, employing a combination of experimental and numerical simulation techniques. Sun [25] examined the impact of support constraints on the stability capacity of large-size and high-strength steel angle members by measuring support rotation stiffness in tested conditions. Additionally, they investigated the behavior of large-size angle steel under eccentric loads to advance the design methodology for large-size angle beam–column members [26]. Numerous researchers have progressively incorporated stainless steel into angle steel applications, mitigating issues related to corrosion. Menezes et al. [27–34] conducted a comprehensive study on the buckling behavior and design methodology of hot-rolled austenitic stainless steel angles. They employed experimental and numerical techniques to examine mechanical performance under fixed-ended and pin-ended conditions. Dobrić et al. [35–37] analyzed cold-formed stainless steel angles, enriching our understanding of relevant mechanical properties and design approaches. Sian [38] explored the material properties and local buckling behavior of extruded WAAM 316L stainless steel angles.

Many of these studies often emphasize altering material or increasing cross-sectional dimensions to enhance the mechanical performance of angle members, aligning with practical engineering requirements. Furthermore, design methodologies primarily revolve around axial load members, with limited attention given to utilizing the plastic properties of angle sections in flexural or beam–column members. In the current steel structure design specifications, the equivalent method often transforms eccentrically compressed angle members into axially compressed members to simplify the design process. However, ensuring that members are not influenced by bending moment under complex external load conditions is challenging, especially for the superstructure, where the wind load will cause the angle member to experience both axial load and bending moment. In addition, a single-angle member bearing capacity cannot match engineering requirements, and it eventually evolves into double-combined sections. It is unavoidable for double-combined section angle members to withstand the bending moment, but this effect is typically disregarded in design, and such a design strategy may pose potential safety hazards. Consequently, effectively harnessing the plastic properties of angle sections in flexural or beam–column applications becomes challenging. Trahair et al. [39,40] examined steel angle section beams' biaxial bending and torsion behavior. Aydin et al. [4,41] investigated the mechanical performance of angle beam–column members. Vayas [3] examined the mechanical performance of angle beam–column members and discovered that angle sections categorized as Class 1 and Class 2 have a specific plastic deformation capability. Liu [42,43] et al. used the finite element method to study the angle behavior after conducting axial and eccentric compression test tests on 28 equal single-angle specimens with the section L51 × 6.4. Hussain proposed a set of finite element calculations regarding steel angle members under biaxial bending [1]. The mechanical performance of angle beam–column members was studied by Chan [44–46] et al., who recommended utilizing the DSM design approach. These studies observed that angle sections exhibit plastic development ability in flexural or beam–column applications, and they believed that the existing specifications are conservative and only provide some empirical formulas without making due provisions regarding the extent of plastic development of angle steel sections, which confused many designers.

Therefore, this study employs a theoretical analysis approach to examine the plastic properties of equal-angle sections subjected to combined axial force and biaxial bending. The fully plastic development property of the angle steel section was investigated. The formula for calculating the plastic development coefficient of angle sections is derived following the methodology outlined in the GB50017 specification. Plastic development coefficients for regular-size (limb width < 220 mm, limb thickness < 16 mm) and large-size angle sections were computed, and design-recommended values were furnished. Thus, appropriate regulations are made for the extent of plastic development of angle steel sections in plastic design, thereby solving the problem of inconsistent design results and even excessive conservatism caused by replacing eccentric compression angle members with axial compression members.

2. Full Plastic Analysis of the Angle Section

2.1. Cross-Section Simplification

The angle section limb thickness t is small relative to the limb width b. For specification GB/T706-2016 [47], the range of width-to-thickness ratios for the regular-size angle section ($b < 220$ mm) is 5.0 to 18.8, whereas for large-size angle steel ($b \geq 220$ mm) is 7.1 to 13.9. To analyze the angle section mechanically better, the angle section is simplified in that the material concentrates on the center line, the two limbs equate to two intersecting straight lines, and the actual rounded corners are substituted with sharp corners [48]. The simplified angle section result is shown in Figure 1. The angle section is assumed to remain planar when subjected to tension, compression, or bending, and the axis remains perpendicular to its deformed state. The material properties of the steel are considered to exhibit ideal elastic–plastic behavior, with the material following Hooke's law in the elastic phase and the entire section being capable of entering the plastic phase. Tension is considered positive, while compression is deemed negative. In the simplified angle section, the limb width is adjusted as $b' = b - t/2$. Consequently, this simplification process leads to a deviation in the section area due to the difference in the limb width value. When the limb width is taken as b for the regular-size angle section, the area calculation, according to Equation (1), results in an average deviation of 3.23% below the nominal value specified in the GB/T706-2016. When b' is utilized, it falls short of the nominal value by 6.91%. In the case of the large-size angle section, when the limb width is taken as b, the section area calculation, as per Equation (1), exceeds the nominal value specified in the GB/T706-2016 by an average of 4.47%. Nevertheless, when b' is employed, it falls short of the nominal value by 8.35%. Moreover, the moment of inertia between the simplified section's strong axis (u-axis) and weak axis (v-axis) is higher than the nominal values in the standard by 0.92% and 0.13%, respectively. In the subsequent theoretical analysis of angle sections, the calculation formulas involved are applicable to both b and b'. This indicates that the errors introduced by using the simplified section for analysis are acceptable. Therefore, in the subsequent analysis, the limb width of the angle steel is considered as b.

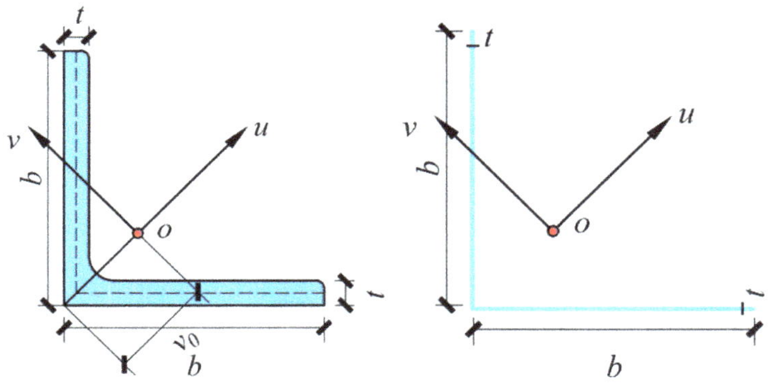

(**a**) (**b**)

Figure 1. Simplification of angle section: (**a**) actual section; (**b**) simplified section.

2.2. Full Plastic Analysis

The simplified angle section is investigated to discuss its mechanical behaviors under axial force N and combined bending moments (M_u and M_v). The overall analysis process of this paper is shown in Figure 2.

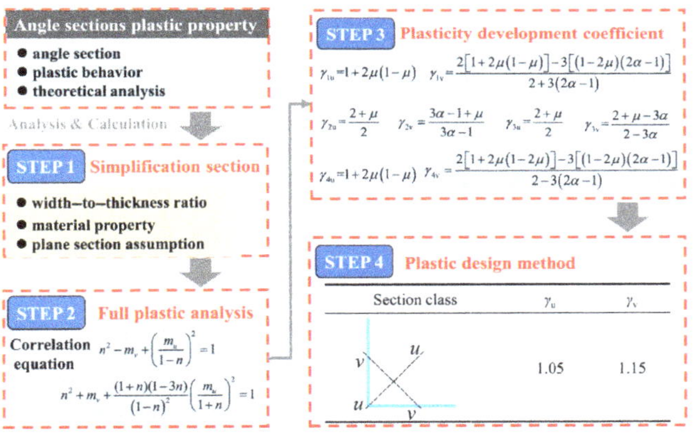

Figure 2. The process of the proposed method.

Based on the position of the neutral axis (NA axis), the stress distribution in the angle section is categorized into four cases: the neutral axis intersects both limbs with the limb back in compression (Figure 3a), the neutral axis intersects the lower limb (Figure 3b), the neutral axis intersects both limbs with the limb back in tension (Figure 3c), and the neutral axis intersects the upper limb (Figure 3d). Different stress scenarios are analyzed for each case, and the section is considered to have reached its ultimate bearing capacity when the full cross-section is yielded. The geometric parameters for the angle section's full plasticity are expressed as follows.

$$A = 2 \cdot b \cdot t \tag{1}$$

$$N_\mathrm{p} = A \cdot f_\mathrm{y} \tag{2}$$

$$M_\mathrm{up} = W_\mathrm{up} \cdot f_\mathrm{y} \tag{3}$$

$$M_\mathrm{vp} = W_\mathrm{vp} \cdot f_\mathrm{y} \tag{4}$$

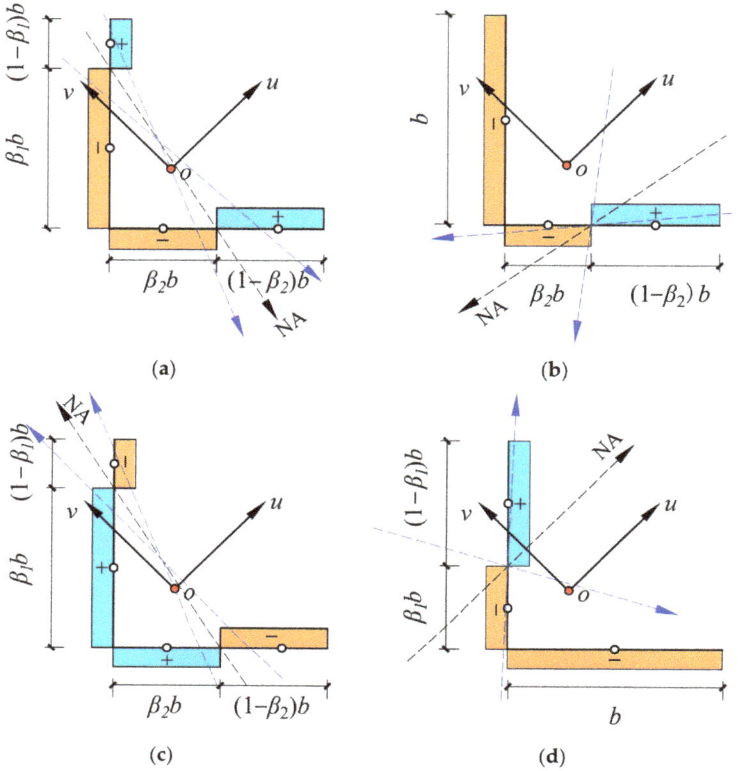

Figure 3. Stress distribution with the different neutral axis positions: (**a**) intersects the two limbs, which limb back in compression; (**b**) intersects the lower limb; (**c**) intersects the two limbs, which limb back in tension; (**d**) intersects the upper limb.

N is the axial force applied to the angle section. M_u is the bending moments applied to the angle section along the strong axis (u-axis). M_v is the bending moments applied to the angle section along the weak axis (v-axis). A is the section area, f_y is the yield strength, N_p is the axial plastic bearing capacity, W_{up} and W_{vp} are the plastic section moduli of the strong and weak axes, respectively, and M_{up} and M_{vp} are the plastic bending moments of the strong and weak axes, respectively.

The stress distribution of the angle section in Figure 3a is analyzed. $\beta_1 b$ and $\beta_2 b$ are the distances from the intersection point of the neutral axis and the two limbs to the limb back, respectively. To satisfy the equilibrium of the force on the angle section, β_1 and β_2 have to satisfy the relationship in Equation (5).

$$\frac{N}{N_p} = 1 - (\beta_1 + \beta_2) \tag{5}$$

The bending moment for the strong axis (u-axis):

$$M_u = btf_y\left[\frac{(1+\beta_1)b}{2\sqrt{2}}(1-\beta_1) - \frac{\beta_1 b}{2\sqrt{2}}\beta_1 + \frac{\beta_2 b}{2\sqrt{2}}\beta_2 - \frac{(1+\beta_2)b}{2\sqrt{2}}(1-\beta_2)\right] \tag{6}$$

Collating the above equation obtains:

$$M_{\mathrm{u}} = \frac{(\beta_2 + \beta_1)(\beta_2 - \beta_1)}{\sqrt{2}} b^2 t f_{\mathrm{y}} \tag{7}$$

The plastic bending moment for the strong axis:

$$M_{\mathrm{up}} = 2 \cdot \frac{b}{2\sqrt{2}} b t f_{\mathrm{y}} = \frac{1}{\sqrt{2}} b^2 t f_{\mathrm{y}} \tag{8}$$

The bending moment for the weak axis (v-axis):

$$M_{\mathrm{v}} = b t f_{\mathrm{y}} \left[-\frac{\beta_1 b}{2\sqrt{2}}(1 - \beta_1) - \frac{(1 - \beta_1) b}{2\sqrt{2}} \beta_1 - \frac{(1 - \beta_2) b}{2\sqrt{2}} \beta_2 - \frac{\beta_2 b}{2\sqrt{2}}(1 - \beta_2) \right] \tag{9}$$

Collating the above equation obtains:

$$M_{\mathrm{v}} = \frac{(\beta_1^2 + \beta_2^2) - (\beta_1 + \beta_2)}{\sqrt{2}} b^2 t f_{\mathrm{y}} \tag{10}$$

The plastic bending moment for the weak axis:

$$M_{\mathrm{vp}} = 4 \cdot \frac{1}{2} \cdot \frac{b}{2\sqrt{2}} \cdot \frac{1}{2} b t f_{\mathrm{y}} = \frac{1}{2\sqrt{2}} b^2 t f_{\mathrm{y}} \tag{11}$$

From Equations (5)–(11), the full plasticity correlation equation for the angle section under the combined action of axial force and biaxial bending moment can be obtained, as shown in Equation (12).

$$\left(\frac{N}{N_{\mathrm{p}}}\right)^2 - \frac{M_{\mathrm{v}}}{M_{\mathrm{vp}}} + \left(\frac{M_{\mathrm{u}}/M_{\mathrm{up}}}{1 - N/N_{\mathrm{p}}}\right)^2 = 1 \tag{12}$$

The stress distribution of the angle steel section in Figure 3b is analyzed. $\beta_2 b$ is the distance from the intersection point of the neutral axis with the lower limb to the limb back, and the parameter β_2 is related to the axial force exerted on the section, shown in Equation (13).

$$\beta_2 = -\frac{N}{N_{\mathrm{p}}} \tag{13}$$

The bending moment for the strong axis:

$$M_{\mathrm{u}} = b t f_{\mathrm{y}} \left[-\frac{b}{2\sqrt{2}} + \frac{\beta_2 b}{2\sqrt{2}} \beta_2 - \frac{(1 + \beta_2) b}{2\sqrt{2}} (1 - \beta_2) \right] \tag{14}$$

Collating the above equation obtains:

$$M_{\mathrm{u}} = \frac{(\beta_2^2 - 1)}{\sqrt{2}} b^2 t f_{\mathrm{y}} \tag{15}$$

The plastic bending moment for the strong axis:

$$M_{\mathrm{up}} = 2 \cdot \frac{b}{2\sqrt{2}} b t f_{\mathrm{y}} = \frac{1}{\sqrt{2}} b^2 t f_{\mathrm{y}} \tag{16}$$

From Equations (13)–(16), the full plasticity correlation equation of the axial force and bending moment about the strong axis can be obtained, as shown in Equation (17).

$$\left(\frac{N}{N_{\mathrm{p}}}\right)^2 - \frac{M_{\mathrm{u}}}{M_{\mathrm{up}}} = 1 \tag{17}$$

The bending moment for the weak axis:

$$M_{\mathrm{v}} = btf_{\mathrm{y}}\left[-\frac{(1-\beta_2)b}{2\sqrt{2}}\beta_2 - \frac{\beta_2 b}{2\sqrt{2}}(1-\beta_2)\right] \tag{18}$$

Collating the above equation obtains:

$$M_{\mathrm{v}} = \frac{-\beta_2(1-\beta_2)}{\sqrt{2}}b^2tf_{\mathrm{y}} \tag{19}$$

The plastic bending moment for the weak axis:

$$M_{\mathrm{vp}} = 4\cdot\frac{1}{2}\cdot\frac{b}{2\sqrt{2}}\cdot\frac{1}{2}btf_{\mathrm{y}} = \frac{1}{2\sqrt{2}}b^2tf_{\mathrm{y}} \tag{20}$$

From Equations (13), (19), and (20), the full plasticity correlation equation of axial forces and bending moments about the weak axis can be obtained as shown in Equation (21).

$$2\left(\frac{N}{N_{\mathrm{p}}}+\frac{N^2}{N_{\mathrm{p}}^2}\right)-\frac{M_{\mathrm{v}}}{M_{\mathrm{vp}}}=0 \tag{21}$$

The same approach can be employed to analyze the stress distribution in the cases depicted in Figure 3c,d, deriving the full plasticity correlation equation for the angle section as shown in Equations (22)–(24).

$$\left(\frac{N}{N_{\mathrm{p}}}\right)^2+\frac{M_{\mathrm{v}}}{M_{\mathrm{vp}}}+\left(\frac{M_{\mathrm{u}}/M_{\mathrm{up}}}{1+N/N_{\mathrm{p}}}\right)^2=1 \tag{22}$$

$$\left(\frac{N}{N_{\mathrm{p}}}\right)^2+\frac{M_{\mathrm{u}}}{M_{\mathrm{up}}}=1 \tag{23}$$

$$2\left(\frac{N}{N_{\mathrm{p}}}+\frac{N^2}{N_{\mathrm{p}}^2}\right)-\frac{M_{\mathrm{v}}}{M_{\mathrm{vp}}}=0 \tag{24}$$

where Equation (22) represents the correlation equation for the stress distribution state in Figure 3c, Equations (23) and (24), conversely, pertain to the correlation equations for the strong and weak axis of the stress distribution state in Figure 3d, respectively.

2.3. Full Plastic Correlation Equations

To conduct a more in-depth analysis of the full plastic correlation equations for the angle section's axial force and biaxial bending moment, we introduce the following dimensionless parameters: $n = N/N_{\mathrm{p}}$, $m_{\mathrm{u}} = M_{\mathrm{u}}/M_{\mathrm{up}}$, and $m_{\mathrm{v}} = M_{\mathrm{v}}/M_{\mathrm{vp}}$. Consequently, we can simplify the fully plastic correlation equations presented in Equations (12), (17), (21)–(24) as follows.

The correlation equation for the stress distribution state depicted in Figure 3a was simplified as Equation (25).

$$n^2 - m_{\mathrm{v}} + \left(\frac{m_{\mathrm{u}}}{1-n}\right)^2 = 1 \tag{25}$$

Similarly, the correlation equation for the stress distribution state in Figure 3b was simplified as Equations (26) and (27).

$$m_{\mathrm{u}} = n^2 - 1 \tag{26}$$

$$m_{\mathrm{v}} = 2\left(n^2 + n\right) \tag{27}$$

Similarly, the correlation equation on the stress distribution state in Figure 3c was simplified as Equation (28).

$$n^2 + m_v + \left(\frac{m_u}{1 + n}\right)^2 = 1 \tag{28}$$

Furthermore, we simplify the correlation equation for the stress distribution state in Figure 3d, resulting in Equation (29) and (30).

$$m_u = 1 - n^2 \tag{29}$$

$$m_v = 2\left(n^2 + n\right) \tag{30}$$

The relationship curves for the fully plastic correlation equations range from (25) to (30) at various values of n, as illustrated in Figure 4. The full plasticity relationship curves exhibit a distinct convex shape. When a fixed value of n is considered, Equations (25) and (28) correspond to the lower and upper segments of the curve—meanwhile, Equations (26), (27), (29), and (30) denote the red angular points indicated in the graph. Notably, these angular points do not completely coincide with the curve itself, thus requiring modifications to the full plasticity relationship curves. In Figure 3b,d, where the neutral axis intersects a single limb of the angle section, although the neutral axis can rotate over a wide range, the lengths of the tensile and compressive zones need to remain constant to satisfy the equilibrium conditions. Therefore, during the rotation process, Figure 3b,d correspond to only one point on the full plasticity relationship cures, specifically the angular points marked in Figure 4. Moreover, in both of these scenarios, the neutral axis and the angle limb are nearly perpendicular, making the stress distribution of the simplified angle section very similar to that of the actual section. Particularly when $v_0 \cong \frac{b}{2\sqrt{2}}$ it can be considered that the simplified angle section centroid is the same as that of the actual section. Hence, these angular points can be used to modify the curve.

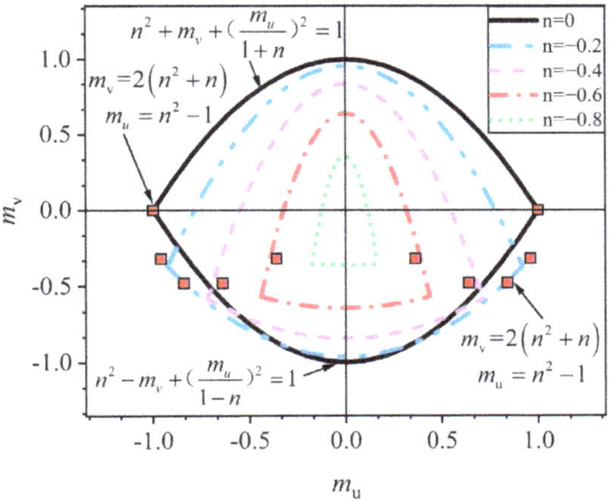

Figure 4. Fully plastic correlation equation curve with different n values.

Introducing correction coefficients ρ_1 and ρ_2, we can rewrite Equations (25) and (28) as follows.

$$n^2 - m_v + \rho_1\left(\frac{m_u}{1 - n}\right)^2 = 1 \tag{31}$$

$$n^2 + m_v + \rho_2\left(\frac{m_u}{1 + n}\right)^2 = 1 \tag{32}$$

By substituting Equations (26) and (27) into Equation (31) and substituting Equations (29) and (30) into Equation (32), we obtain:

$$\rho_1 = 1 \tag{33}$$

$$\rho_2 = \frac{(1 + n)(1 - 3n)}{(1 - n)^2} \tag{34}$$

The calculation results of ρ_1 and ρ_2 are substituted in Equations (29) and (30) to obtain Equations (35) and (36) and depict the full plasticity relationships of angle sections under different n values as shown in Figure 5.

$$n^2 - m_v + \left(\frac{m_u}{1 - n}\right)^2 = 1 \tag{35}$$

$$n^2 + m_v + \frac{(1 + n)(1 - 3n)}{(1 - n)^2}\left(\frac{m_u}{1 + n}\right)^2 = 1 \tag{36}$$

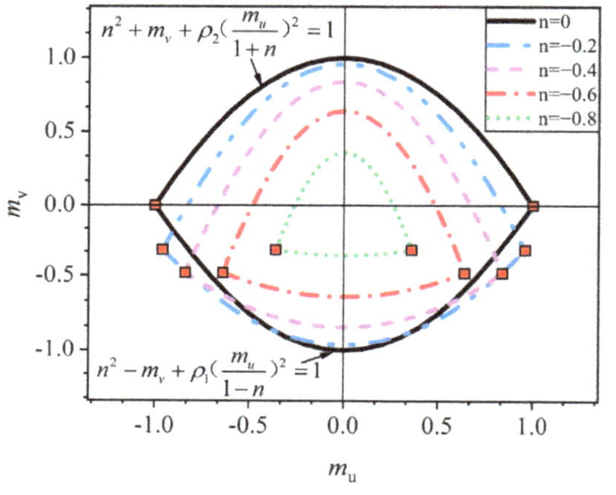

Figure 5. Fully plastic correlation equation modified curve with different n values.

From Figure 5, it can be observed that the relationship curves remain convex, with Equation (35) representing the lower portion of the curve and Equation (36) representing the upper portion. Moreover, both curves encompass the angular points, indicating that Equations (35) and (36) can serve as the full plasticity correlation equation for the angle section under the combined action of axial force and biaxial bending moment. In standard GB50017, to facilitate the design and consider unfavorable factors, the strength design formula for members subjected to axial force and biaxial bending moment, whether in tension or compression, is presented in a linear form, as shown in Equation (37).

$$\frac{N}{N_p} \pm \frac{M_x}{M_{px}} \pm \frac{M_y}{M_{py}} \leq 1 \tag{37}$$

where, $N_p = A_n f_y$, $M_{px} = \gamma_x W_{nx} f_y$, $M_{py} = \gamma_y W_{ny} f_y$. A_n is the net cross-section area, W_{nx} and W_{ny} are the x-axis and y-axis net cross-section moduli, and γ_x and γ_y are the x-axis and y-axis cross-section plasticity development coefficients.

Comparing the relationship curves of the derived full plasticity correlation equations for the angle section with the relationship curves of the strength design formula in the standard GB50017, as shown in Figure 6. It can be observed that the full plasticity equation results in a convex curve, whereas the standard design formula follows a linear relationship.

Furthermore, the standard design formula is relatively conservative compared with the full plasticity equation. This conservative formula arises from the fact that the standard employs a linear simplified formula, utilizing the section plastic development coefficient γ to restrict the section plastic development depth. This approach ensures that members can utilize the plastic properties of the section while maintaining an adequate margin of safety, fully leveraging the material's performance. Therefore, for structural design safety considerations, it is also possible to refer to the linear relationships in the standard and discuss the values of the section plastic development coefficient γ, which reflects the plastic performance of angle sections.

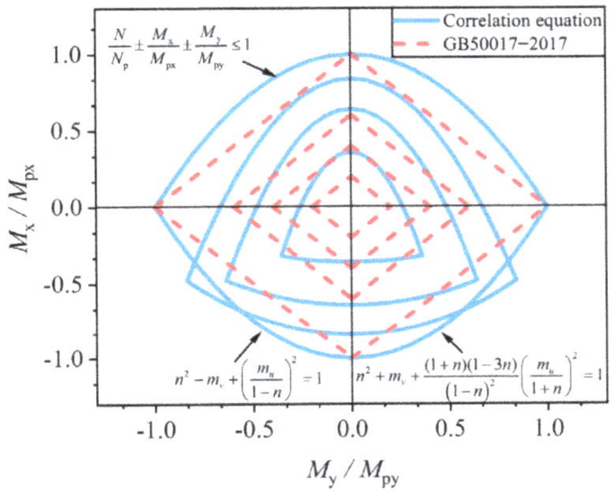

Figure 6. Comparison of full plasticity correlation equation and standard design formulas.

3. Plasticity Development Coefficient

3.1. Section Shape Coefficient

The plastic section modulus W_p divided by the elastic section modulus W_e is denoted by the form factor γ_F, which represents the full plastic capacity of the section. The section shape coefficients of the angle section strong axis γ_{Fu} and the weak axis γ_{Fv} for both regular size (limb width of 100 mm and 150 mm) and large-size (limb width of 220 mm and 250 mm), with different width-to-thickness ratios are calculated, as per the GBT706-2016 standard. The results are depicted in Figure 7.

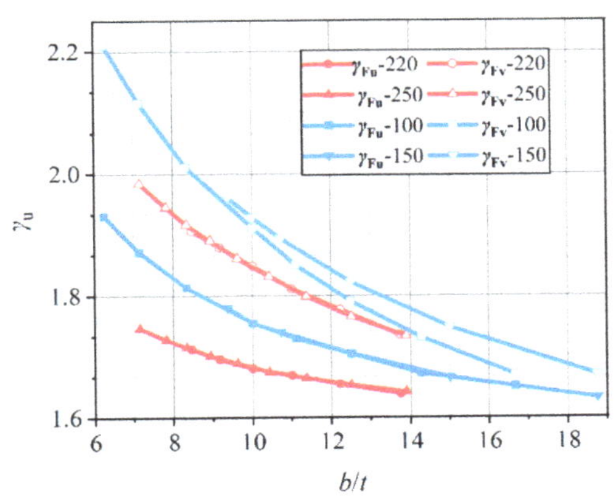

Figure 7. Comparison of section shape coefficient.

It can be observed that for a regular-size section angle with a limb width of 100 mm, the strong axis section shape coefficient γ_{Fu} ranges from 1.65 to 1.93, while the weak axis γ_{Fv} ranges from 1.67 to 2.21. For a limb width of 150 mm, the strong axis γ_{Fu} ranges from 1.63 to 1.78, and the weak axis γ_{Fv} ranges from 1.67 to 1.96. Large-size angle section with a limb width of 220 mm, the strong axis γ_{Fu} ranging from 1.64 to 1.71, and the weak axis γ_{Fv} ranging from 1.73 to 1.91. Similarly, in a large-size angle section with a limb width of 250 mm, the strong axis γ_{Fu} ranges from 1.64 to 1.75, and the weak axis γ_{Fv} ranges from 1.73 to 1.98. This reveals that for regular-size and large-size angles, the section shape coefficient of the weak axis is greater than that of the strong axis, which is consistent with the behavior of I-section biaxial symmetrically sections. In general, the limb width exerts a relatively modest impact on the section shape coefficient of angle sections, especially notable for large-size angle sections, where the trends in section shape coefficient exhibit remarkable consistency across various limb widths. The section shape coefficients are inversely associated with their width-to-thickness ratio; an escalation in this ratio results in a gradual reduction in the section shape coefficients. This implies that angle sections possessing higher width-to-thickness ratios display diminished plasticity. Additionally, the section shape coefficients for large-size angle sections are smaller than those of regular-size sections, signifying that regular-size angle sections exhibit pronounced plasticity.

3.2. Plastic Development Coefficient

The section's plastic development coefficient γ is closely linked to its section shape, plastic development depth μh, and stress state. In this section, we investigate the method for calculating the plastic development coefficient of angle sections, drawing inspiration from the derivation approach for the plastic development coefficient of H-section in GB50017. During the analysis, the stress state is categorized into four scenarios: stress state 1, where both the limb tip and limb back simultaneously experience plastic compressive stress and plastic tensile stress; stress state 2, where the limb tip undergoes plastic compressive stress; stress state 3, where the limb back undergoes plastic tensile stress; stress state 4, where both the limb tip and limb back simultaneously experience plastic tensile stress and plastic compressive stress. The stress distribution states were visualized in Figure 8, where σ_1 represents the tensile stress at the limb back in stress state 2, σ_2 represents the tensile stress at the limb tip in stress state 3, α represents the limb width ratio, defined as the ratio of the distance from the intersection point (between the weak axis and a single limb) to the limb back to the limb width. f_y signifies the yield stress, and σ_0 denotes the equivalent stress.

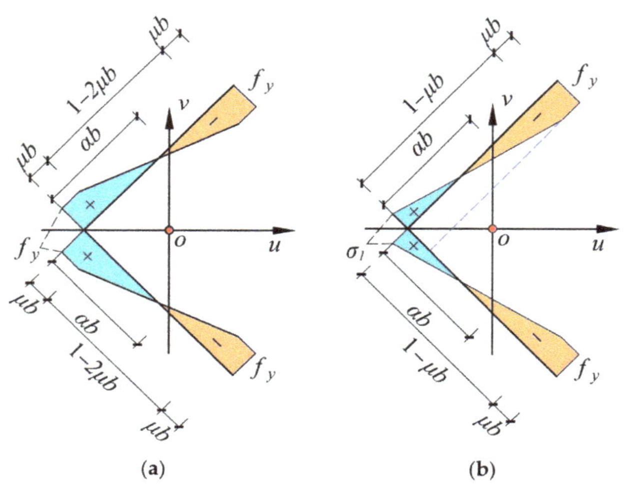

Figure 8. *Cont.*

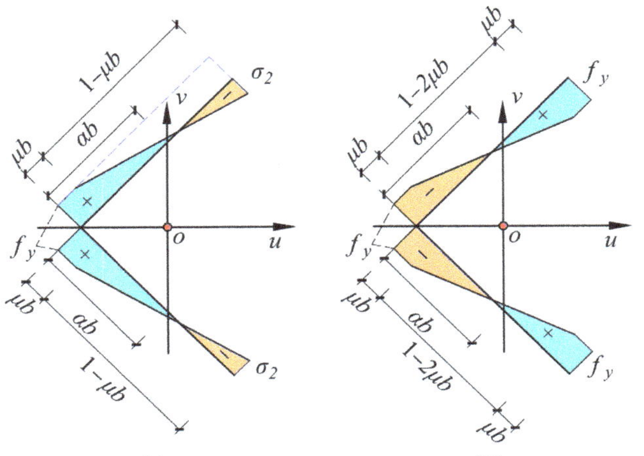

Figure 8. Different stress states: (**a**) stress state 1; (**b**) stress state 2; (**c**) stress state 3; (**d**) stress state 4.

The different stress states are analyzed to determine the calculation method for the plastic development coefficient. For the strong axis of stress state 1, based on the stress equilibrium conditions, we can obtain:

$$N = 0 \tag{38}$$

$$\begin{aligned} M_{1u} = & \left[\mu b\left(b - \frac{\mu b}{2}\right) - \mu b\frac{\mu b}{2}\right]\frac{1}{\sqrt{2}}btf_y \cdot 2 \\ & + \left\{\frac{(1-2\mu)b}{2}\left[b - \mu b - \frac{(1-2\mu)b}{3}\right]\frac{1}{\sqrt{2}}\right. \\ & \left. - \frac{(1-2\mu)b}{2}\left[\mu b + \frac{(1-2\mu)b}{3}\right]\frac{1}{\sqrt{2}}\right\}btf_y \cdot 2 \end{aligned} \tag{39}$$

Collating the above equation obtains:

$$M_{1u} = \frac{1 + 2\mu(1-\mu)}{3\sqrt{2}}b^2tf \tag{40}$$

When $\mu = 0$, obtains:

$$M_{1ue} = \frac{1}{3\sqrt{2}}b^2tf \tag{41}$$

Therefore,

$$\gamma_{1u} = \frac{M_{1u}}{M_{1ue}} = 1 + 2\mu(1-\mu) \tag{42}$$

For the weak axis, according to the stress equilibrium conditions, it can be obtained that:

$$N = 0 \tag{43}$$

$$\begin{aligned} M_{1v} = & \left\{\mu b\left[(1-\alpha)b - \frac{\mu b}{2}\right]\frac{1}{\sqrt{2}} + \mu b\left[\alpha b - \frac{\mu b}{2}\right]\frac{1}{\sqrt{2}}\right\}btf_y \cdot 2 \\ & + \left\{\frac{(1-2\mu)b}{2}\left[(1-\alpha b) - \mu b - \frac{(1-2\mu)b}{3} - \left(\alpha b - \frac{b}{2}\right)\right]\frac{1}{\sqrt{2}}\right. \\ & \left. + \frac{(1-2\mu)b}{2}\left[\alpha b - \mu b - \frac{(1-2\mu)b}{3}\right]\frac{1}{\sqrt{2}}\right\}btf_y \cdot 2 \end{aligned} \tag{44}$$

Collating the above equation obtains:

$$M_{1v} = \frac{2[1 + 2\mu(1-\mu)] - 3(1-2\mu)(2\alpha-1)}{6\sqrt{2}}b^2tf_y \tag{45}$$

When $\mu = 0$, obtains:

$$M_{1\text{ve}} = \frac{2 - 3(2\alpha - 1)}{6\sqrt{2}} b^2 t f_y \tag{46}$$

Therefore,

$$\gamma_{1\text{v}} = \frac{M_{1v}}{M_{1\text{ve}}} = \frac{2[1 + 2\mu(1 - \mu)] - 3[(1 - 2\mu)(2\alpha - 1)]}{2 + 3(2\alpha - 1)} \tag{47}$$

For the strong axis of stress state 2, we can obtain that:

$$N = 2bt\sigma_0 = 2btf_y - \frac{(1 - \mu)\text{b}}{2} t(f_y + \sigma_1) \cdot 2 \tag{48}$$

$$M_{2\text{u}} = \frac{(1 - \mu)b}{2} t\left[b - \frac{(1 - \mu)b}{3}\right] \frac{1}{\sqrt{2}} (f_y + \sigma_1) \cdot 2 \tag{49}$$

Then:

$$(f_y + \sigma_1) = \left(\frac{2}{1 - \mu}\right)(f_y - \sigma_0) \tag{50}$$

$$\begin{aligned} M_{2\text{u}} &= \left(\tfrac{1 - \mu}{2} \cdot \tfrac{2 + \mu}{3\sqrt{2}}\right)\left(\tfrac{2}{1 - \mu}\right) b^2 t (f_y + \sigma_1) \cdot 2 \\ &= \tfrac{2(2 + \mu)}{3\sqrt{2}} b^2 t (f_y - \sigma_0) \end{aligned} \tag{51}$$

When $\mu = 0$, obtains:

$$M_{2\text{ue}} = \frac{4}{3\sqrt{2}} b^2 t (f_y - \sigma_0) \tag{52}$$

Therefore:

$$\gamma_{2\text{u}} = \frac{M_{2\text{u}}}{M_{2\text{ue}}} = \frac{2 + \mu}{2} \tag{53}$$

For the weak axis:

$$N = 2bt\sigma_0 = 2btf_y - \frac{(1 - \mu)\text{b}}{2} t(f_y + \sigma_1) \cdot 2 \tag{54}$$

$$M_{2\text{v}} = \frac{(1 - \mu)b}{2} t\left[\alpha b - \frac{(1 - \mu)b}{3}\right] \frac{1}{\sqrt{2}} (f_y + \sigma_1) \cdot 2 \tag{55}$$

Then:

$$(f_y + \sigma_1) = \left(\frac{2}{1 - \mu}\right)(f_y - \sigma_0) \tag{56}$$

$$\begin{aligned} M_{2\text{v}} &= \left(\tfrac{1 - \mu}{2} \cdot \tfrac{3\alpha - 1 + \mu}{3\sqrt{2}}\right)\left(\tfrac{2}{1 - \mu}\right) b^2 t (f_y + \sigma_1) \cdot 2 \\ &= \tfrac{2(3\alpha - 1 + \mu)}{3\sqrt{2}} b^2 t (f_y - \sigma_0) \end{aligned} \tag{57}$$

When $\mu = 0$, obtains:

$$M_{2\text{ve}} = \frac{2(3\alpha - 1)}{3\sqrt{2}} b^2 t (f_y - \sigma_0) \tag{58}$$

Therefore,

$$\gamma_{2\text{v}} = \frac{M_{2v}}{M_{2\text{ve}}} = \frac{3\alpha - 1 + \mu}{3\alpha - 1} \tag{59}$$

For the strong axis of stress state 3, we can obtain that:

$$N = 2bt\sigma_0 = \frac{(1 - \mu)\text{b}}{2} t(f_y + \sigma_2) \cdot 2 - 2btf_y \tag{60}$$

$$M_{3\text{u}} = \frac{(1 - \mu)b}{2} t\left[b - \frac{(1 - \mu)b}{3}\right] \frac{1}{\sqrt{2}} (f_y + \sigma_2) \cdot 2 \tag{61}$$

Then:

$$(f_y + \sigma_2) = \left(\frac{2}{1 - \mu}\right)(f_y - \sigma_0) \tag{62}$$

$$\begin{aligned} M_{3u} &= \left(\frac{1-\mu}{2} \cdot \frac{2+\mu}{3\sqrt{2}}\right)\left(\frac{2}{1-\mu}\right)b^2t(f_y - \sigma_0) \cdot 2 \\ &= \frac{2(2+\mu)}{3\sqrt{2}}b^2t(f_y - \sigma_0) \end{aligned} \tag{63}$$

When $\mu = 0$, one obtains:

$$M_{3ue} = \frac{4}{3\sqrt{2}}b^2t(f_y - \sigma_0) \tag{64}$$

Therefore,

$$\gamma_{3u} = \frac{M_{3u}}{M_{3ue}} = \frac{2 + \mu}{2} \tag{65}$$

For the weak axis:

$$N = 2bt\sigma_0 = \frac{(1 - \mu)\mathrm{b}}{2}t(f_y + \sigma_2) \cdot 2 - 2btf_y \tag{66}$$

$$M_{3v} = \frac{(1 - \mu)b}{2}t\left[(1 - \alpha)b - \frac{(1 - \mu)b}{3}\right]\frac{1}{\sqrt{2}}(f_y + \sigma_2) \cdot 2 \tag{67}$$

Then:

$$(f_y + \sigma_2) = \left(\frac{2}{1 - \mu}\right)(f_y - \sigma_0) \tag{68}$$

$$\begin{aligned} M_{3v} &= \left(\frac{1-\mu}{2} \cdot \frac{2+\mu-3\alpha}{3\sqrt{2}}\right)\left(\frac{2}{1-\mu}\right)b^2t(f_y + \sigma_2) \cdot 2 \\ &= \frac{2(2+\mu-3\alpha)}{3\sqrt{2}}b^2t(f_y - \sigma_0) \end{aligned} \tag{69}$$

When $\mu = 0$, one obtains:

$$M_{3ve} = \frac{2(2 - 3\alpha)}{3\sqrt{2}}b^2t(f_y - \sigma_0) \tag{70}$$

Therefore,

$$\gamma_{3v} = \frac{M_{3v}}{M_{3ve}} = \frac{2 + \mu - 3\alpha}{2 - 3\alpha} \tag{71}$$

For the strong axis of stress state 4, we can obtain that:

$$N = 0 \tag{72}$$

$$\begin{aligned} M_{4u} &= \left[\mu b\left(b - \frac{\mu b}{2}\right) - \mu b\frac{\mu b}{2}\right]\frac{1}{\sqrt{2}}btf_y \cdot 2 \\ &+ \left\{\frac{(1-2\mu)b}{2}\left[b - \mu b - \frac{(1-2\mu)b}{3}\right]\frac{1}{\sqrt{2}} \right. \\ &\left. - \frac{(1-2\mu)b}{2}\left[\mu b + \frac{(1-2\mu)b}{3}\right]\frac{1}{\sqrt{2}}\right\}btf_y \cdot 2 \end{aligned} \tag{73}$$

Collating the above equation obtains:

$$M_{4u} = \frac{1 + 2\mu(1 - \mu)}{3\sqrt{2}}b^2tf \tag{74}$$

When $\mu = 0$, one obtains:

$$M_{4ue} = \frac{1}{3\sqrt{2}}b^2tf \tag{75}$$

Therefore,

$$\gamma_{4u} = \frac{M_{4u}}{M_{4ue}} = 1 + 2\mu(1 - \mu) \tag{76}$$

For the weak axis:

$$N = 0 \tag{77}$$

$$\begin{aligned} M_{4\mathrm{v}} = & \left\{\mu b\left[(1-\alpha)b-\tfrac{\mu b}{2}\right]\tfrac{1}{\sqrt{2}}+\mu b\left[\alpha b-\tfrac{\mu b}{2}\right]\tfrac{1}{\sqrt{2}}\right\}btf_y\cdot 2 \\ & +\left\{\tfrac{(1-2\mu)b}{2}\left[(1-\alpha b)-\mu b-\tfrac{(1-2\mu)b}{3}\right]\tfrac{1}{\sqrt{2}}\right. \\ & \left.+\tfrac{(1-2\mu)b}{2}\left[\alpha b-\mu b-\tfrac{(1-2\mu)b}{3}-\left(\alpha b-\tfrac{\mu b}{2}\right)\right]\tfrac{1}{\sqrt{2}}\right\}btf_y\cdot 2 \end{aligned} \tag{78}$$

Collating the above equation obtains:

$$M_{4\mathrm{v}} = \left\{\frac{2[1+2\mu(1-2\mu)]-3[(1-2\mu)(2\alpha-1)]}{6\sqrt{2}}\right\}b^2tf_y \tag{79}$$

When $\mu = 0$, one obtains:

$$M_{4\mathrm{ve}} = \frac{2-3(2\alpha-1)}{6\sqrt{2}}b^2tf_y \tag{80}$$

Therefore,

$$\gamma_{4\mathrm{v}} = \frac{M_{4v}}{M_{4\mathrm{ve}}} = \frac{2[1+2\mu(1-2\mu)]-3[(1-2\mu)(2\alpha-1)]}{2-3(2\alpha-1)} \tag{81}$$

Based on the preceding analysis, it can be inferred that the plastic development coefficients γ_{u} and γ_{v} of angle sections in various stress states for the strong and weak axis are dependent on μ and α. Consequently, Tables 1 and 2 present the values of γ_{u} and γ_{v} for angle sections with varying μ values (0.125, 0.15, 0.20, 0.25) and different α values. The α values in Table 1 are obtained from regular-size angle sections L150 × 8, L150 × 12, and L150 × 16, with respective α values of 0.532, 0.553, and 0.575. The α values in Table 2 are derived from large-size angle sections L250 × 18, L250 × 26, and L250 × 35, yielding α values of 0.547, 0.572, and 0.598, respectively.

Table 1. Plastic development coefficient for regular-size angle sections.

Stress State	μ	0.125			0.15			0.20			0.25		
	α	**0.532**	**0.553**	**0.575**	**0.532**	**0.553**	**0.575**	**0.532**	**0.553**	**0.575**	**0.532**	**0.553**	**0.575**
1	γ_{1u}		1.219			1.255			1.320			1.375	
	γ_{1v}	1.269	1.308	1.354	1.314	1.361	1.415	1.396	1.457	1.528	1.468	1.542	1.628
2	γ_{2u}		1.063			1.075			1.100			1.125	
	γ_{2v}	1.210	1.189	1.173	1.252	1.227	1.207	1.336	1.303	1.276	1.419	1.379	1.345
3	γ_{3u}		1.063			1.075			1.100			1.125	
	γ_{3v}	1.309	1.368	1.453	1.371	1.441	1.543	1.495	1.588	1.725	1.619	1.735	1.906
4	γ_{4u}		1.219			1.255			1.320			1.375	
	γ_{4v}	1.269	1.308	1.354	1.314	1.361	1.415	1.396	1.457	1.528	1.468	1.542	1.628

Table 2. Plastic development coefficient for large-size angle sections.

Stress State	μ	0.125			0.15			0.20			0.25		
	α	**0.547**	**0.572**	**0.598**	**0.547**	**0.572**	**0.598**	**0.547**	**0.572**	**0.598**	**0.547**	**0.572**	**0.598**
1	γ_{1u}		1.219			1.255			1.320			1.375	
	γ_{1v}	1.296	1.348	1.415	1.347	1.408	1.487	1.439	1.518	1.622	1.519	1.616	1.741
2	γ_{2u}		1.063			1.075			1.100			1.125	
	γ_{2v}	1.195	1.175	1.157	1.234	1.209	1.189	1.312	1.279	1.252	1.390	1.349	1.314
3	γ_{3u}		1.063			1.075			1.100			1.125	
	γ_{3v}	1.349	1.440	1.610	1.419	1.528	1.732	1.558	1.704	1.977	1.698	1.880	2.221
4	γ_{4u}		1.219			1.255			1.320			1.375	
	γ_{4v}	1.296	1.348	1.415	1.347	1.296	1.348	1.415	1.347	1.296	1.348	1.415	1.347

The results demonstrate that as the section's plastic development depth increases, both γ_u and γ_v values also rise. Furthermore, the weak axis value γ_v surpasses the strong axis γ_u, signifying superior plastic development capability along the weak axis, thereby affirming the earlier findings presented in this paper. The strong axis γ_u values for large-size angle sections are the same as those for regular-size angle sections. However, when considering various plastic development depths, the weak axis γ_v values for large-size angle sections exhibit average deviations of 3.94% (μ = 0.125), −1.72% (μ = 0.15), 7.32% (μ = 0.20), and 3.94% (μ = 0.25) in comparison to regular-size angle sections. This suggests no substantial disparity in the plastic development coefficients between regular-size and large-size angle sections by the analysis method presented in this paper. For ease of design and safety margin considerations, the plastic development coefficients for the strong and weak axis of the angle section are adopted as γ_u = 1.05 and γ_v = 1.15, respectively. This choice aligns with the Chinese steel structure design codes stipulating that the plastic development depth should not exceed 1/8 of the section height. Notably, these selected values are smaller than the minimum calculated values in Tables 1 and 2.

Substituting the plastic development coefficients of the angle section into the strength design formula for biaxial bending specimens in the GB50017 and comparing it with the H-section and box-section, as shown in Figure 9. It can be observed that the strength design formula for the angle section follows a similar form to that of the H-section and box-section, and the plastic development capability of the angle section falls between that of the H-section and box-section.

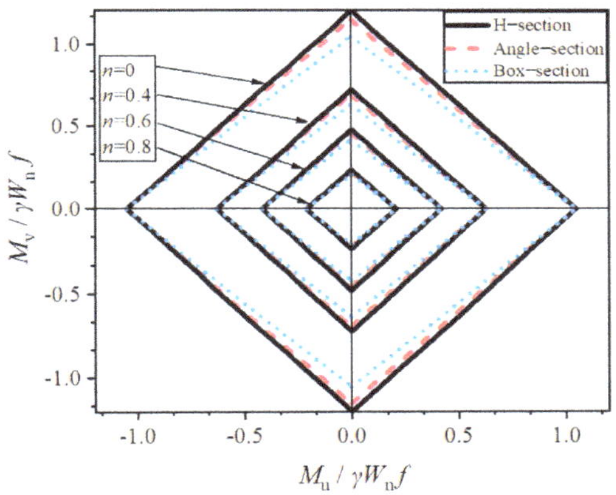

Figure 9. Comparison of strength design formulas for different section forms.

4. Angle Section Classification Discussion

In this study, we found that the plastic development capacity of the angle section is intimately connected with the component plates' width-to-thickness ratio and its plastic development depth, which is consistent with the conclusions of the references [3,4,39–41]. Moreover, Trahair believes that in comparison with flat bar sections, the plastic moment for angle sections is 1.5 times the first yield moment. However, in the standard GB50017, the plastic development depth of steel components is limited, not exceeding 1/8 of the section height, to ensure the plastic design's safety margin. Therefore, based on the calculation method of the plastic development coefficient in this paper, the value of the plastic development coefficient for angle sections is discussed. The GB50017 and Eurocode 3 establish distinct grades for various section types when designing flexural and beam–column members, delineating them according to their width-to-thickness ratio to represent the member's plastic development capacity. Nonetheless, it is noteworthy that these codes do not explicitly address angle sections. Trahair [49], based on the corresponding design standards, discussed the section grades classification for the angle section using local buckling coefficients. They considered the single-leg plates of the angle section as the extensions of the hot-rolled I-section flange. The Eurocode 3 also adopted a similar approach and classified the angle section into four categories based on the slenderness ratio λ_T of the constituent plates, which include plastic ($\lambda_T < \lambda_p$), compact ($\lambda_p < \lambda_T \leq \lambda_c$), semi-compact ($\lambda c < \lambda_T \leq \lambda_y$), and slender ($\lambda_y < \lambda_T$).

Furthermore, the slenderness ratio λ_T for the strong and weak axes differ because of variations in local buckling coefficients. In the context of angle section classification, the plastic and compact types can be subject to plastic design, while the semi-compact type can attain the first-order yield moment. However, the slender type, constrained by local stability limitations, necessitates elastic design, and the section's area must be determined based on the effective area. Following the classification method outlined in this reference, angle sections with Q420 strength grades, encompassing both regular-size (with limb widths of 100 mm and 150 mm) and large-size (with limb widths of 220 mm and 250 mm), are categorized and presented in Tables 3 and 4. Tables 3 and 4 reveal that, for regular-size angle sections, when assessed along the strong axis, all sections, with the exception of L100 × 7, L100 × 6, L150 × 10, and L100 × 8, are eligible for plastic design. When considering the weak axis, in addition to adhering to the criteria for the strong axis, sections L100 × 8 and L150 × 12 are also ineligible for plastic design. Concerning large-size angle sections, when assessed along the strong axis, all sections, with the exception of L220 × 16 and L250 × 18, meet the criteria for plastic design. In contrast, when evaluated along the weak axis, similar to the strong axis criteria, sections L220 × 18 and L250 × 20 are also ineligible for plastic design. This observation highlights that, when appraised along the strong axis, a larger number of angle sections qualify for plastic design in comparison to the weak axis. Furthermore, no angle sections are assigned to this category within the classification of slender sections, indicating that this classification method may be somewhat stringent.

Table 3. Section classification of regular-size angle section.

Axis	Plastic λ_p	Compact λ_c	Semi-Compact λ_y	Slender
Strong	12	16	26	>26
	L100 × 16, L100 × 14 L100 × 12, L100 × 10 L150 × 16, L150 × 15	L100 × 9, L100 × 8 L150 × 14, L150 × 12	L100 × 7, L100 × 6 L150 × 10, L100 × 8	/
Weak	10	14	23	>23
	L100 × 16, L100 × 14 L100 × 12	L100 × 10, L100 × 9 L150 × 16, L150 × 15 L150 × 14	L100 × 8, L100 × 7 L100 × 6, L150 × 12 L150 × 10, L100 × 8	/

Table 4. Section classification of large-size angle.

Axis	Plastic λ_p	Compact λ_c	Semi-Compact λ_y	Slender
Strong	12	16	26	>26
	L220 × 26, L220 × 24 L220 × 22, L250 × 35 L250 × 32, L250 × 30 L250 × 28, L250 × 26	L220 × 20, L220 × 18 L250 × 24, L250 × 22 L250 × 20	L220 × 16 L250 × 18	/
Weak	10	14	23	>23
	L220 × 26, L250 × 35 L250 × 32, L250 × 30	L220 × 24, L220 × 22 L220 × 20, L250 × 28 L250 × 26, L250 × 24 L250 × 22	L220 × 16 L220 × 18 L250 × 18 L250 × 20	/

For the sake of design convenience, this paper treats the single-leg plates of angle sections as extensions of hot-rolled I-section flanges and adopts the section classification method outlined in GB50017. Both regular-size and large-size angle sections are classified into five grades, as presented in Tables 5 and 6. According to the standard, plastic design is appropriate when the width-to-thickness ratio grade of the section plates falls within S1, S2, or S3. Conversely, elastic design should be employed when the width-to-thickness ratio grade is S4 or S5. These guidelines are equally applicable to angle sections, aligning with the stipulations in the standard.

Table 5. Slenderness ratio limit for classification of regular-size angle section.

Section Class	S1	S2	S3	S4	S5
b/t	$9\varepsilon_k$	$11\varepsilon_k$	$13\varepsilon_k$	$15\varepsilon_k$	20
Angle section	L100 × 16 L100 × 14	L150 × 16 L100 × 12	L100 × 10 L150 × 15 L150 × 14	L100 × 9 L100 × 8 L150 × 12	L100 × 6 L100 × 7 L150 × 10 L150 × 8

Table 6. Slenderness ratio limit for classification of large-size angle section.

Section Class	Class S1	Class S2	Class S3	Class S4	Class S5
b/t	$9\varepsilon_k$	$11\varepsilon_k$	$13\varepsilon_k$	$15\varepsilon_k$	20
Angle section	L250 × 35 L250 × 32	L220 × 26 L250 × 30 L250 × 28	L220 × 24 L220 × 22 L220 × 20 L250 × 26 L250 × 24	L220 × 18 L250 × 22 L250 × 20	L220 × 16 L250 × 18

ε_k is the correction coefficient of the steel grade. $\varepsilon_k = \sqrt{\frac{235}{f_y}}$.

In the process of classification, the method used did not distinguish angle sections based on their different principal axis but directly categorized them based on the width-to-thickness ratio. A comparison between Tables 3 and 4, and Tables 5 and 6 reveals that, with the exception of the inclusion of regular-size angle section L100 × 9 and large-size angle section L250 × 22, which require elastic design, all other section classifications correspond to plastic design. In contrast, the classification method in reference [49] divides angle sections into four grades, whether regular-size or large-size, with no sections allocated to the slender section grades. However, following the classification method outlined in the GB50017 standard, angle sections are classified into five grades, encompassing both regular-size and large-size angle sections in all five grades. This suggests that the section grade classification for the angle section in this paper is relatively cautious compared with the reference [49]. This implies that, for the angle section, adopting the GB50017 standard's classification method is a more prudent approach, offering increased safety margins for design and ensuring a safer and more reliable design. Therefore, when performing plastic design for angle sections, for section width-to-thickness ratio grades S1, S2, and S3, the section plastic development coefficients γ_u and γ_v should be taken as 1.05 and 1.15, respectively, i.e., the plastic development depth of the angle section is taken as 1/8 of the section height. When the section width-to-thickness ratio grade is S4 or S5, a value of 1.0 should be used.

5. Conclusions

This paper explored the plastic behavior of angle sections under the combined action of axial force and biaxial bending, derived the correlation equations for axial force and moment, analyzed the affecting factors of the angle sections' plastic development capacity, introduced a method for computing the plastic development coefficients of angle sections, and summarized the key findings as follows:

1. The axial force and moment full plasticity equation results are nonlinear, whereas the specification design formula follows a linear relationship. Furthermore, the specification design formula is relatively conservative compared with the full plasticity

equation. This conservative arises from the fact that the standard employs a linear simplified formula, utilizing the section plastic development coefficient to restrict the section plastic development depth.

2. The section shape coefficients are inversely associated with their width-to-thickness ratio; an escalation in this ratio results in a gradual reduction in the section shape coefficients. This implies that angle sections possessing higher width-to-thickness ratios display diminished plasticity. Additionally, the section shape coefficients for large-size angle sections are smaller than those of regular-size sections, signifying that regular-size angle sections exhibit pronounced plasticity.
3. The angle sections' plastic development coefficients are related to plastic development depth. Higher values of development depths are associated with larger values of γ_u and γ_v. The plastic development capability of angle sections is also between that of H-sections and box-sections, and they exhibit better plastic development capacity along the weak axis than the strong axis.
4. For regular-size and large-size angle sections, the plastic development coefficients for plastic design can be set to the same values. This choice aligns with the steel structure design code GB50017, stipulating that the plastic development depth should be taken as 1/8 of the section height, with strong and weak axes taken as $\gamma_u = 1.05$ and $\gamma_v = 1.15$, respectively.

Author Contributions: Conceptualization, Q.C.; Formal analysis, Y.S.; Methodology, Y.S.; Writing—original draft, D.S. and S.S.; Validation, Y.L. and Y.Y.; Writing—review and editing, Y.S. and D.S. All authors have read and agreed to the published version of the manuscript.

Funding: This research was funded by the Nanyang Institute of Technology Ph.D. Research Initiation Fund Project (Grant No. NGBJ-2023-34), the Henan Province Science and Technology Plan Project (Grant No. 242102320216), and the Henan Province Key Research Project Plan for Higher Education Institution (Grant No. 24A560016).

Data Availability Statement: The data presented in this study are available on request from the corresponding author. The data are not publicly available as research is continuing.

Conflicts of Interest: The authors declare that they have no known competing financial interests or personal relationships that could have appeared to influence the work reported in this paper.

References

1. Hussain, A.; Liu, Y.P.; Chan, S.L. Finite Element Modeling and Design of Single Angle Member Under Bi-axial Bending. *Structures* **2018**, *16*, 373–389. [CrossRef]
2. Charalampakis, A.; Koumousis, V. Ultimate strength analysis of composite sections under biaxial bending and axial load. *Adv. Eng. Softw.* **2008**, *39*, 923–936. [CrossRef]
3. Vayas, I.; Charalampakis, A.; Koumousis, V. Inelastic resistance of angle sections subjected to biaxial bending and normal forces. *Steel Constr. Des. Res.* **2009**, *2*, 138–146. [CrossRef]
4. Charalampakis, A.E. Full plastic capacity of equal angle sections under biaxial bending and normal force. *Eng. Struct.* **2011**, *33*, 2085–2090. [CrossRef]
5. *BS EN 1993-1-1*; Eurocode3: Design of Steel Structures: Part 1-1: General Rules and Rules for Buildings. British Standards Institution (BSI): London, Britain, 2005.
6. *ANSI/AISC 360-16*; Specification for Structural steel Buildings, ANSI/AISC 360-10. American Institute of Steel Construction (AISC): Chicago, IL, USA, 2016.
7. *GB50017-2017*; Standard for Design of Steel Structures. China Architecture & Building Press: Beijing, China, 2017.
8. Jiang, W.-Q.; Liu, Y.-P.; Chan, S.-L.; Wang, Z.-Q. Direct analysis of an ultrahigh-voltage lattice transmission tower considering joint effects. *J. Struct. Eng.* **2017**, *143*, 4017009. [CrossRef]
9. Cho, S.; Chan, S.L. Second-order analysis and design of angle trusses, Part II: Plastic analysis and design. *Eng. Struct.* **2008**, *30*, 626–631. [CrossRef]
10. Li, B.; Cao, P.; Zhang, D.; Guo, Y. Experimental research on behavior of Q420 dual-angle steel with cruciform section under dynamic compression. *J. Vibroeng.* **2017**, *19*, 2031–2042. [CrossRef]
11. Qu, S.; Zhang, B.; Guo, Y.; Sun, Q.; Wang, Y. Ultimate strength of pinned-end dual-angle cross combined section columns under axial compression. *Thin Wall Struct.* **2020**, *157*, 107062. [CrossRef]

12. Ban, H.Y.; Shi, G.; Shi, Y.J.; Wang, Y.Q. Residual Stress Tests of High-Strength Steel Equal Angles. *J. Struct. Eng.* **2012**, *138*, 1446–1454. [CrossRef]
13. Shi, G.; Zhou, W.J.; Bai, Y.; Liu, Z. Local buckling of steel equal angle members with normal and high strengths. *Int. J. Steel Struct.* **2014**, *14*, 447–455. [CrossRef]
14. Ban, H.Y.; Shi, G.; Shi, Y.J.; Wang, Y.Q. Column Buckling Tests of 420 Mpa High Strength Steel Single Equal Angles. *Int. J. Struct. Stab. Dyn.* **2013**, *13*, 1250069. [CrossRef]
15. Cao, X.L.; Xu, Y.; Hao, J.P. Experimental analysis on ultimate bearing capacity of Q460 high-strength angle members under axial compression. *J. Civ. Environ. Eng.* **2018**, *40*, 146–152.
16. Bezas, M.-Z.; Demonceau, J.-F.; Vayas, I.; Jaspart, J.-P. Experimental and numerical investigations on large angle high strength steel columns. *Thin Wall Struct.* **2021**, *159*, 107287. [CrossRef]
17. Wang, F.; Liang, Y.; Zhao, O.; Young, B. Pin-ended press-braked S960 ultra-high strength steel angle section columns: Testing, numerical modelling and design. *Eng. Struct.* **2021**, *228*, 111418. [CrossRef]
18. Zhang, L.; Liang, Y.; Zhao, O. Flexural-torsional buckling behaviour and resistances of fixed-ended press-braked S690 high strength steel angle section columns. *Eng. Struct.* **2020**, *223*, 111180. [CrossRef]
19. Zhang, L.; Wang, F.; Liang, Y.; Zhao, O. Press-braked S690 high strength steel equal-leg angle and plain channel section stub columns: Testing, numerical simulation and design. *Eng. Struct.* **2019**, *201*, 109764. [CrossRef]
20. Du, R.Z. Applications of Q420 large section angle steel in UHV power transmission line. *Power Syst. Clean Energy* **2011**, *27*, 30–34.
21. Zhang, Z.F.; Li, Q.H.; Xing, H.J.; Tang, C.J.; Yuan, X. Application Study of 300 mm Large Size Angle Steel in +/− 1100 kV Ultra-high Voltage Tower Transmission Tower. *Aer. Adv. Eng. Res.* **2016**, *30*, 1304–1311.
22. Cao, K.; Guo, Y.J.; Zeng, D.W. Buckling behavior of large-section and 420 MPa high-strength angle steel columns. *J. Constr. Steel Res.* **2015**, *111*, 11–20. [CrossRef]
23. Chen, H.; Zeng, D.; Cao, K.; Guo, Y. Buckling behavior and design of large-section angle steel columns under axial compression. *J. Build Eng.* **2023**, *78*, 107578. [CrossRef]
24. Chen, H.; Chen, H.; Cao, K.; Guo, Y. Study on Axial Compression Stability of Q345 Large-Section Angle Steel Columns. *Buildings* **2023**, *13*, 1030. [CrossRef]
25. Sun, Y.; Guo, Y.J.; Chen, H.Y.; Cao, K. Stability of Large-Size and High-Strength Steel Angle Sections Affected by Support Constraint. *Int. J. Steel Struct.* **2021**, *21*, 85–99. [CrossRef]
26. Sun, Y.; Song, D.; Sun, S.; Guo, Y. Behavior of large-size and high-strength steel angle subjected to eccentric load. *Structures* **2023**, *57*, 105161. [CrossRef]
27. de Menezes, A.A.; Vellasco, P.C.d.S.; de Lima, L.R.; da Silva, A.T. Experimental and numerical investigation of austenitic stainless steel hot-rolled angles under compression. *J. Constr. Steel Res.* **2019**, *152*, 42–56. [CrossRef]
28. Liang, Y.; Jeyapragasam, V.V.K.; Zhang, L.; Zhao, O. Flexural-torsional buckling behaviour of fixed-ended hot-rolled austenitic stainless steel equal-leg angle section columns. *J. Constr. Steel Res.* **2019**, *154*, 43–54. [CrossRef]
29. Sun, Y.; Liu, Z.; Liang, Y.; Zhao, O. Experimental and numerical investigations of hot-rolled austenitic stainless steel equal-leg angle sections. *Thin Wall Struct.* **2019**, *144*, 106225. [CrossRef]
30. Sarquis, F.R.; de Lima, L.R.; Vellasco, P.C.G.d.S.; Rodrigues, M. Experimental and numerical investigation of hot-rolled stainless steel equal leg angles under compression. *Thin Wall Struct.* **2020**, *151*, 106742. [CrossRef]
31. Zhang, L.; Liang, Y.; Zhao, O. Experimental and numerical investigations of pin-ended hot-rolled stainless steel angle section columns failing by flexural buckling. *Thin Wall Struct.* **2020**, *156*, 106977. [CrossRef]
32. Dobrić, J.; Filipović, A.; Baddoo, N.; Buđevac, D.; Rossi, B. Design criteria for pin-ended hot-rolled and laser-welded stainless steel equal-leg angle columns. *Thin Wall Struct.* **2021**, *167*, 108175. [CrossRef]
33. Filipović, A.; Dobrić, J.; Baddoo, N.; Može, P. Experimental response of hot-rolled stainless steel angle columns. *Thin Wall Struct.* **2021**, *163*, 107659. [CrossRef]
34. Behzadi-Sofiani, B.; Gardner, L.; Wadee, M.A. Testing, numerical analysis and design of stainless steel equal-leg angle section beams. *Structures* **2022**, *37*, 977–1001. [CrossRef]
35. Dobrić, J.; Filipović, A.; Baddoo, N.; Marković, Z.; Buđevac, D. Design procedures for cold-formed stainless steel equal-leg angle columns. *Thin Wall Struct.* **2021**, *159*, 107210. [CrossRef]
36. Dobrić, J.; Filipović, A.; Marković, Z.; Baddoo, N. Structural response to axial testing of cold-formed stainless steel angle columns. *Thin Wall Struct.* **2020**, *156*, 106986. [CrossRef]
37. Zhang, L.; Tan, K.H.; Zhao, O. Experimental and numerical studies of fixed-ended cold-formed stainless steel equal-leg angle section columns. *Eng. Struct.* **2019**, *184*, 134–144. [CrossRef]
38. Evans, S.I.; Xu, F.; Wang, J. Material properties and local stability of WAAM stainless steel equal angle sections. *Eng. Struct.* **2023**, *287*, 116160. [CrossRef]
39. Trahair, N. Biaxial bending of steel angle section beams. *J. Struct. Eng.* **2004**, *130*, 554–561. [CrossRef]
40. Trahair, N. Biaxial bending and torsion of steel equal angle section beams. *J. Struct. Eng.* **2007**, *133*, 78–84. [CrossRef]
41. Aydın, R.; Doğan, M. Elastic, full plastic and lateral torsional buckling analysis of steel single-angle section beams subjected to biaxial bending. *J. Constr. Steel Res.* **2007**, *63*, 13–23. [CrossRef]
42. Liu, Y.; Hui, L.B. Finite element study of steel single angle beam-columns. *Eng. Struct.* **2010**, *32*, 2087–2095. [CrossRef]

43. Aydin, R.; Dogan, M.; Gunaydin, A.; Gokdemir, H.; Kirac, N. Experimental and code based study of beam-column behaviour of equal leg single-angles. *J. Constr. Steel Res.* **2011**, *67*, 780–789. [CrossRef]
44. Chan, S.L.; Liu, Y.P.; Liu, S.W. A New Codified Design Theory of Second-order Direct Analysis for Steel and Composite Structures—From Research to Practice. *Structures* **2017**, *9*, 105–111. [CrossRef]
45. Dinis, P.B.; Camotim, D.; Landesmann, A.; Martins, A.D. Improving the Direct Strength Method prediction of column flexural-torsional failure loads. *Thin Wall Struct.* **2020**, *148*, 106461. [CrossRef]
46. Dinis, P.B.; Camotim, D. A novel DSM-based approach for the rational design of fixed-ended and pin-ended short-to-intermediate thin-walled angle columns. *Thin Wall Struct.* **2015**, *87*, 158–182. [CrossRef]
47. *GB/T706-2016*; Hot Rolled Section Steel. China Standard Press: Beijing, China, 2016.
48. Deng, C.; Zhang, C.; Zhou, J.; Chen, Y. Plate assembly elasto-plastic interactive buckling analyses of H Section steel members under compression and bending. *J. Tongji Univ. (Nat. Sci.)* **2016**, *44*, 1307–1315. [CrossRef]
49. Trahair, N. Moment capacities of steel angle sections. *J. Struct. Eng.* **2002**, *128*, 1387–1393. [CrossRef]

buildings

MDPI

Article

An Experimental and Numerical Study on the Lateral Stiffness Limits of Straddle-Type Monorail Tour-Transit Systems

Hong Zhang [1], Pengjiao Wang [2], Qin Li [3], Junhui Jin [4], Shiqi Wei [2], Fengqi Guo [2,*], Cheng Feng [3] and Qun Deng [3]

[1] College of Mechanical and Intelligent Manufacturing, Central South University of Forestry and Technology, Changsha 410075, China; t20040415@csuft.edu.cn
[2] School of Civil Engineering, Central South University, Changsha 410075, China; w2592570@gmail.com (P.W.); 224811107@csu.edu.cn (S.W.)
[3] 3rd Construction Co., Ltd. of China Construction 5th Engineering Bureau, Changsha 410004, China; zngcb2023@126.com (Q.L.); 19174948228@163.com (C.F.); accdengqun@163.com (Q.D.)
[4] Zhuzhou CRRC Special Equipment Technology Co., Ltd., Zhuzhou 412001, China; jinjunhui0514@163.com
* Correspondence: fengqiguo@csu.edu.cn

Abstract: The development of the straddle-type monorail tour-transit system (MTTS) has received keen attention. Regarding the unspecified regulations on the lateral stiffness limit of steel substructures, which is essential for the design of MTTSs, this work presents a comprehensive assessment of the issue. Firstly, a wind–vehicle–bridge coupling model was established, integrating multibody dynamics and finite element methods. This model was then validated against field test results, considering measured track irregularities and simulated wind loadings as the excitations. Afterwards, a trend analysis and a variance-based sensitivity analysis was performed to investigate the effect of various factors on the dynamic response of the MTTS. Results indicate that the pier height significantly impacts the lateral displacement of the pier top, accounting for 87% of the first-order sensitivity index and 75% of the total sensitivity index. In comparison, the lateral stiffness of track beams contributes over 70% to the maximum responses at the mid-span. Based on this, two responses, the lateral displacement of the pier top and the lateral deflection–span ratio of the track beam, were utilized as evaluation indicators. With the analysis of indicators in terms of lateral acceleration, Sperling index, and lateral wheel force, the limited values for the two indicators were determined as 8.04 mm and L/4200, respectively. These findings can serve as valuable references for future research and designs in this field.

Keywords: monorail tour-transit system; lateral stiffness; finite element modelling; multi-body dynamics; dynamic response; sensitivity analysis

1. Introduction

The straddle-type monorail tour-transit system (MTTS) has developed rapidly in recent years. At present, there are more than 30 MTTS projects under construction or operation across the world [1]. Two main subsystems can be easily defined in an MTTS project, the monorail train and the steel bridge. Compared to conventional monorail systems utilized in urban transportation, the configuration of monorail trains in an MTTS is not much different, but there are significant diversities in the bridge structures employed. That is, the MTTS uses lightweight steel box girders as the supporting track beam and steel columns as the substructures, which is easy for construction, has good ornamental characteristics, and is very suitable for scenic areas. Extensive studies have been conducted on steel bridges of a more complex type, such as steel–concrete composite beams and girders with corrugated webs [2–4], regarding overall structural performance and critical responses of the main structural components [5]. Nevertheless, the general configuration of the supporting system in an MTTS is relatively simple. The two main structural components, track beams and columns, of the bridge in an MTTS are basically connected through a flange joints system; a finger band (expansion joint) is used between adjacent track beams to eliminate potential effects due to temperature change. In addition, flat webs are employed in monorails to ensure the stable operation of steering and stabilizing wheels. With these configurations, the weight of the track beam of an MTTS is light, and the overall stiffness is relatively low. Hence, the dynamic response of the track beam of an MTTS may be more significant than that in a conventional monorail system under the same conditions.

For the design of track beams, more attention has been paid to their overall structural performance under practical loads, such as wind and live loads. At present, related designs of track beams does not consider some long-term issues, such as fatigue and material degradation [6]. In particular, the lateral stiffness is an essential factor affecting the smoothness and comfort of train operation and hence, the related limited values must be reasonably determined. However, the design code for the MTTS, i.e., the "Large-scale amusement device safety code (GB 8408-2018) [7]", does not have specific regulations in this respect. Of course, provisions in relevant codes (e.g., design code for urban rail transit bridges (GB/T 51234-2017) [8] and monorails in urban transit systems (GB 50458-2022) [9]) can be used for reference. Nevertheless, practical use has proven that these codes are not completely applicable because of the structural diversity in various transit systems and the design concepts of different codes. Therefore, it is important to carry out targeted research by considering the coupling effects of practical loads, thereby improving relevant design standards to promote the development of modern tourism systems.

In general, existing studies on the dynamic responses of transportation systems were conducted through numerical simulations, e.g., using the finite element method (FEM) and multi-body dynamics (MBD). For example, Gao et al. developed a co-simulation model using FEM and MBD software to study the effects of vehicle speed and bridge pier height on a rail system [10]. Liu studied the dynamic response of a T-girder continuous rigid frame bridge; by using computational fluid dynamics (CFD) software, the aerodynamic parameters and their effects on the bridge under passing vehicles were simulated [11]. Furthermore, many studies have been able to consider the coupling effect of various loadings. For example, Miao et al. [12] established a refined space vibration model of a wind–vehicle–bridge system with time-varying characteristics to investigate the dynamic features of bridges. Huang et al. [13] utilized a co-simulation method to analyze the mechanical properties of bridges under wind–vehicle–bridge coupling actions. Chen et al. [14] and Wang et al. [15] studied the aerodynamic performance of a vehicle–bridge system under the action of a crosswind. Zou et al. [16] developed a joint model to analyze the dynamic response of a wind–monorail–car–suspended bridge system under wind loadings. In addition, laboratory tests on the coupling system were reported. Luo et al. [17] measured the aerodynamic coefficients of relevant bridge and vehicle models under the action of crosswind through a wind tunnel test. Jiang et al. [18] focused on the most unfavourable load that long-span bridges bore when vehicles and subways were in operation at the same time, and relevant models were established for dynamic response studies. Guo et al. [19] performed wind tunnel tests and numerical simulations to investigate the dynamic characteristics of bridges; the wind–car–bridge coupling model was also developed via the FEM and CFD. It should be noted that these studies are dominantly focused on rail transportations; therefore, results in general monorail systems are still limited, not to mention relevant studies on MTTSs. Nevertheless, the methodologies of general transportation systems can be used as a great reference.

In terms of the lateral stiffness limit, a perspective more inclined towards practical application, research has made great progress on general transit systems. In general, the main body of a vehicle can be simulated or constructed by using MBD. Then, a targeted dynamic evaluation system can be proposed based on the results of the dynamic responses of the vehicle and bridges [20]. Xiang et al. [21] investigated the lateral stiffness limit of a suspension bridge through a combination of a wind tunnel test and a numerical simulation. Zhou et al. [22] investigated the effects of the height of the bridge pier and the lateral stiffness of the bridge beam on the dynamic response of a vehicle–bridge system.

Zhu et al. [23] developed multiple suspension bridge models and investigated various methods to improve the structural stiffness of railway bridges. Liu et al. [24] proposed an analytical model to quantitatively analyze the influence and mechanism of several important parameters on the mapped deformation of the rail, including the lateral beam deformation and the lateral stiffness of fasteners. Wang et al. [25] performed numerical simulations of the coupled vibration of 12 track–beam systems with different stiffnesses. In the study, track irregularities and wind-induced lateral loads were considered. Yan et al. [26] investigated the design value of the torsional stiffness reduction factor for the serviceability limit states. Carden et al. [27] proposed a simplified calculation method based on the elastic stress and stiffness of bridge beams, and the results were found comparable with those obtained by the FEM.

To summarize, existing studies on the dynamic response of monorail systems, including MTTSs, and effects of the lateral stiffness of a monorail bridge, i.e., track beams and piers, are limited. Research has reported field tests and numerical simulations of the monorail systems considering track irregularities [28–31], wind loadings [5], and ground motions [28]. Moreover, the usage of the FEM and MBD on monorails has been verified, providing an effective method for investigating the dynamic effects of an MTTS. To the best of our knowledge, no existing studies have focused on the determination of lateral limited values of MTTSs.

On the basis of a comprehensive analysis of the existing research and considering the unique structural configurations of MTTS, this work employs FEM and MBD software to establish a wind–vehicle–bridge coupling model of an MTTS. The effects of variables such as vehicle load and track beam lateral stiffness on vehicle running stability, acceleration, and the lateral force of guide wheels and stabilizing wheels are investigated. Through a numerical simulation and parametric analysis, the lateral deflection–span ratio limit and the lateral displacement limit of the pier top were determined based on the riding comfort standard, i.e., the mid-span acceleration of the monorail track beam and the lateral force limit of the monorail train wheels. The results of the study can improve the problem of unclear limits of lateral stiffness of MTTSs, which can provide a reference for related research and design.

2. Methodologies

2.1. Wind–Vehicle–Bridge Coupling Mode

2.1.1. Modelling Track Beams

For easy demonstration, a one-connection five-span curve bridge in the Dajueshan project in Jiangxi, China is taken as an example. The curve radius of the track beam is 30 m, the span of a single girder is 10 m, and the bridge pier is a single column with a pier height of 2 m. A solid element C3D8R with 6 degrees of freedom in ABAQUS [32] (FEM software) was chosen to model the monorail substructures (track beams and piers). In a typical span, one side of a track beam body was fixed to the flange on the pier top, and the longitudinal freedom at the other side was released. The connection between the steel column and the ground was simplified and set as a fixed boundary. It should also be noted that the entire analysis was under the elastic stage. The cross-section of track beams, pier column elevation, and FE model of the track beam structure are shown in Figure 1.

2.1.2. Modelling Monorail Trains

Owing to the complexity of the vehicle system, necessary simplifications should be performed to efficiently simulate their mechanical characteristics, e.g., using MBD. In this work, the mass, geometric characteristics, and dynamic characteristics of each component of the monorail train were considered. According to the principle of MBD, main bodies, such as the car body and bogies, are modelled as rigid components, which are then connected by springs and dampers. The main components of a bogie are the frame, bolster, spring, and other shock damping equipment; the power drive device; and three types of wheels, the travelling wheels, steering wheels, and stabilizing wheels. To improve the accuracy of the model, springs and other shock damping equipment are simulated via the force element in SIMPACK [33].

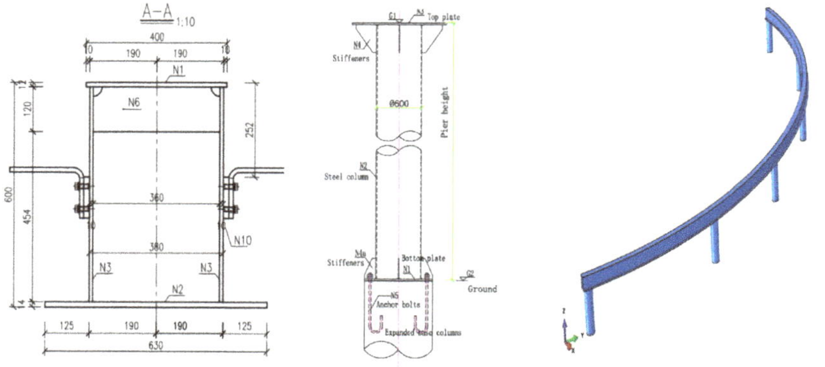

(a) (b) (c)

Figure 1. Representative design diagrams and numerical model of monorail substructures. (**a**) Cross-section of the track beam; (**b**) design diagram of the pier; and (**c**) finite element model.

The modelling of the monorail train adopts the substructure technique in SIMPACK. Specifically, a bogie is firstly constructed and transferred as a substructure; then, two identical substructures (bogie) can be used to model one carriage, along with the car body. Similarly, the carriage can be transformed into a substructure and be used to model the entire train. The main modelling parameters were provided by the Zhuzhou CRRC Special Equipment Technology Co., Ltd. The developed MBD models of a 4-carriage monorail system are shown in Figure 2.

(a) (b)

Figure 2. Onsite view of the Dajueshan MTTS project in Jiangxi province, China. (**a**) A monorail train running on the track beam of steel bridges. (**b**) MBD model of a typical monorail train; each carriage has two identical bogies surrounding the track beam in practise.

2.1.3. Coupling Model

During the actual operation of an MTTS, the track beam will be disturbed by the pulsation wind component in the crosswind, which will generate a lateral rocking motion, thereby reducing the ride comfort and running stability of monorail trains. The static wind force experienced by an MTTS in a natural wind field can usually be decomposed into three parts, namely aerodynamic drag (FD), aerodynamic lift (FL), and the aerodynamic moment (FM). These three coefficients of the MTTS were determined by numerical simulations via the FLUENT [34] software with CFD grid technology. That is, one boundary side (e.g., left) was set as the velocity inlet and the other side was the pressure outlet; symmetric boundary conditions were utilized for the upper and lower boundaries. The track beam and the vehicle surface were defined as smooth walls [35,36]. Adaptive mesh size was used to enhance calculation efficiency. The SST $k-\omega$ turbulence model was used; the SIMPLE algorithm was employed to solve the model, and the calculation step is 10^{-4} s. Results for subsequent analysis are shown in Table 1. In this work, the effect of fluctuating winds was simulated by using the Kamal spectrum, which can consider the height factor in a given wind field. For simplification, the harmonic synthesis method was used for wind field simulation.

Table 1. Aerodynamic coefficients of vehicles and bridges.

Type	FD	FL	FM
Vehicle	1.53	0.68	−0.03
Bridge	1.12	−0.15	−0.04

In addition, previous studies have reported the results of track irregularity and related dynamic responses of MTTSs through field measurement and numerical simulations, and the spectrum of longitudinal and horizontal level irregularity was estimated [29,30]. According to the mathematics of the spectrum, observing the power spectral density versus frequency curve, the higher the spectrum line, the worse the track irregularity condition and vice versa. Based on the measured result in several projects, an evaluation system for the spectrum of track irregularity has been proposed. This system contains spectra of six levels to meet various actual conditions of the track lines in practical projects, as shown Figure 3. In this work, level C was chosen because it matches the actual states of track lines in most of the operating projects. In addition, since a combination with wind loadings was used, taking level C is more reasonable and practical.

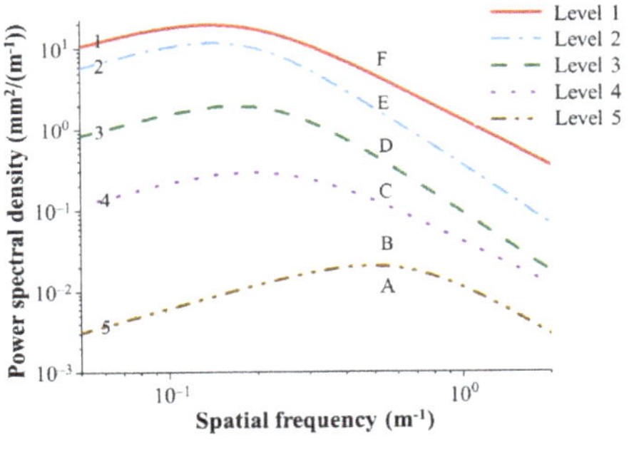

Figure 3. Spectrum curves of different level track irregularities. Note that original data were collected from field measurements. Fourier transform and curve fitting was used to obtain the spectrum curve; the higher the spectral line, the worse the irregularity condition and hence the red line (level F) indicates the worst status of a track beam condition.

In the track beam model, the wheel–rail forces are transferred, and the interaction between the vehicle and the track is simulated via substructure analysis, fixed master nodes, an interpolation algorithm, elastic deformation, and dynamic interaction. Similar modelling methods have been validated in existing studies [5,28,30,31]. The simulation results of the Kamal spectrum are obtained via the harmonic synthesis method, and the wind load time history is directly applied to the car body and track beam via force elements. After defining the spectral formula, the track irregularity can be retrieved in the form of excitation type in SIMPACK.

2.2. Model Validation

2.2.1. Test Site Information

To validate the effectiveness of the developed coupling model, an on-site dynamic response test was conducted in the MTTS test line located in Zhuzhou City, China. The construction standard of the test line is the same as that of the actual MTTS project. In other words, the configuration of the trains and track girders is consistent with the actual project. During the test of track beams, the focus was placed on the vertical displacement and lateral and vertical acceleration of the mid-span cross-section. The location of measuring points was determined according to the most unfavourable locations obtained from finite element analysis, with consideration of the installation space of equipment. The measuring points of vehicle body acceleration are arranged at the seat bottoms of the middle compartment and the rear car. Equipment such as linear variable differential transform and vibration pickups were used to capture the dynamic signals. General configurations of the on-site test are shown in Figure 4.

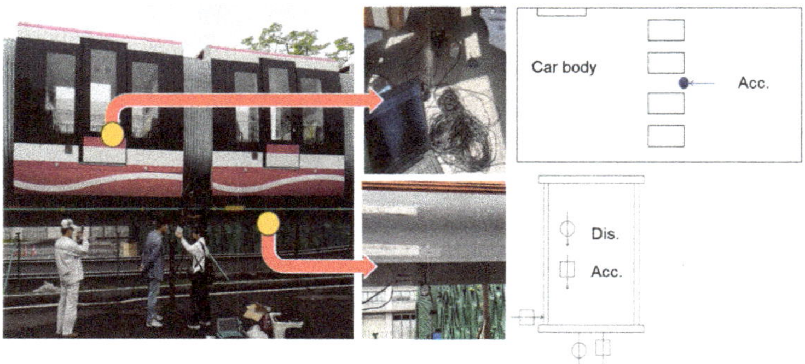

Figure 4. Layout diagram of the measuring points. For the car body, vehicle accelerations (Acc.) were tested; for the track beam, vertical displacement (Dis.) and vertical and lateral Acc. were tested.

2.2.2. Validation

Before further simulation analysis, the vertical displacement, lateral and vertical acceleration, and lateral and vertical acceleration parameters of the car body were predefined in SIMPACK so that the corresponding postprocessing results could be directly viewed during simulation. The analysis results of the co-simulation model are obtained by observing the dynamic response at the corresponding positions, such as the acceleration and deflection of the lateral vibration of the track beam, as shown in Figure 5. The validation was mainly conducted at the mid-span of track beams. This is because the positions are in the typical and representative area where the most unfavourable loadings would occur.

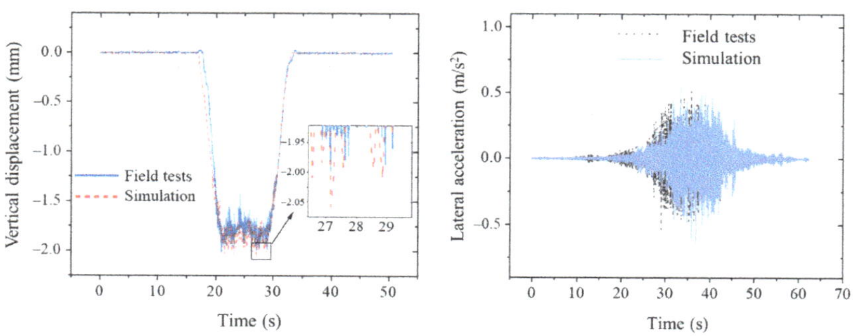

(**a**) (**b**)

Figure 5. Comparison of tested and simulation results at the mid-span of a track beam. (**a**) Vertical displacement and (**b**) lateral acceleration.

An analysis of Figure 5 reveals that the simulation results and test results are in good agreement. The simulated vertical displacement of the monorail beam mid-span was slightly larger than the actual measured value, which was mainly because the model did not consider the influence of some stiffeners. The analysis results prove that the co-simulation model constructed in this study via ABAQUS and SIMPACK has high accuracy and can be applied to subsequent analysis.

3. Results and Analysis

3.1. Influencing Factors

In the dynamic analysis of vehicle–bridge coupling, with reference to relevant codes, various indexes that represent the dynamic response of the MTTS were selected. The evaluation indexes of the dynamic response of track beams are (i) horizontal and vertical deflection at the beam mid-span; (ii) acceleration at the beam mid-span; and (iii) lateral displacement at the pier top. The evaluation indexes of the dynamic response of monorail trains are (i) the wheel unloading rate; (ii) acceleration of the car body; (iii) riding quality; and (iv) the index of tyre stress. It should be noted that the last index is only applicable in monorail systems utilizing rubber tyres.

Main factors influencing the dynamic response of the MTTS are the speed of the tourist vehicles, vehicle loadings, the pier height, and the lateral stiffness of the monorail bridge. To lay a solid basis for subsequent analyses, the effects of these four influencing factors are firstly analyzed in the following sections.

3.1.1. Effects of Travelling Speed

At present, the MTTS generally serves as an amusement facility, and the maximum speed is 40 km/h under current specifications. In this work, the maximum speed used was 50 km/h, about 1.2 times the maximum operating speed, which is recommended by the GB/T 51234-2017 [8]. A total of 10 intervals of vehicle speed were set as follows: 5, 10, . . . , 50 km/h, so that a comprehensive analysis on the dynamic response can be performed. The model described in Section 2 is then used, with consideration of train loading (maximum capacity), wind loading, and track irregularities. Dynamic responses of the track beam and monorail train are shown in Figure 6.

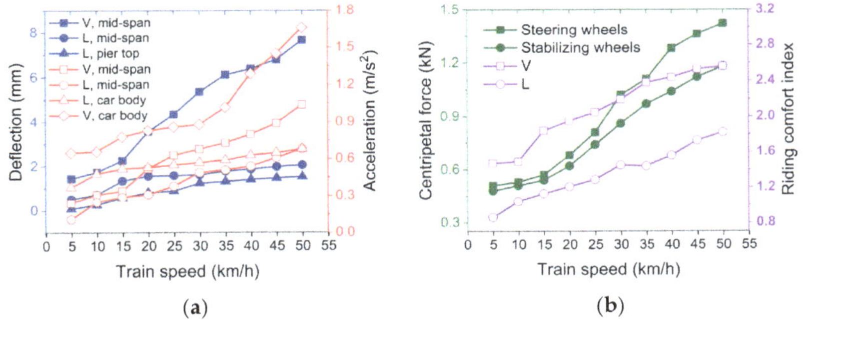

Figure 6. Response results corresponding to different vehicle speeds. (**a**) Displacement and acceleration response and (**b**) centripetal force and the calculated riding comfort index. Note V and L denote vertical and lateral, respectively.

Figure 6 shows that each kind of dynamic response continuously increases with vehicle speed. Compared with the other responses, vertical deflection at the mid-span and the vertical acceleration of the car body varied significantly with the train speed, indicating that the train speed had the most significant effects on the two. In addition, a significant change in various dynamic responses can be found under a train speed from 20 to 30 km/h. Moreover, in terms of deflection and acceleration, vertical responses are more sensitive to the variation in train speed compared to lateral responses, which is also reflected in the curves of the riding comfort index in Figure 6b. It should be noted that the steering wheels and stabilizing wheels are located at the webs of the track beam during monorail operation. Hence, similar patterns of centripetal force can be observed.

3.1.2. Effects of Vehicle Loading

To study the effect of vehicle loading on the dynamic response of the MTTS, three different working conditions were used as follows: 1, no load; 2, full load (3.8 tons); and 3, overloaded (5.3 tons). The train speed commonly used in an MTTS is divided into three gears as follows: slow, 10 km/h; medium, 20 km/h; and fast, 30 km/h. The rest of the conditions are the same as those described in Section 3.1.1. The dynamic responses of the monorail beam and the vehicle are shown in Figure 7.

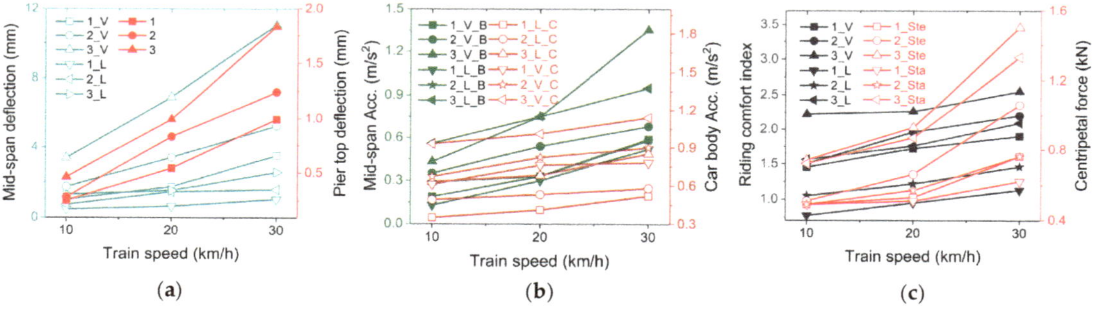

Figure 7. Results under various loading conditions and train speed. (**a**) Deflection of bridge components; (**b**) accelerations; and (**c**) centripetal force and the calculated riding comfort index. Note that 1, 2, and 3 denote the loading conditions; B and C denote the bridge and car body, respectively; and Ste and Sta denote steering and stabilizing wheels, respectively.

Figure 7 shows that various types of dynamic responses increase as the vehicle load continuously increases, which is consistent to practical experiences. Specifically, the weight of the vehicle significantly affects the lateral displacement of the pier top, with a greater impact on the vertical deflection at the mid-span than in the lateral direction. Furthermore, this weight has a more pronounced effect on the vertical acceleration response and vertical stability index compared to the lateral direction.

3.1.3. Effect of Pier Height

In this section, six pier heights of the monorail bridge, i.e., 2 m, 4 m, 6 m, 8 m, 10 m, and 15 m, were used to analyze the dynamic response of the MTTS. The designed maximum train speed (40 km/h) was used in simulations. The rest of the conditions are the same as those described in Section 3.1.1. The dynamic responses of the monorail beam and vehicles are shown in Figure 8.

As shown in Figure 8, similar patterns can be found in various dynamic responses versus pier height curves, indicating that the induced dynamic responses generally increase with the increasing pier height. The effect of the pier height on the variation in the lateral displacement of the pier top was more significant than that of the other responses. The variation in the lateral riding comfort index is also significant. Nevertheless, the overall lateral comfort index is less than that of the vertical index.

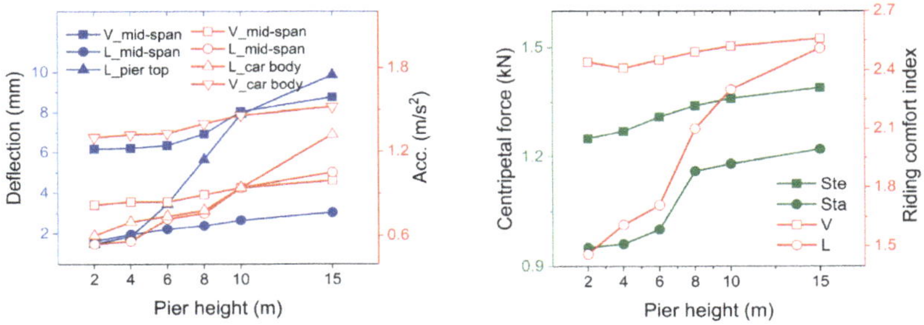

(a) (b)

Figure 8. Results under various pier heights. (**a**) Displacement and acceleration response and (**b**) centripetal force and the calculated riding comfort index.

3.1.4. Effects of Beam Stiffness

By changing the width and stiffness of the monorail beam, the lateral stiffness of the monorail beam can be altered and hence, the effects of lateral stiffness on the track beam can be further investigated. To this end, this work adopted the method of changing the transverse moment of inertia of track beams. The specific method is as follows: the lateral stiffness of the beam EI was multiplied by a reduction factor Rh, where Rh was set to values of 0.3, 0.5, 0.7, 1.0, 1.2, 1.5, and 1.8 to ensure representativeness and coverage of the lateral stiffness of practical track beams. The remaining conditions are as described in Section 3.1.1. Results are shown in Figure 9.

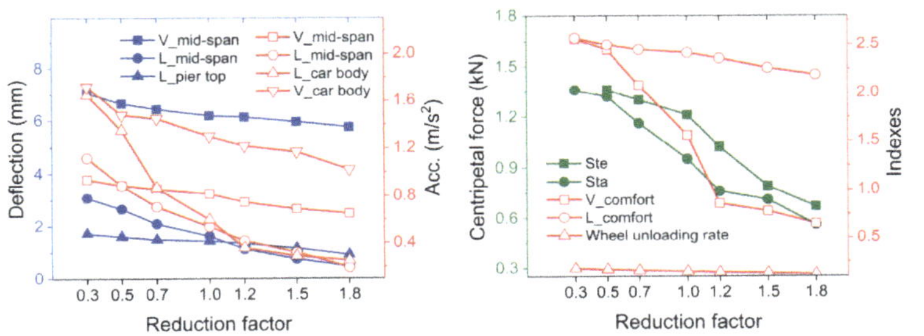

(a) (b)

Figure 9. Response results corresponding to different reduction factors. (**a**) Displacement and acceleration responses and (**b**) centripetal force, riding comfort index, and wheel unloading rate.

As shown in Figure 9, all kinds of dynamic responses decrease with increasing beam lateral stiffness. The variations in the mid-span lateral deflection and the vehicle body lateral acceleration with the girder lateral stiffness were more significant than the other responses were, and the two kinds of wheel vertical forces and stationarity indicators also varied significantly with the girder lateral stiffness.

3.2. Sensitivity Analysis

3.2.1. Variance-Based Methods

In the variance-based sensitivity analysis, the ratio of the parameter influencing factors to the output variance was calculated to assess the sensitivity of the parameter to the output. In this work, the Sobol index analysis method combined with the Saltelli method [37] was used to perform sensitivity analysis. This method uses quasi-random number sequences to construct two random number matrices A and B, then constructs matrix C_i on the basis of A and B and inputs the parameters of matrices A and B and each row of C_i into the model calculation to obtain three matrices. Output vectors are u_A, u_B, *and* u_C, from which the first-order sensitivity coefficient S_i and the overall sensitivity coefficient S_{Ti} of the i-th parameter can be calculated. The method is expressed as follows:

$$S_i = \frac{\frac{1}{N}\sum_{j=1}^{N} u_B(j)\left(u_{C_i}(j) - u_A(j)\right)}{E\left((u_A)^2\right) - (E(u_A))^2} \tag{1}$$

$$S_{T_i} = \frac{\frac{1}{2N}\sum_{j=1}^{N} \left(u_A(j) - u_{C_i}(j)\right)^2}{E\left((u_A)^2\right) - (E(u_A))^2} \tag{2}$$

3.2.2. Sensitivity Analysis of the Lateral Response

In the sensitivity analysis of the horizontal response, the four influencing factors mentioned above were set as sensitivity input parameters. A program was compiled based on the Saltelli method. The sensitivity indicators of each parameter were obtained after integrating the determined parameters, as shown in Figure 10.

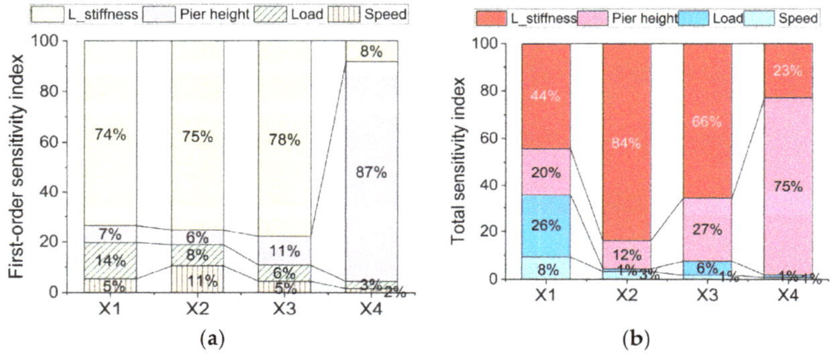

Figure 10. Results of sensitivity indexes. (**a**) First-order level and (**b**) total level. Note that X1 represents the vehicle lateral acceleration, X2 represents the lateral displacement at the mid-span, X3 represents the lateral acceleration at the mid-span, and X4 represents the lateral displacement at the pier top.

As shown in Figure 10, the height of the monorail girder pier and the lateral stiffness of the beam have the greatest impacts on the lateral response of the overall structure. Among them, the pier height significantly contributes to the lateral displacement of the pier top, accounting for 87% of the first-order sensitivity index and 75% of the total sensitivity index. The beam lateral stiffness contributes significantly to the car body lateral acceleration and monorail beam mid-span. The maximum lateral acceleration and the maximum lateral displacement at the mid-span significantly contribute to 74%, 75%, and 78% of the first-order sensitivity indicators and account for 44%, 84%, and 66% of the total sensitivity indicators, respectively.

3.3. Determination of Lateral Stiffness Index

For the evaluation index of bridge lateral stiffness, the width–span ratio, lateral deflection–span ratio, natural frequency, and lateral amplitude are usually chosen. Compared with other indicators, the lateral deflection-to-span ratio can fully reflect the interaction between the vehicle and the bridge system. It is also closely related to the magnitude and distribution pattern of the horizontal force acting on the bridge. Therefore, it can effectively reflect the interaction between the vehicle and the bridge system, considering the magnitude, distribution pattern, and phase distribution of the lateral force. In fact, the deflection–span ratio has also been used in bridges that are sensitive to lateral stiffness, such as long cable-stayed bridges [38,39]. Accordingly, the lateral deflection–span ratio was used in the present study to evaluate the lateral stiffness of bridges. As an important structure in travel and transportation systems, the lateral deflection at the top was chosen as the evaluation scale for the lateral stiffness of the steel pipe columns. This was also determined by the expected most unfavourable serving condition, as presented in studies on the dynamic responses of bridge piers [40].

To determine the limit value of these two parameters, a comprehensive parametric study was conducted to obtain their relationship with the dynamic performance of the MTTS. In this work, three indicators of dynamic performance, including the mid-span lateral acceleration of track beams, the Sperling index (riding comfort [41]) of train operation, and the lateral wheel force of trains, were investigated because of their deterministic nature. With reference to related specifications and the consideration of structural configurations of the MTTS, the evaluation limit of the mid-span lateral acceleration was set to 0.98 m/s^2 (0.1 g); the excellent level of the Sperling index, i.e., 2.5, was utilized to ensure the riding comfort of monorail trains. In addition, the evaluation limit value of the lateral wheel force was set to 1.37 kN according to the technical conditions of the polyurethane wheel tyre provided by the manufacturer.

3.3.1. Lateral Acceleration Perspective

Based on the proposed model, the maximum lateral deflection of the track beam (in the mid-span) and the pier (at the top), as well as the indicator of dynamic performance, lateral acceleration, can be calculated. In the subsequent analysis, a 10 m long track beam was used, and wind loads and track irregularity of level C were employed. To obtain the required limit values, the pier height was changed in each individual calculation to obtain dynamic responses of different levels, where the higher the pier, the stronger the response.

Figure 11a,b illustrate the relationship between the lateral deflection and the lateral acceleration at the pier top and mid-span, respectively. As it shows, the lateral response of different structural components in the MTTS has a similar variation pattern under the change in pier height. Curve fitting was conducted for the data in Figure 11; the determination coefficient R^2 was used to quantify the fitting quality. In this case, the coefficients of the pier top and the mid-span are 0.96 and 0.98, respectively, indicating excellent fitting performance. Point interpolation was then performed.

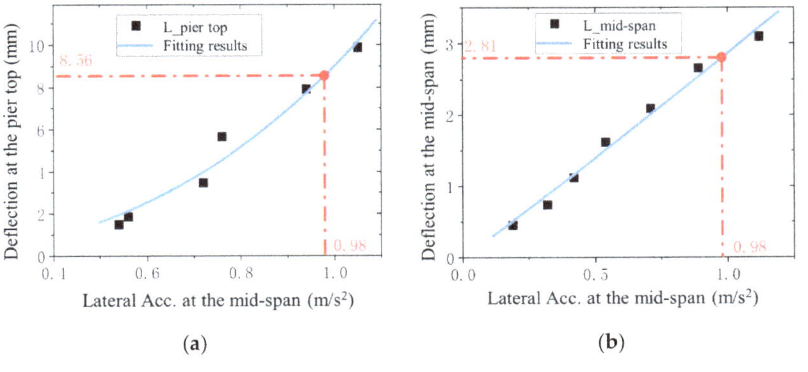

Figure 11. Fitting results of the maximum lateral deflection versus lateral acceleration of the track beam. (**a**) Lateral deflection at the pier top and (**b**) lateral deflection at the mid-span.

Based on the predefined threshold value of 0.98 m/s^2, the limit value of lateral deflection at the pier top and mid-span can be determined as 8.56 mm and 2.81 mm, respectively. For 10 m long track beams, the deflection–span ratio is 2.81 mm over 10 m (1/3559); for convenient design, the limit value of the lateral deflection–span ratio is approximately taken as L/3600.

3.3.2. Lateral Riding Comfort Perspective

Similarly, this section will investigate the lateral stiffness of MTTS components in terms of the riding comfort of monorail train operation. By following the approach in Section 3.3.1, the relationship between the riding comfort index and the maximum lateral deflection, as well as the fitting result, are plotted in Figure 12. As it shows, the lateral deflection increases a little bit faster than the Sperling index. The determination coefficient of the pier top and the mid-span are 0.96 and 0.97, respectively.

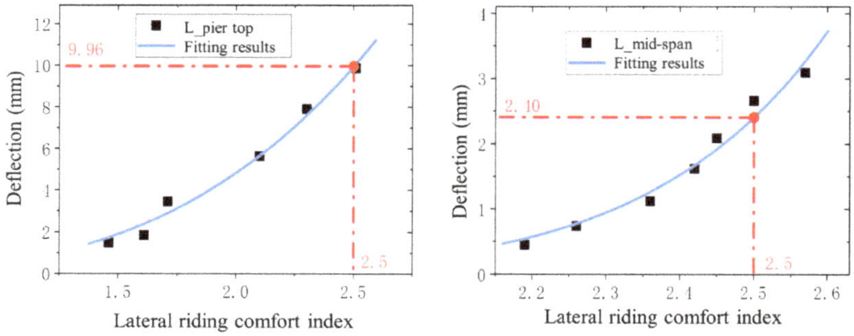

(**a**) (**b**)

Figure 12. Fitting results of the maximum lateral deflection versus lateral riding comfort index. (**a**) Lateral deflection at the pier top and (**b**) lateral deflection at the mid-span.

Based on the predefined threshold value of 2.5, the limit value of lateral deflection at the pier top and mid-span can be determined as 9.96 mm and 2.40 mm, respectively. In this case, the deflection–span ratio is about 1/4167; for convenient design, the limit value is approximately taken as L/4200.

3.3.3. Lateral (Centripetal) Wheel Force Perspective

This section will investigate the lateral stiffness limit by considering the lateral force of rubber wheels. The relationship between the centripetal force and the maximum lateral deflection, as well as the fitting result, are plotted in Figure 13. A similar pattern is observed in Figure 13, where the centripetal force rises with the increase in pier height. The determination coefficient of the pier top and the mid-span are 0.98 and 0.96, respectively.

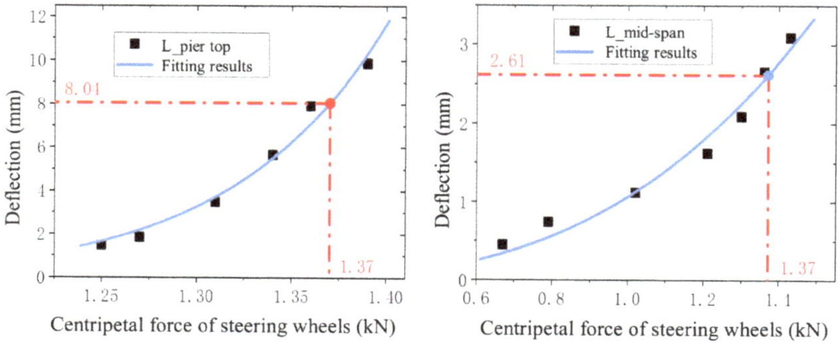

(**a**) (**b**)

Figure 13. Fitting results of the maximum lateral deflection versus centripetal force of lateral wheels. (**a**) Lateral deflection at the pier top and (**b**) lateral deflection at the mid-span.

Based on the predefined threshold value of 1.37 kN, the limit value of lateral deflection at the pier top and mid-span can be determined as 8.04 mm and 2.64 mm, respectively. In this case, the deflection–span ratio is about 1/3831; for convenient design, the limit value is approximately taken as L/3900.

3.3.4. Limits of the Lateral Displacement and Lateral Deflection–Span Ratio of the Pier Top

In the foregoing text, the interpolation method was used to select the values of the fitting function, and the corresponding data are summarized in Table 2.

Table 2. Summary of the lateral deflection and deflection–span ratio limits at the pier top.

Judging Criteria	Lateral Acc.	Lateral Riding Comfort	Wheel Lateral Force Limit
Lateral deflection at the pier top (mm)	8.56	9.96	8.04
Lateral deflection–span ratio	L/3600	L/4200	L/3900

Based on the data in Table 2, and for the sake of a relatively conservative design, the minimum value of 8.04 mm in the table is taken as the lateral displacement of the pier top, and the minimum value of L/4200 in the table is taken as the lateral deflection–span ratio limit of the MTTS.

4. Discussion

To investigate the difference between the lateral stiffness limit of the MTTS and other specifications, the provisions in similar specifications are compared in Table 3. An analysis of Table 3 shows that among the lateral stiffness limits proposed in this paper, the lateral deflection–span ratio is close to the "Code for the Design of Urban Rail Transit Bridges (GB/T 51234-2017) [8]" and lower than the code of "Design Standards for Straddle-Type Monorail Transit (GB 50458-2022) [9]". The lateral displacement of the pier top was much lower than those of the other two specifications, which reflects the structural characteristics of the MTTS, which are different from those of rail transportation; therefore, special attention should be given to the design.

Table 3. Comparison of the lateral stiffness limits with specifications (non-mandatory for MTTS).

Specifications	GB 50458-2022 [9]	GB/T 51234-2017 [8]	MTTS
Lateral deflection at the pier Top (mm)	30	25	8.04
Lateral deflection–span ratio	L/800	L/4000	L/4200

Limitations and Future Work

1. The simulation model established in this paper has room for improvement. For example, a more detailed measurement of the in situ wind field can be carried out to consider the wind-induced effects better. This may help realize the coupled simulation of extreme load effects and dynamic effects of large-span track beams. In addition, more effects of track irregularity can be further included, which can be achieved by employing spectra of various levels and considering the irregularity-induced effect under complex loading conditions, such as ground motion. Of course, a more complicated model generally means a more time-consuming solving process. Moreover, more practical design preferences should be integrated to draw more meaningful conclusions and to promote potential applications.
2. This work focuses on the lateral response of an MTTS. However, supplying the vertical response of an MTTS may be useful for effectively capturing the dynamic performance of the MTTS. In addition, a more targeted evaluation system and standard limit values can be proposed. The selection of the lateral stiffness index and dynamic performance

indicator mainly relies on referring to existing specifications. The possibility of using a more precise indicator for the MTTS should be further studied.

5. Conclusions

The main conclusions of this paper are as follows:

(i) A wind–train–bridge coupling model validated by field measurements was developed to investigate the effects of multiple factors on the dynamic response of an MTTS. During the changes in vehicle speed and vehicle weight, the vertical response increase in the MTTS is greater than the horizontal response increase. Changes in pier height have a significant impact on the lateral response of monorail beams. Moreover, it is concluded that the variation in lateral stiffness can distinctly affect the dynamic responses.

(ii) A variance-based sensitivity analysis was performed on various influencing factors. Results revealed that the pier height and the lateral stiffness significantly affected the lateral response of the overall structure. Among them, the pier height significantly contributed to the lateral displacement of the pier top, accounting for 87% of the first-order sensitivity index and 75% of the total sensitivity index. The lateral acceleration and the maximum lateral displacement at the mid-span significantly contributed to 74%, 75%, and 78% of the first-order sensitivity indexes and accounted for 44%, 84%, and 66% of the total sensitivity indexes, respectively.

(iii) Based on the obtained results, the lateral responses of the MTTS were evaluated in terms of index of beam acceleration, riding comfort of the monorail train, and wheel force on the bogie. Finally, two lateral limited values, i.e., a lateral displacement limit of 8.04 mm for the pier top and a lateral deflection–span ratio limit of L/4200 were determined.

Author Contributions: Methodology, H.Z., P.W. and S.W.; software and formal analysis, P.W. and S.W.; data curation, P.W., J.J. and S.W.; writing—original draft preparation, H.Z. and S.W.; writing—review and editing, P.W. and F.G.; supervision and project administration, F.G.; engineering application, Q.L., J.J., C.F. and Q.D. All authors have read and agreed to the published version of the manuscript.

Funding: This research was funded by the National Key Research and Development Program during the 14th Five-Year Plan period of China, grant number 2022YFC3004304.

Data Availability Statement: The data presented in this study may be available upon reasonable request.

Conflicts of Interest: Authors Qin Li, Cheng Feng, and Qun Deng were employed by the 3rd Construction Co., Ltd. of the China Construction 5th Engineering Bureau. Author Junhui Jin was employed by Zhuzhou CRRC Special Equipment Technology Co., Ltd. The remaining authors declare that the research was conducted in the absence of any commercial or financial relationships that could be construed as potential conflicts of interest.

References

1. Guo, F.; Chen, K.; Gu, F.; Wang, H.; Wen, T. Reviews on current situation and development of straddle-type monorail tour transit system in China. *J. Cent. South Univ. Sci. Technol.* **2021**, *52*, 4540–4551.
2. Lee, J.; Kim, H.; Lee, K.; Kang, Y.J. Effect of load combinations on distortional behaviors of simple-span steel box girder bridges. *Metals* **2021**, *11*, 1238. [CrossRef]
3. Cheng, J.; Yao, H. Simplified method for predicting the deflections of composite box girders. *Eng. Struct.* **2016**, *128*, 256–264. [CrossRef]
4. Zhou, M.; Liu, Z.; Zhang, J.; An, L.; He, Z. Equivalent computational models and deflection calculation methods of box girders with corrugated steel webs. *Eng. Struct.* **2016**, *127*, 615–634. [CrossRef]
5. Shamsi, M.; Ghanbari, A. Nonlinear dynamic analysis of Qom monorail bridge considering soil-pile-bridge-train interaction. *Transp. Geotech.* **2020**, *22*, 100309. [CrossRef]
6. Nettis, A.; Nettis, A.; Ruggieri, S.; Uva, G. Corrosion-induced fragility of existing prestressed concrete girder bridges under traffic loads. *Eng. Struct.* **2024**, *314*, 118302. [CrossRef]
7. *GB 8408-2018*; Large-Scale Amusement Device Safety Code. State Administration of Work Supervision: Beijing, China, 2018.
8. *GB/T 51234-2017*; Code for Design of Urban Rail Transit Bridge. Ministry of Housing and Urban-Rural Development of the People's Republic of China: Beijing, China, 2017.

9. *GB 50458-2022*; Standard for Design of Straddle Monorail Transit. Ministry of Housing and Urban-Rural Development of the People's Republic of China: Beijing, China, 2022.
10. Gao, Q.; Cui, K.; Li, Z.; Li, Y. Numerical investigation of the dynamic performance and riding comfort of a straddle-type monorail subjected to moving trains. *Appl. Sci.* **2020**, *10*, 5258. [CrossRef]
11. Liu, M. Study on Coupling Vibration of Windmill Bridge for T-beam Continuous Rigid Frame Bridge in High Altitude Area. Master's Thesis, Chongqing Jiaotong University, Chongqing, China, 2019.
12. Miao, Y. Studies on Coupled Vibration of Rail-cum-Road Cable-stayed Bridge with a Super Span under Wind Loads. Master's Thesis, Central South University, Changsha, China, 2014.
13. Huang, G.; Ren, N.; Zheng, H. Structural dynamics of bridges under the coupling effect of windmills and bridges. *Eur. J. Comput. Mech.* **2023**, *31*, 601–620. [CrossRef]
14. Chen, J.; Li, R.; Xu, J.; Li, P. Study on the aerodynamic performances and the operational safety of the vehicle-bridge system under the crosswind. *Mech. Based Des. Struct. Mach.* **2024**, *2*, 1–23. [CrossRef]
15. Wang, S.; Wan, X.; Guo, M.; Giao, H.; Zhang, N.; Ye, Q. Nonlinear dynamic analysis of the wind–train–bridge system of a long-span railway suspension truss bridge. *Buildings* **2023**, *13*, 277. [CrossRef]
16. Zou, Y.; Liu, Z.; Shi, K.; He, X. Analysis of effects of aerodynamic interference on dynamic response of suspended monorail wind–vehicle–bridge system using joint simulation approach. *Structures* **2022**, *45*, 179–198. [CrossRef]
17. Luo, S. The Research on the Aerodynamic Characteristics of Road Vehicles on Bridge Deck and Its Driving Safety. Master's Thesis, Hunan University of Science and Technology, Xiangtan, China, 2023.
18. Jiang, Y. Study on Coupled Vibration and Lateral Stiffness of Steel Girder Cable-Stayed Long-Span Road-Rail Bridge. Master's Thesis, Chongqing Jiaotong University, Chongqing, China, 2023.
19. Guo, X.; Zou, X. Influence of wind guide railing parameters on wind-vehicle-bridge coupling vibration of steel truss bridge. *J. Railw. Sci. Eng.* **2024**, *21*, 1068–1078.
20. Sugimura, H.; Takeda, M.; Takei, M.; Yamaoka, H.; Ogata, T. Development of HEV engine start-shock prediction technique combining motor generator system control and multi-body dynamics (MBD) models. *SAE Int. J. Passeng. Cars-Mech. Syst.* **2013**, *6*, 1363–1370. [CrossRef]
21. Xiang, H.; Tao, Y.; Chen, X.; Hu, H.; Li, Y. Study on Transverse Deflection-span Ratio Limit of Long-span Railway Suspension Bridge Considering Wind Barrier. *J. China Railw. Soc.* **2023**, *45*, 62–69.
22. Zhou, C.; Xu, X.; Zheng, X.; Chen, X. Study on the Influence of Pier Stiffness on Vehicle-bridge Coupled Vibration of Long-span Continuous Girder Bridge of Railway. *High Speed Railw. Technol.* **2022**, *13*, 23–29.
23. Zhu, Z.; Jiang, Z. Analysis of dynamic characteristics of large-span suspension bridges with extended spans of stiffening girders. *J. Railw. Sci. Eng.* **2022**, *19*, 1014–1023.
24. Liu, L.; Jiang, L.; Liu, X.; Zhou, W.; Nie, L.; Shao, G. Research on analytical model of transverse deformation and rail surface deformation of high-speed railway bridge based on Ritz method. *J. Cent. South Univ. Nat. Sci. Ed.* **2021**, *52*, 4349–4360.
25. Wang, M.; Song, Z.; Teng, N. Research on Transverse Stiffness of High Speed Maglev Guideway Girder Based on Coupled Vibration. *Railw. Eng.* **2021**, *61*, 1–5.
26. Yan, Y.; Zhan, J.; Zhang, N.; Shen, K. Influence of Pier Transverse Displacement and Foundation Stiffness Change on High Speed Running Safety. *Railw. Eng.* **2020**, *60*, 8–12.
27. Carden, L.P.; Buckle, I.G.; Itani, A.M. Transverse displacement capacity and stiffness of steel plate girder bridge superstructures for seismic loads. *J. Constr. Steel Res.* **2007**, *63*, 1546–1559. [CrossRef]
28. Naeimi, M.; Tatari, M.; Esmaeilzadeh, A.; Mehrali, M. Dynamic interaction of the monorail–bridge system using a combined finite element multibody-based model. *Proc. Inst. Mech. Eng. Part K J. Multi-Body Dyn.* **2015**, *229*, 132–151. [CrossRef]
29. Guo, F.; Wang, P. Measurement and analysis of the longitudinal level irregularity of the track beam in monorail tour-transit systems. *Sci. Rep.* **2022**, *12*, 19219. [CrossRef] [PubMed]
30. Wang, P.; Guo, F.; Zhang, H.; Jin, J.; Liao, Q.; Yan, Y. Acquiring the High-Precision Spectrum of Track Irregularity by Integrating Inclination in Chord Methods: Mathematics, Simulation, and a Case Study. *Mathematics* **2024**, *12*, 2227–7390. [CrossRef]
31. Jiang, Y.; Wu, P.; Zeng, J.; Qu, S.; Wang, X.; Wang, S. Simplified and relatively precise back-calculation method for the pavement excitation of the monorail. *Int. J. Pavement Eng.* **2021**, *22*, 480–497. [CrossRef]
32. ABAQUS. *ABAQUS User's Manual, Version 2016*; Dassault Systemes: Velizy-Villacoublay, France, 2016. Available online: http://62.108.178.35:2080/v2016/index.html (accessed on 20 September 2024).
33. *SIMPACK, Version 2021x*; Dassault Systemes: Velizy-Villacoublay, France, 2021. Available online: https://blog.3ds.com/brands/simulia/simpack-r2021x-multibody-system-simulation-updates/ (accessed on 20 September 2024).
34. ANSYS. *ANSYS Fluent 17.0 Documentation*; ANSYS Inc.: Lebanon, NH, USA, 2016.
35. Bao, Y.L.; Xiang, H.Y.; Li, Y.L.; Yu, C.J.; Wang, Y.C. Study of wind–vehicle–bridge system of suspended monorail during the meeting of two trains. *Adv. Struct. Eng.* **2019**, *22*, 1988–1997. [CrossRef]
36. Zou, Y.; Liu, Z.; Shi, K.; Ou, S.; He, X.; Deng, H.; Zhou, S. Experimental study of aerodynamic interference effects for a suspended monorail vehicle–bridge system using a wireless acquisition system. *Sensors* **2021**, *21*, 5841. [CrossRef]
37. Saltelli, A.; Tarantola, S.; Chan, K. A quantitative model-independent method for global sensitivity analysis of model output. *Technometrics* **1999**, *41*, 39–56. [CrossRef]

38. Qin, S.; Han, S.; Li, S. In-situ testing and finite element model updating of a long-span cable-stayed bridge with ballastless track. *Structures* **2022**, *45*, 1412–1423. [CrossRef]
39. Moore, J.C.; Glenncross-Grant, R.; Mahini, S.S.; Patterson, R. Regional timber bridge girder reliability: Structural health monitoring and reliability strategies. *Adv. Struct. Eng.* **2012**, *15*, 793–806. [CrossRef]
40. Tubaldi, E.; Tassotti, L.; Dall'Asta, A.; Dezi, L. Seismic response analysis of slender bridge piers. *Earthq. Eng. Struct. Dyn.* **2014**, *43*, 1503–1519. [CrossRef]
41. *GB/T 5599-2019*; Specification for Dynamic Performance Assessment and Testing Verification of Rolling Stock. State Administration for Market Regulation; Standardization Administration of PRC: Beijing, China, 2020.

buildings

MDPI

Article

Stability Analysis of Curved Beams Based on First-Order Shear Deformation Theory and Moving Least-Squares Approximation

Yuxiao Li [1], Yajing Liao [1], Zhen Xie [1] and Linxin Peng [1,2,3,4,*]

1 College of Civil Engineering and Architecture, Guangxi University, Nanning 530000, China
2 State Key Laboratory of Featured Metal Materials and Life-Cycle Safety for Composite Structures, Guangxi University, Nanning 530000, China
3 Key Laboratory of Disaster Prevention and Structural Safety of Ministry of Education, Guangxi University, Nanning 530000, China
4 Guangxi Key Laboratory of Disaster Prevention and Structural Safety, Guangxi University, Nanning 530000, China
* Correspondence: penglx@gxu.edu.cn

Abstract: Based on the first-order shear deformation theory (FSDT) and moving least-squares approximation (MLS), a new meshfree method that considers the effects of geometric nonlinearity and the pre- and post-buckling behaviors of curved beams is proposed. An incremental equilibrium equation is established with the Updated Lagrangian (UL) formulation under the von Karman deflection theory. The proposed method is applied to several numerical examples, and the results are compared with those from previous studies to demonstrate its convergence and accuracy. The pre- and post-buckling behaviors of the curved beam with different parameters, such as vector span ratios, bending forms, inclusion angles, boundary conditions, slenderness ratios, and axial shear stiffness ratios, are also investigated. The effects of the parameters on the buckling response are demonstrated. The proposed method can be extended to the study of double nonlinearities of curved beams in the future. This extension will provide a more scientific reference basis for the structural selection of curved girder structures in practical engineering.

Keywords: mesh-free method; curved beam; nonlinear buckling; first-order shear deformation theory; updated Lagrangian method

1. Introduction

Curved beams have been widely applied in civil engineering, aerospace engineering, and machinery engineering, and they have attracted much research interest from scholars who have aimed to explore the intrinsic mechanical properties and behaviors of these structures [1–3]. However, the presence of initial curvature in curved beams brings complexity to the study and makes the beams more susceptible to bending and torsional instabilities compared to straight beams. Efforts have been made to overcome these difficulties. In the 19th century, Saint-Venant [4] pioneered the study of torsional problems in curved bars with circular cross sections. In the 1950s and 1960s, Timoshenko and Gere [5] carried out an in-depth study of curved members and analyzed the in-plane and out-of-plane stability of arches with narrow rectangular cross sections subjected to uniform radial loads and moments at both ends. Власов [6] modified the expression of internal forces of straight beams based on the effect of curvature and established basic equations for stability analysis of curved beams. Subsequently, many scholars further investigated the stability of curved beams based on the work of Власов [7–10]. As the span continued to rise, large deformation issues in curved beams occurred [11–15]. Therefore, it became necessary to include nonlinear factors in analyses. Currently, the commonly used methods for solving nonlinear stability problems of curved beams fall into two categories, i.e., analytic/semi-analytic methods and numerical ones.

The analytical solution facilitates the discussion on the stability of curved beams and provides more guidance for the design of curved beams. Conway and Lo [16,17] investigated the elastic buckling behaviors of a one-dimensional curved beam subjected to both bending moments and concentrated loads by using linear, nonlinear (elastic) extensible, and nonlinear inexpandable theories, respectively. Pi et al. [18] considered the effect of nonlinear terms and derived the in-plane buckling load of flat arches under radially distributed loads. They [19,20] also analyzed the nonlinear buckling behavior of flat arch structures under torsion spring boundaries.

However, analytical methods for studying the stability of curved beams have limitations in dealing with complex loads and boundary conditions. With the advancement of computer hardware and computing techniques, numerical methods, such as the finite difference method, the finite strip method, and the finite element method (FEM), have emerged. Among the numerical methods, the FEM can accurately analyze complex structures with flexible modeling capabilities. Noor et al. [21] conducted an analysis on the geometrically nonlinear behavior of deep arches using hybrid curved beam elements, considering transverse shear deformation and tension–bending coupling. Akoussah et al. [22] investigated the shear self-locking phenomenon in curved beam elements. Elias and Chen [23] developed a nonlinear curved beam element to obtain the buckling load and load-displacement curves of flat and deep arches. Yang and Kuo [24,25] introduced a nodal geometric stiffness matrix to handle large rotational deformations and simulated curved beam structures with straight beam elements. Litewka and Rakowski [26] developed a two-node curved beam element with triangular series shape functions. Hashemian and Hosseini [27] predicted the nonlinear behavior of curved beams by combining isoparametric analysis with an arc-length expansion method.

Nevertheless, the FEM also has disadvantages. Its solutions are based on meshes, and mesh distortion is unavoidable when dealing with large deformation problems. In recent years, meshfree or meshless methods [28–38] have emerged as alternatives to FEMs, such as the smoothed particle hydrodynamics (SPH) [28] method, the element-free Galerkin (EFG) [29,30] method, and the reproducing kernel particle method (RKPM) [31], which utilize discrete points or nodes to represent the problem domain and adjust their density and distribution according to the problem complexity, which can improve computational accuracy and efficiency. Compared with finite element methods, meshless methods offer greater flexibility and adaptability. They can effectively handle complex geometries, material inhomogeneities, multiscale phenomena, large deformations, and problems related to localized damage or destruction. Many scholars have explored the application of meshless methods to solve nonlinear problems in curved beams, taking advantage of the aforementioned benefits. Specifically, Yang [32], Kou [33], Dodagoudar [34], and Chinh [35] et al. have studied functionally graded beams, cracked beams, beams on elastic foundations, and multi-unit mezzanine beams, respectively. Cheng et al. [36] carried out a geometric nonlinear analysis of elastic beams by the point interpolation meshless method. Wang et al. [37] proposed an unlocked meshless curved beam formulation based on stabilizing compliant nodal integrals. Yeh et al. [38] proposed a three-dimensional finite deformation theory, solving curved beam geometrically nonlinear problems in a strong form.

1.1. FSDT

The stability analysis of curved beams relies on a series of mature shear deformation theories, and the most widely used theories are the classical linear plate theory (CLPT), the FSDT [39], and higher-order shear deformation theory (HSDT). The CLPT neglects the influence of transverse shear deformation when the curved beam is stressed. However, the FSDT assumes that even if the normal line perpendicular to the middle surface of the curved beam maintains a straight line during the deformation process, it will also tilt due to the action of transverse shear and will no longer strictly maintain the vertical relationship with the middle surface but show a certain deflection angle. Improvements in this theory, by taking into account the transverse shear deformation, improve the accuracy of predicting the complex deformation behavior of the curved beam, especially when dealing with the complex load of the curved beam. Therefore, the FSDT is used to research the stability of curved beams in this paper.

1.2. The Novelty of This Work

It can be found that there are few meshless methods developed for the analysis of nonlinear behaviors of curved beams, which motivates this study. The objective of this paper is to propose a mesh-free method to trace the pre-buckling as well as the post-buckling nonlinear equilibrium paths for curved beams. Firstly, an equation governing the nonlinear buckling of curved beams is derived based on the FSDT, the MLS, and the von Karman's large deflection theory. Subsequently, the nonlinear equilibrium equation is solved by the arc-length method, and the results are compared with those from the literature to verify the validity of the proposed method. Finally, the influence of geometric parameters on the nonlinear behaviors of curved beams is investigated. This study provides a theoretical basis and technical guidance for the optimization design of curved beam structures.

2. Updated Lagrangian Method

The nonlinear equilibrium paths of curved beams in the pre-buckling as well as the post-buckling phases are controlled by the large deflection equation. The coordinate systems commonly used in large deformation analysis are the Lagrangian and Eulerian coordinate systems, respectively. In this paper, the Updated Lagrangian (UL) [40] method is used to establish the governing equation for the large deflection of curved beams.

All variables in the UL method use the configuration of moment t as the reference configuration, so the equilibrium configuration at moment $t + \Delta t$ is

$$\int \delta \boldsymbol{E}^{\mathrm{T}} \boldsymbol{S} \mathrm{d}V = \int \delta \boldsymbol{u}^{\mathrm{T}} \boldsymbol{p} \mathrm{d}V + \int \delta \boldsymbol{u}^{\mathrm{T}} \boldsymbol{q} \mathrm{d}A \tag{1}$$

where V and A are the volume and surface area of the structure at t, $\boldsymbol{E}$ and $\boldsymbol{S}$ are the Cauchy stress and Green strain at $t + \Delta t$, and $\boldsymbol{p}$ and $\boldsymbol{q}$ are the volumetric force and surface force at $t + \Delta t$, respectively. In UL, the strains and the displacements at moment t are considered to be zero for time steps t to $t + \Delta t$, so the stress, strain, and displacement at time $t + \Delta t$ can all be expressed as:

$$\boldsymbol{S} = \boldsymbol{\tau} + \Delta \boldsymbol{S} \tag{2}$$

$$\boldsymbol{E} = \Delta \boldsymbol{E} \tag{3}$$

$$\boldsymbol{u} = \Delta \boldsymbol{u} \tag{4}$$

where $\boldsymbol{\tau}$ is the Cauchy stress. $\Delta \boldsymbol{S}$, $\Delta \boldsymbol{E}$ and $\Delta \boldsymbol{u}$ are the stress, strain, and displacement increments, respectively, for the configuration of moment $t + \Delta t$.

According to the definition of Green strain, Equation (1) is transformed to

$$\int \boldsymbol{B}_{\mathrm{G}} \boldsymbol{S} \mathrm{d}V + \boldsymbol{K}_{\sigma} \boldsymbol{u} = \boldsymbol{R} - \boldsymbol{R}_{\sigma} \tag{5}$$

where $\boldsymbol{B}_{\mathrm{G}}$ is the displacement–strain transformation matrix composed of the linear part $\boldsymbol{B}_{\mathrm{L}}$ and the nonlinear part $\boldsymbol{B}_{\mathrm{NL}}$. $\boldsymbol{K}_{\sigma}$ and $\boldsymbol{R}_{\sigma}$ are the initial stress stiffness matrix and structural internal forces, respectively. They are both related to the Cauchy stress of the configuration of moment t. $\boldsymbol{u}$ is the displacement increment; and $\boldsymbol{R}$ is the external load vector. Expressions of these matrices and vectors will be given in Section 3.

The linearization of Equation (5) by the intrinsic and geometrical correlation gives

$$(\boldsymbol{K}_{L} + \boldsymbol{K}_{\sigma}) \boldsymbol{u} = \boldsymbol{R} - \boldsymbol{R}_{\sigma} \tag{6}$$

Equation (6) can be rewritten as

$$\boldsymbol{K}_{\mathrm{T}}\boldsymbol{u} = \boldsymbol{\Phi} \tag{7}$$

where $\boldsymbol{K}_{\mathrm{T}}$ is the tangent stiffness matrix and $\boldsymbol{\Phi}$ is the unbalance force vector.

3. MLS Approximation

According to the MLS, a function $u(\eta)$ in a domain Ω can be approximated by $u^h(\eta)$ in the sub-domain Ω_η [41], and

$$u^h(\eta) = \sum_{i=1}^{m} p_i(\eta)a_i(\eta) = \boldsymbol{p}^{\mathrm{T}}(\eta)\boldsymbol{a}(\eta) \tag{8}$$

where $p_i(\eta)$ are the monomial basis functions; $a_i(\eta)$ are the corresponding unknown coefficients; h is a factor that measures the domain of influence of the nodes; and m is the number of basis functions. According to ref. [37], the basis functions in curved coordinate system are defined as

$$\boldsymbol{p}(\eta) = \begin{bmatrix} 1 & \eta & \sin\left(\frac{\eta}{R}\right) & \cos\left(\frac{\eta}{R}\right) \end{bmatrix} \tag{9}$$

The unknown coefficients $a_i(\eta)$ can be determined by minimizing a weighted discrete L_2 norm

$$J = \sum_{I=1}^{n} \varpi(\eta - \eta_I)\left[\boldsymbol{p}(\eta_I)^{\mathrm{T}}\boldsymbol{a}(\eta) - u_I\right]^2 \tag{10}$$

where $\varpi(\eta - \eta_I)$ and $\varpi(\eta)$ are the weight functions associated with node I, n is the number of nodes in Ω_η that make the weight function $\varpi(\eta) > 0$, and u_I are the nodal parameters. The minimization of J in Equation (10) with respect to $\boldsymbol{a}(\eta)$ leads to the following set of linear equations

$$\boldsymbol{H}(\eta)\boldsymbol{a}(\eta) = \boldsymbol{B}(\eta)\boldsymbol{u} \tag{11}$$

where

$$\boldsymbol{u}^{\mathrm{T}} = \begin{bmatrix} u_1 & u_2 & u_3 & \cdots & u_n \end{bmatrix}^{\mathrm{T}} \tag{12}$$

$$\boldsymbol{H}(\eta) = \sum_{I=1}^{n} \varpi(\eta - \eta_I)\boldsymbol{p}^{\mathrm{T}}(\eta_I)\boldsymbol{p}(\eta_I) \tag{13}$$

$$\boldsymbol{B}(\eta) = [\varpi(\eta - \eta_1)\boldsymbol{p}(\eta_1), \ldots, \varpi(\eta - \eta_n)\boldsymbol{p}(\eta_n)] \tag{14}$$

The coefficients $\boldsymbol{a}(\eta)$ are derived from Equation (11)

$$\boldsymbol{a}(\eta) = \boldsymbol{H}^{-1}(\eta)\boldsymbol{B}(\eta_I)\boldsymbol{u} \tag{15}$$

ϖ is the cubic spline weight function

$$\varpi(r) = \begin{cases} \frac{2}{3} - 4r^2 + 4r^3 & r \le 1/2 \\ \frac{4}{3} - 4r + 4r^2 - \frac{4}{3}r^3 & 1/2 < r \le 1 \\ 0 & r > 1 \end{cases} \tag{16}$$

where r is the normalized distance between point η and node η_I. Circular support is used in this paper, so $r = [(x - x_I)^2 + (y - y_I)^2]^{0.5}/d_{mI}$, where x_I, y_I are the coordinates of node I and d_{mI} is the radius of the support domain.

By substituting Equation (15) into Equation (8), $u^h(\eta)$ can be expressed in a standard form as

$$u(\eta) = \sum_{I=1}^{n} N_I(\eta)u_I \tag{17}$$

where the shape function $N_I(\eta)$ is given by

$$N_I(\eta) = \boldsymbol{p}^{\mathrm{T}}(\eta)\boldsymbol{H}^{-1}(\eta)\boldsymbol{B}_I(\eta) \tag{18}$$

From Equation (14), the following can be expressed as

$$B(\eta) = p^{\mathrm{T}}(\eta_I)\varpi(\eta - \eta_I) \tag{19}$$

Thus, Equation (18) can be rewritten as

$$N_I(\eta) = p^{\mathrm{T}}(\eta)H^{-1}(\eta_I)p^{\mathrm{T}}(\eta_I)\varpi(\eta - \eta_I) \tag{20}$$

4. The Meshfree Formulation for a Curved Beam

A meshfree model of a curved beam with a rectangular cross section of length L and a radius of R is shown in Figure 1. The curved beam is discretized by a set of nodes. The degree of freedom of every node is (u, w, and θ). u is the vertical displacement, w is the horizontal displacement, and θ is the angular displacement of the node.

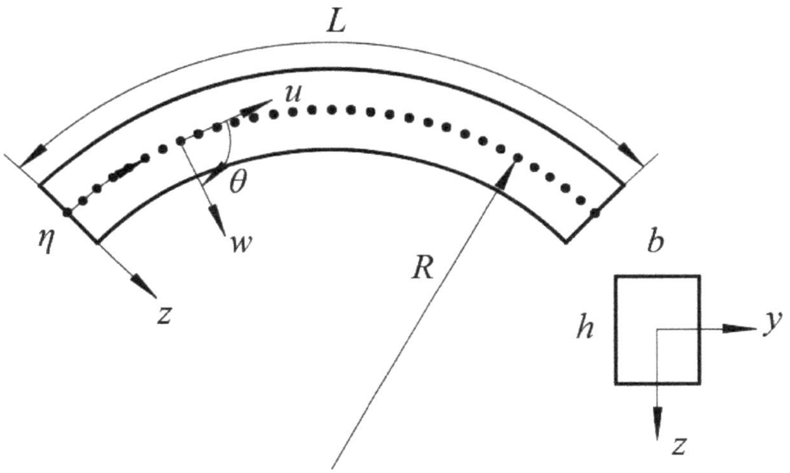

Figure 1. Meshfree model of the curved beam.

4.1. Displacement Approximation

4.1.1. Displacement Field

According to the FSDT [39] and the MLS, the displacement of a curved beam is expressed as

$$\begin{cases} u(\eta, z) = u_0(\eta) - z\theta(\eta) = \sum\limits_{I=1}^{n} N_I(\eta)u_I - z\sum\limits_{I=1}^{n} N_I(\eta)\theta_I \\ w(\eta, z) = w(\eta) = \sum\limits_{I=1}^{n} N_I(\eta)w_I \end{cases} \tag{21}$$

Equation (21) can be written in a matrix form

$$U = \begin{bmatrix} u \\ w \end{bmatrix} = \sum_{I=1}^{n} N_I a_I = Na \tag{22}$$

where

$$N_I = \begin{bmatrix} N_I & 0 & -zN_I \\ 0 & N_I & 0 \end{bmatrix} \tag{23}$$

$$a_I = [u_I \; w_I \; \theta_I]^{\mathrm{T}} \tag{24}$$

$$N = [N_1 \; N_2 \; N_3 \cdots N_I] \tag{25}$$

$$\boldsymbol{a} = \left[\boldsymbol{a}_1\ \boldsymbol{a}_2\ \boldsymbol{a}_3 \cdots \boldsymbol{a}_I\right]^{\mathrm{T}} \tag{26}$$

According to Equation (22), the displacement increments can be denoted by

$$\Delta U = \boldsymbol{N}\Delta\boldsymbol{a} \tag{27}$$

where

$$\Delta\boldsymbol{a}_I = \begin{bmatrix}\Delta u_I & \Delta w_I & \Delta\theta_I\end{bmatrix}^{\mathrm{T}} \tag{28}$$

4.1.2. Displacement and Strain Relation

Based on the von Karman large deflection theory, the relationship between the strain increment and displacement of a curved beam can be expressed as

$$\left.\begin{aligned} \Delta\varepsilon &= \tfrac{\partial\Delta u}{\partial\eta} - \tfrac{\Delta w}{R} + \tfrac{1}{2}\left(\tfrac{\partial\Delta u}{\partial\eta}\right)^2 + \tfrac{1}{2}\left(\tfrac{\partial\Delta w}{\partial\eta}\right)^2 \\ \Delta\gamma &= \tfrac{\partial\Delta w}{\partial\eta} - \Delta\theta + \tfrac{\Delta u}{R} \\ \Delta\kappa &= -z\tfrac{\partial\Delta\theta}{\partial\eta} \end{aligned}\right\} \tag{29}$$

Decomposing the strain increment into linear strain and nonlinear strain, we have

$$\Delta\varepsilon_i = \Delta\varepsilon_i^{\mathrm{L}} + \Delta\varepsilon_i^{\mathrm{N}} \tag{30}$$

where

$$\Delta\varepsilon_i^{\mathrm{L}} = \begin{bmatrix} \frac{\partial\Delta u}{\partial\eta} - \frac{\Delta w}{R} \\ \frac{\partial\Delta w}{\partial\eta} - \Delta\theta + \frac{\Delta u}{R} \\ -z\frac{\partial\Delta\theta}{\partial\eta} \end{bmatrix} = \begin{bmatrix} \frac{\partial N}{\partial\eta} & -\frac{N}{R} & 0 \\ \frac{N}{R} & \frac{\partial N}{\partial\eta} & -N \\ 0 & 0 & -z\frac{\partial N}{\partial\eta} \end{bmatrix} \begin{bmatrix}\Delta u \\ \Delta w \\ \Delta\theta\end{bmatrix} \tag{31}$$

$$\begin{aligned} \Delta\varepsilon_i^{\mathrm{N}} &= \left[\tfrac{1}{2}\left(\tfrac{\partial\Delta u}{\partial\eta}\right)^2 + \tfrac{1}{2}\left(\tfrac{\partial\Delta w}{\partial\eta}\right)^2\right] \\ &= \tfrac{1}{2}\begin{bmatrix} \frac{\partial\Delta u}{\partial\eta} & \frac{\partial\Delta w}{\partial\eta} \end{bmatrix} \begin{bmatrix} \frac{\partial N}{\partial\eta} & 0 & 0 \\ 0 & \frac{\partial N}{\partial\eta} & 0 \end{bmatrix} \begin{bmatrix}\Delta u \\ \Delta w \\ \Delta\theta\end{bmatrix} \end{aligned} \tag{32}$$

In a matrix form,

$$\Delta\varepsilon_i^{\mathrm{L}} = \boldsymbol{B}_{\mathrm{L}}\Delta\boldsymbol{a} \tag{33}$$

and

$$\Delta\varepsilon_i^{\mathrm{N}} = \frac{1}{2}\Delta\boldsymbol{C}\Delta\boldsymbol{\theta}_i = \frac{1}{2}\Delta\boldsymbol{C}\boldsymbol{G}\Delta\boldsymbol{a} = \boldsymbol{B}_{\mathrm{N}}\Delta\boldsymbol{a} \tag{34}$$

where

$$\Delta\boldsymbol{C} = \begin{bmatrix} \frac{\partial\Delta u}{\partial\eta} & \frac{\partial\Delta w}{\partial\eta} \end{bmatrix} \tag{35}$$

$$\boldsymbol{G} = \begin{bmatrix} \frac{\partial N}{\partial\eta} & 0 & 0 \\ 0 & \frac{\partial N}{\partial\eta} & 0 \end{bmatrix} \tag{36}$$

$$\boldsymbol{B}_{\mathrm{L}} = \begin{bmatrix} \frac{\partial N}{\partial\eta} & -\frac{N}{R} & 0 \\ \frac{N}{R} & \frac{\partial N}{\partial\eta} & -N \\ 0 & 0 & -z\frac{\partial N}{\partial\eta} \end{bmatrix} \tag{37}$$

$$\boldsymbol{B}_{\mathrm{N}} = \frac{1}{2}\begin{bmatrix} \frac{\partial N}{\partial\eta}\frac{\partial\Delta u}{\partial\eta} & \frac{\partial N}{\partial\eta}\frac{\partial\Delta w}{\partial\eta} & 0 \end{bmatrix} \tag{38}$$

$$\Delta\boldsymbol{\theta}_i = \begin{bmatrix} \frac{\partial\Delta u}{\partial\eta} \\ \frac{\partial\Delta w}{\partial\eta} \end{bmatrix} \tag{39}$$

$$\Delta\boldsymbol{a} = \begin{bmatrix}\Delta u & \Delta w & \Delta\theta\end{bmatrix}^{\mathrm{T}} \tag{40}$$

4.2. Governing Equations for Nonlinear Buckling of Curved Beams

The virtual work is denoted by

$$\varphi = \int \delta\overline{\varepsilon}^{\mathrm{T}}\overline{\sigma}\mathrm{d}V - \int \delta\overline{u}^{\mathrm{T}}\overline{p}\mathrm{d}V - \int \delta\overline{u}^{\mathrm{T}}\overline{q}\mathrm{d}A \tag{41}$$

where V and A are the volume and surface area of the beam at moment t, respectively. $\bar{\varepsilon}$, $\bar{\sigma}$ are the stress and strain, respectively. $\bar{p}$ and $\bar{q}$ are the volumetric force and surface forces, respectively.

$$\overline{\varepsilon} = \Delta\varepsilon \tag{42}$$

$$\overline{\sigma} = \tau + \Delta\sigma \tag{43}$$

where τ is the Cauchy stress. Yin [42] obtained the $\Delta\sigma$ formula based on the relationship between Cauchy stress and Kirchhoff stress according to the UL.

$$\Delta\sigma = D_{\mathrm{ep}}\Delta\varepsilon - 2M\Delta\varepsilon \tag{44}$$

where D_{ep} is the elastic–plastic matrix. Because material nonlinearity is not considered in this article, D_{ep} is degraded to the elastic matrix D. Thus, Equation (44) can be written as:

$$\Delta\sigma = D\Delta\varepsilon - 2M\,\Delta\varepsilon \tag{45}$$

$$D = \begin{bmatrix} E & 0 & 0 \\ 0 & \kappa GA & 0 \\ 0 & 0 & E \end{bmatrix} \tag{46}$$

where E is the Young's modulus, G is the shear modulus, κ is the shear coefficient, and A is the cross sectional area of the curved beam. M is a matrix related to the stress and

$$M = \begin{bmatrix} \sigma_{11}I & \sigma_{12}I & \sigma_{13}I \\ \sigma_{12}I & \sigma_{22}I & \sigma_{23}I \\ \sigma_{13}I & \sigma_{23}I & \sigma_{33}I \end{bmatrix} \tag{47}$$

I is the identity matrix.

By substituting Equation (43) into Equation (41), we have

$$\int \delta\overline{\varepsilon}^{\mathrm{T}}\overline{\sigma}\mathrm{d}V = \int \delta\Delta\overline{\varepsilon}^{\mathrm{T}}\sigma\mathrm{d}V + \int \delta\Delta\overline{\varepsilon}^{\mathrm{T}}\Delta\sigma\mathrm{d}V \tag{48}$$

where

$$\delta\Delta\overline{\varepsilon} = \delta\left(\Delta\varepsilon^{\mathrm{L}} + \Delta\varepsilon^{\mathrm{N}}\right) \tag{49}$$

Equations (31) and (34) give us

$$\delta\Delta\varepsilon^{\mathrm{L}} = B_{\mathrm{L}}\delta\Delta a \tag{50}$$

$$\delta\Delta\varepsilon^{\mathrm{N}} = B_{\mathrm{N}}\delta\Delta a \tag{51}$$

By substituting Equations (50) and (51) into Equation (49), we obtain

$$\delta\Delta\overline{\varepsilon} = B_{\mathrm{L}}\delta\Delta a + B_{\mathrm{N}}\delta\Delta a = (B_{\mathrm{L}} + B_{\mathrm{N}})\delta\Delta a = B\delta\Delta a \tag{52}$$

So, the first item on the right side of Equation (48) is expressed as

$$\begin{aligned} \textstyle\int \delta\Delta\overline{\varepsilon}^{\mathrm{T}}\sigma\mathrm{d}V &= \delta\Delta a^{\mathrm{T}} \textstyle\int (B_{\mathrm{L}} + B_{\mathrm{N}})^{\mathrm{T}}\sigma\mathrm{d}V \\ &= \delta\Delta a^{\mathrm{T}}(F_{\sigma} + K_{\sigma}\Delta a) \end{aligned} \tag{53}$$

where $\boldsymbol{F}_\sigma$ is the structural internal force and $\boldsymbol{K}_\sigma$ is the initial stress matrix:

$$\boldsymbol{F}_\sigma = \int \boldsymbol{B}_\mathrm{L}^\mathrm{T}\sigma \mathrm{d}V \tag{54}$$

$$\boldsymbol{K}_\sigma = \int \boldsymbol{G}^\mathrm{T}\boldsymbol{M}\boldsymbol{G}^\mathrm{T}\mathrm{d}V \tag{55}$$

According to Equations (45) and (52), the second item on the right side of Equation (48) can be written as

$$\int \delta\Delta\overline{\varepsilon}^\mathrm{T}\Delta\sigma\mathrm{d}V = \delta\Delta\boldsymbol{a}^\mathrm{T}(\boldsymbol{K}_1 - \boldsymbol{K}_2)\Delta\boldsymbol{a} \tag{56}$$

where

$$\boldsymbol{K}_1 = \int \boldsymbol{B}^\mathrm{T}\boldsymbol{D}\boldsymbol{B}\mathrm{d}V \tag{57}$$

$$\boldsymbol{K}_2 = 2\int \boldsymbol{B}^\mathrm{T}\boldsymbol{M}\boldsymbol{B}\mathrm{d}V \tag{58}$$

By substituting Equations (53) and (56) into Equation (41), we have:

$$\boldsymbol{\varphi} = \delta\Delta\boldsymbol{a}^\mathrm{T}\boldsymbol{K}_\sigma\Delta\boldsymbol{a} + \delta\Delta\boldsymbol{a}^\mathrm{T}(\boldsymbol{K}_1 - \boldsymbol{K}_2)\Delta\boldsymbol{a} + \delta\Delta\boldsymbol{a}^\mathrm{T}\boldsymbol{F}_\sigma - \delta\Delta\boldsymbol{a}^\mathrm{T}\boldsymbol{F} \tag{59}$$

where $\boldsymbol{F}$ is the vector of the external load

$$\boldsymbol{F} = \int \boldsymbol{N}^\mathrm{T}\overline{\boldsymbol{p}}\mathrm{d}V + \int \boldsymbol{N}^\mathrm{T}\overline{\boldsymbol{q}}\mathrm{d}A \tag{60}$$

According to the principle of the virtual work,

$$\boldsymbol{\varphi} = \boldsymbol{K}_\sigma\Delta\boldsymbol{a} + (\boldsymbol{K}_1 - \boldsymbol{K}_2)\Delta\boldsymbol{a} + \boldsymbol{F}_\sigma - \boldsymbol{F} = 0 \tag{61}$$

Assuming

$$\boldsymbol{K}_\mathrm{T} = \boldsymbol{K}_\sigma + \boldsymbol{K}_1 - \boldsymbol{K}_2 \tag{62}$$

Equation (61) can be written as

$$\boldsymbol{\varphi} = \boldsymbol{K}_\mathrm{T}\Delta\boldsymbol{a} + \boldsymbol{F}_\sigma - \boldsymbol{F} = \boldsymbol{K}_\mathrm{T}\Delta\boldsymbol{a} - \boldsymbol{\Phi} = 0 \tag{63}$$

where $\boldsymbol{K}_\mathrm{T}$ is the tangent stiffness matrix, which represents the tangent slope corresponding to a certain node in the load displacement curve. $\boldsymbol{\Phi}$ is an unbalanced force, and

$$\boldsymbol{\Phi} = \boldsymbol{F} - \boldsymbol{F}_\sigma \tag{64}$$

So, Equation (63) can be written as

$$\boldsymbol{K}_\mathrm{T}\Delta\boldsymbol{a} = \boldsymbol{\Phi} \tag{65}$$

However, due to the fact that the shape function obtained from Equation (20) does not satisfy the Kronecker delta's properties, and the displacement of Equation (63) is a nodal parameter rather than the actual displacement of the node, the governing Equation (61) of the curved beam cannot be directly solved. The full transformation method proposed by Chen et al. [43] is used in this article to modify the governing Equation (61) [34].

4.3. Nonlinear Buckling Analysis of Curved Beams

Equation (65) is a nonlinear equation that needs to be solved by an iterative method.

$$\boldsymbol{K}_\mathrm{T}^{(n)}\Delta\boldsymbol{a}_n = \boldsymbol{\Phi}_n \tag{66}$$

$$\boldsymbol{\Phi}_n = \boldsymbol{P} - \boldsymbol{F}_\sigma^{(n)} \tag{67}$$

where $K_T^{(n)}$ is the tangent stiffness matrix at the n-th iteration, where its value is derived from the displacement at the end of the (n − 1)-th iteration; $\boldsymbol{\Phi}_n$ is the unbalanced force at the n-th iteration; $\boldsymbol{P}$ is the total load applied to the structure; and $F_\sigma^{(n)}$ is the internal force of the structure at the nth iteration, where its value is computed from the stress at the end of the (n − 1)-th iteration.

The arc length method is typically employed to conduct iterative computation for solving nonlinear buckling equilibrium path problems [43,44] to extend the original n-dimensional equations to n + 1 dimensions by introducing load coefficients and constraint equations and form a hypersurface in the solution space. It can be used to solve the displacement equilibrium, the displacement, and the load factor at the intersection of the hypersurface. Figure 2 shows the iteration of the arc length method.

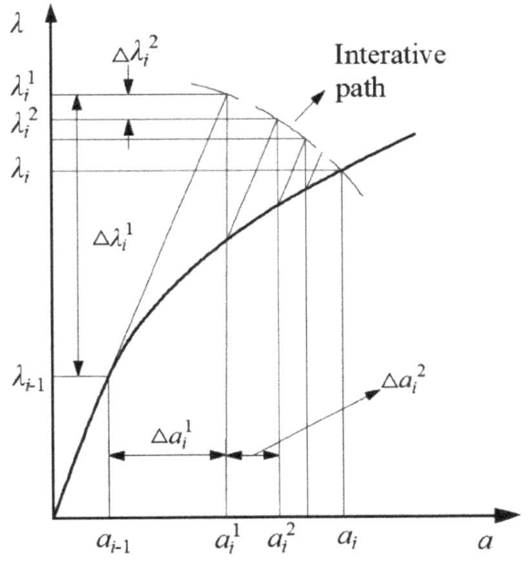

Figure 2. Schematic diagram of the arc length method.

According to the arc length method, a load parameter λ is introduced and embedded in Equation (64) to obtain a system of equilibrium equations with the parameter λ

$$\boldsymbol{\Phi} = \lambda \boldsymbol{F} - \boldsymbol{F}_\sigma \tag{68}$$

Substituting Equation (68) into Equation (63) yields

$$\boldsymbol{K}_T \Delta \boldsymbol{a} = \lambda \boldsymbol{F} - \boldsymbol{F}_\sigma \tag{69}$$

For Equation (68), because a new parameter is added, another equation needs to be created in order to be solved. The equation can be obtained from the schematic diagram of the arc length method in Figure 2, as follows:

$$\Delta \boldsymbol{a}^{\mathrm{T}} \Delta \boldsymbol{a} + \beta i \Delta \lambda^2 \boldsymbol{F}^{\mathrm{T}} \boldsymbol{F} = l^2 \tag{70}$$

where βi is the scaling parameter for loading terms. The spherical arc length method [45–47] is used in this article and βi = 1, so Equation (70) can be re-written as

$$\Delta \boldsymbol{a}^{\mathrm{T}} \Delta \boldsymbol{a} + \Delta \lambda^2 \boldsymbol{F}^{\mathrm{T}} \boldsymbol{F} = l^2 \tag{71}$$

Figure 2 shows the calculation process of the arc length method. The figure represents the iteration process of the j-th iteration step of the i-th load increment step, and the tangent lines in the figure represent the tangent stiffness matrix $\boldsymbol{K}_{\mathrm{T}}$, which can be taken from the first iteration of each load increment step and used for the remaining iteration steps within the current incremental step so that the tangents within the same load incremental step are shown to be parallel in Figure 2. $\Delta \boldsymbol{a}_i^j$ denotes the displacement increment in the j-th iteration step of the i-th load increment step; then, the displacement relationship can be expressed as

$$\boldsymbol{a}_i = \boldsymbol{a}_{i-1} + d\boldsymbol{a}_i \tag{72}$$

where $\boldsymbol{a}_i$ is the total displacement at the end of the i-th load increment step, $\boldsymbol{a}_{i-1}$ is the total displacement at the end of the $(i-1)$-th load increment step, and $\mathrm{d}\boldsymbol{a}_i$ denotes the total displacement increment within the i-th load increment step. The displacement increment at the j-th iteration step can be expressed as

$$\mathrm{d}\boldsymbol{a}_i^j = \sum \Delta \boldsymbol{a}_i^j \tag{73}$$

For ease of expression, assume

$$\boldsymbol{r} = \begin{bmatrix} \lambda \boldsymbol{F} \\ \Delta \boldsymbol{a} \end{bmatrix} \tag{74}$$

Therefore, Equation (71) can be written as

$$(\|\boldsymbol{r}\|_2)^2 = l^2 \tag{75}$$

In the j-th iteration step,

$$\boldsymbol{r}_i^j = \boldsymbol{r}_i^{j-1} + \boldsymbol{l}_i^{j-1} \tag{76}$$

where $\boldsymbol{l}_i$ is the vector form of the arc length, and substituting Equation (76) into Equation (75) yields

$$\left(\boldsymbol{r}_i^{j-1} + \boldsymbol{l}_i^j\right)^{\mathrm{T}} \left(\boldsymbol{r}_i^{j-1} + \boldsymbol{l}_i^j\right) = \left(l_i^{j-1}\right)^2 \tag{77}$$

For the j-th iteration step

$$\left(\boldsymbol{r}_i^{j-1}\right)^{\mathrm{T}} \boldsymbol{r}_i^{j-1} = \left(l_i^{j-1}\right)^2 \tag{78}$$

$$\left(\boldsymbol{l}_i^j\right)^{\mathrm{T}} \boldsymbol{l}_i^j = \left(\Delta \boldsymbol{a}_i^j\right)^{\mathrm{T}} \Delta \boldsymbol{a}_i^j + \left(\Delta \lambda_i^j\right)^2 \boldsymbol{F}^{\mathrm{T}} \boldsymbol{F} \tag{79}$$

Therefore, Equation (77) can be written as

$$\left(\Delta \boldsymbol{a}_i^j\right)^{\mathrm{T}} \left(\Delta \boldsymbol{a}_i^j + 2d\boldsymbol{a}_i^{j-1} + \Delta \lambda_i^j \boldsymbol{F}^{\mathrm{T}} \left(\Delta \lambda_i^j + 2\lambda_i^{j-1}\right) \boldsymbol{F}\right) = 0 \tag{80}$$

where $\Delta \boldsymbol{a}_i^j$ can be decomposed, and

$$\Delta \boldsymbol{a}_i^j = \Delta \lambda_i^j \left[\Delta \boldsymbol{a}_i^j\right]_{\mathrm{I}} + \left[\Delta \boldsymbol{a}_i^j\right]_{\mathrm{II}} \tag{81}$$

Substituting Equation (81) into Equation (80) yields a quadratic equation

$$b1\left(\Delta \lambda_i^i\right)^2 + b2\Delta \lambda_i^i + b3 = 0 \tag{82}$$

where

$$b1 = 1 + \left[\Delta \boldsymbol{a}_i^j\right]_{\mathrm{I}}^T \left[\Delta \boldsymbol{a}_i^j\right]_{\mathrm{I}} \tag{83}$$

$$b2 = \lambda_i^{j-1} + \left[\Delta \boldsymbol{a}_i^j\right]_{\mathrm{I}}^{\mathrm{T}} \left(\left[\Delta \boldsymbol{a}_i^j\right]_{\mathrm{II}} + d\boldsymbol{a}_i^{j-1}\right) \tag{84}$$

$$b3 = \left[\Delta \boldsymbol{a}_i^j\right]_{\mathrm{II}}^{\mathrm{T}} \left(\left[\Delta \boldsymbol{a}_i^j\right]_{\mathrm{II}} + 2\mathrm{d}\boldsymbol{a}_i^{j-1}\right) \tag{85}$$

Equation (82) is solved to obtain $\Delta\lambda_i^j$. The process of solving the quadratic equation is slightly cumbersome, so the tangent plane method can be used to simplify the equation.

By making

$$\left(\boldsymbol{r}_i^{j-1}\right)^{\mathrm{T}} \boldsymbol{l}_i^j = 0 \tag{86}$$

we obtain

$$\left(\boldsymbol{a}_i^{j-1}\right)^{\mathrm{T}} \Delta \boldsymbol{a}_i^j + \lambda_i^{j-1} \Delta\lambda_i^j \boldsymbol{F}^{\mathrm{T}} \boldsymbol{F} = 0 \tag{87}$$

The expression for $\Delta\lambda_i^j$ can be obtained from Equation (87):

$$\Delta\lambda_i^j = -\frac{\left(d\boldsymbol{a}_i^{j-1}\right)^{\mathrm{T}} \left[\Delta \boldsymbol{a}_i^j\right]_{\mathrm{II}}}{d\boldsymbol{a}_i^{j-1} \left[\Delta \boldsymbol{a}_i^j\right]_{\mathrm{I}} + \lambda_i^{j-1} \boldsymbol{F}^{\mathrm{T}} \boldsymbol{F}} \tag{88}$$

We solve Equation (88) for $\left[\Delta \boldsymbol{a}_i^j\right]_{\mathrm{I}}$ and $\left[\Delta \boldsymbol{a}_i^j\right]_{\mathrm{II}}$.

Again, by discretizing the external loads, Equation (64) can be rewritten as

$$\widetilde{\boldsymbol{\Phi}}_i^j = \Delta\lambda_i^j \widehat{\boldsymbol{F}} + \boldsymbol{F}_i^{j-1} - (\boldsymbol{F}_\sigma)_i^{j-1} = \Delta\lambda_i^j \widehat{\boldsymbol{F}} + \boldsymbol{R}_i^{j-1} \tag{89}$$

Substituting Equation (89) into Equation (66) yields

$$\boldsymbol{K}_{\mathrm{T}}^{j-1} \Delta \boldsymbol{a}_i^j = \Delta\lambda_i^j \widehat{\boldsymbol{F}} + \boldsymbol{R}_i^{j-1} \tag{90}$$

By combining Equation (90) and Equation (81), we obtain

$$\left[\Delta \boldsymbol{a}_i^j\right]_{\mathrm{I}} = \left[\boldsymbol{K}_{\mathrm{T}}^{j-1}\right]^{-1} \widehat{\boldsymbol{F}} \tag{91}$$

$$\left[\Delta \boldsymbol{a}_i^j\right]_{\mathrm{II}} = \left[\boldsymbol{K}_{\mathrm{T}}^{j-1}\right]^{-1} \boldsymbol{R}_i^{j-1} \tag{92}$$

Arc length in the arc length method is an important factor affecting calculation accuracy and convergence. In general,

$$l_i = \sqrt[n]{\frac{N_{i-1}}{N}} l_{i-1} \tag{93}$$

where N_{i-1} is the number of iterations that the $(i-1)$-th load increment undergoes, and its size is j. If the value of N_{i-1} is larger than 10, then it should take 10 [47]; N is a parameter related to the number of iterations, and its value varies according to different methods; l_{i-1} is the arc length of the $(i-1)$-th load increment step; and n varies for different methods, generally ranging from 2 to 6.

Before calculating $\Delta\lambda_i^j$, it is necessary to calculate its initial value

$$\Delta\lambda_i^1 = \sqrt{\frac{(l_i)^2}{\boldsymbol{F}^{\mathrm{T}} \boldsymbol{F} + \left(\Delta \boldsymbol{a}_i^1\right)^{\mathrm{T}} \Delta \boldsymbol{a}_i^1}} \tag{94}$$

Based on Equation (94), a flowchart for the calculation of the j-th iteration step of the i-th load increment step of the arc length method is shown in Figure 3.

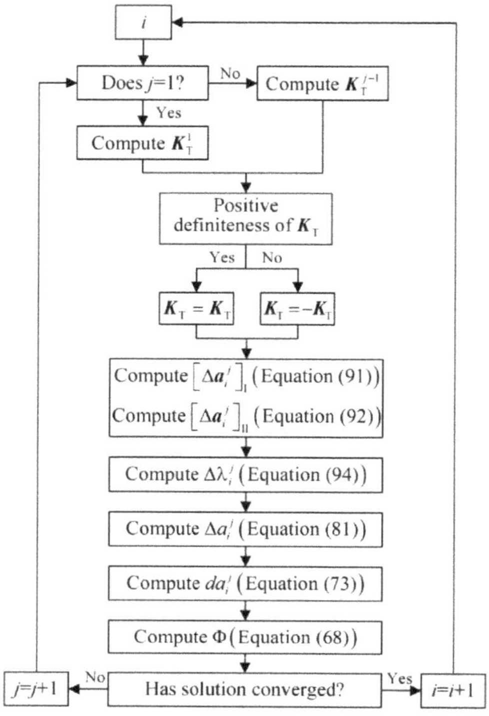

Figure 3. Flowchart for the solution procedure of the arc length method.

5. Results and Discussion

5.1. Convergence and Validation Studies

5.1.1. Convergence Study

Example 1 studies the convergence of the proposed method and investigates how the number of nodes and support size d_{mI} ($d_{mI} = scale \times C_I$, and C_I is the distance between two adjacent nodes) influence the convergence. A clamped circle arch (curved beam) is studied (Figure 4). The dimensions of the circular arch R = 3.384 m, xL = 0.8636 m, and θ =14.66°—and the Young's modulus and the shear modulus of the material of the arch are $E = 6.8948 \times 10^4$ N/mm^2, and $G = 2.0891 \times 10^4$ N/mm^2, respectively, which are consistent with those in Ref. [48] The cross section of the curved beam is rectangular with a height of h = 25.4 mm and a width of b = 4.76 mm.

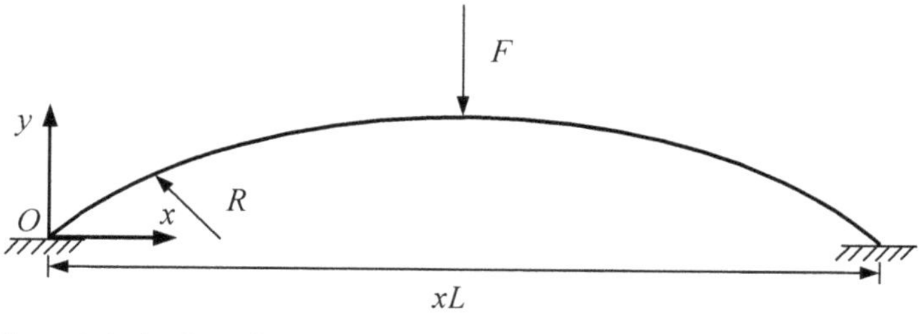

Figure 4. A circular arch (curve beam) under concentrated load.

The buckling loads calculated by the proposed method under different *scales* and nodes are given in Figure 5 and compared with those from ref. [48]. The nodes are evenly distributed. It can be found that in the figure, when the *scale* that denotes the size of the influence domain is set to 3, the calculated buckling load closely matches the literature solution, regardless of the number of nodes. However, for the *scales* of 3.5 and 4, as well as a smaller number of nodes, deviations from the literature solution are observed. But with an increased number of nodes, the calculated buckling load gradually converges towards the literature solution 151 N. Therefore, after considering computational efficiency and accuracy, *scale* 3 and 25 nodes are chosen for the nonlinear buckling study of curved beams in this section.

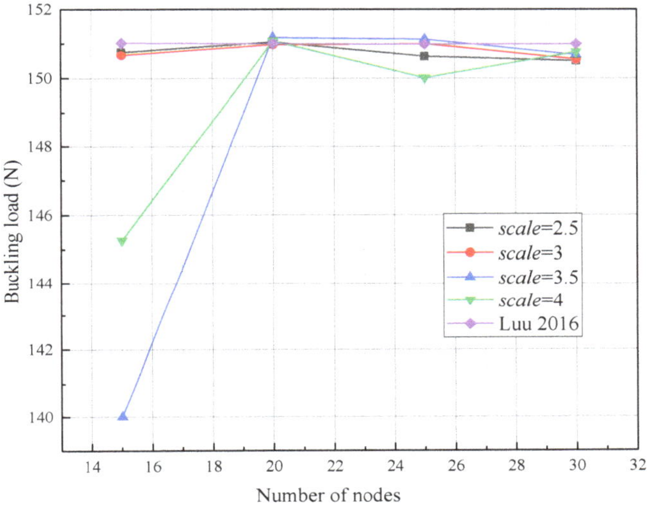

Figure 5. Convergence analysis on nonlinear buckling load of circular arch [48].

5.1.2. Validation Study

Example 2 computes load–displacement curve of example 1 with the proposed method. The curve is shown in Figure 6 and compared with those given by ref. [45], the 3D finite element method, and an experiment [48]. The curves are very close to one another.

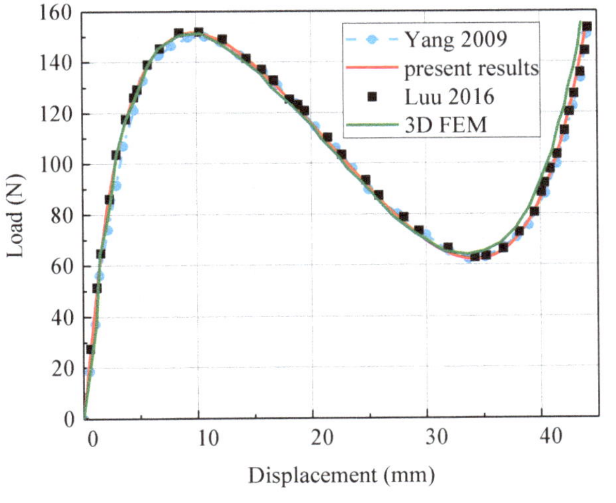

Figure 6. Mid-span load–deflection curve of the circular arch [45,48].

Example 3 investigates a simply supported parabolic curved beam with a rise to span ratio of 0.1 and a length to slenderness ratio of 40, subjected to a uniform load of q_0 (Figure 7). The Young's modulus E and the shear modulus G of the material have the following relation: $E = G \times 3$. The dimensionless deflection is defined as w/i. The dimensionless load is represented by $\frac{q_0 x L^3}{EA}$. i is the radius of gyration, and $i = (h^2 + b^2)/6h$ (h and b are defined in Example 1).

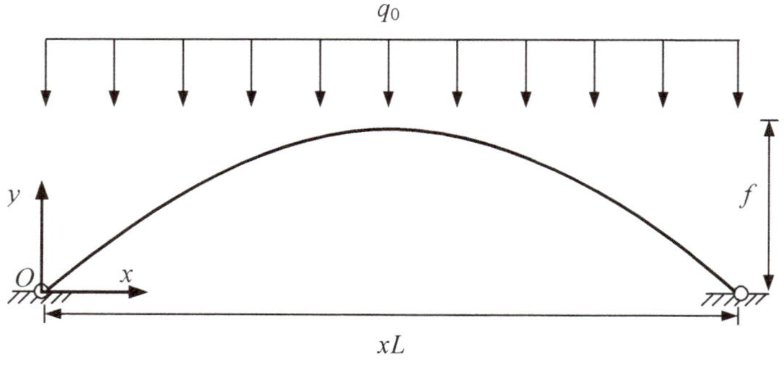

Figure 7. A parabolic curved beam under uniform load.

The load–displacement curve calculated by the proposed method is shown in Figure 8 and compared with that given by ref. [48]. The figure demonstrates that the proposed method has high accuracy in analyzing nonlinear buckling behaviors of curved beams.

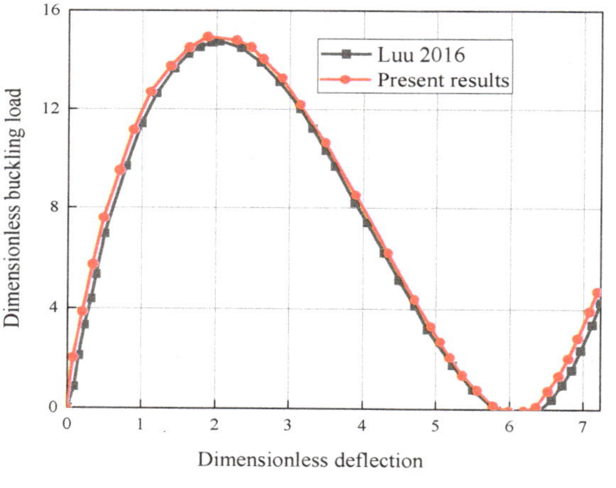

Figure 8. Mid-span load–displacement curve of curved beam under uniformly distributed load [48].

5.2. Parameter Study

5.2.1. Included Angle

Example 4 investigates the effect of different included angles on the nonlinear buckling of curved beams. A clamped circular arch (curved beam) with a radius curvature of $R = L/\varphi$ and a span diameter of $L = 33.33$ m is subjected to a radial uniform load of q_0 (Figure 9). The Young's modulus, shear modulus, and Poisson's ratio are $E = 300$ Mpa, $G = E/3$, and $v = 0.3$, respectively. The cross section of the curved beam is rectangular, with a height of $h = 1$ m and a width of $b = 1$ m.

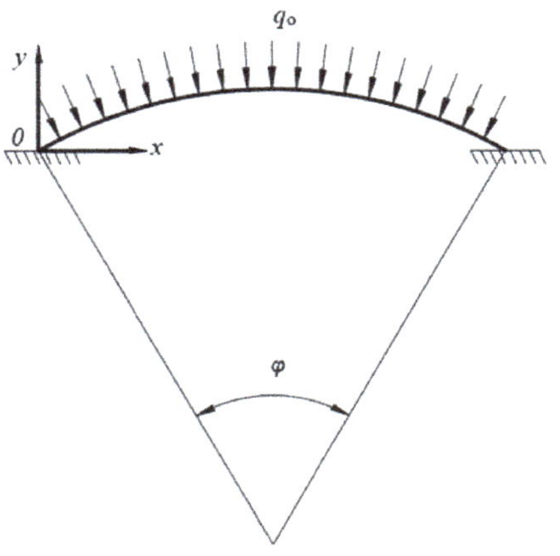

Figure 9. The circular arc curved beam with an included angle of φ.

The mid-span load–deflection curves for the curved beam with the five cases of included angles, i.e., 20°, 30°, 40°, 50°, and 60°, are calculated by the proposed method and shown in Figure 10. When the included angle φ = 20°, the load–deflection response of the structure shows that there is continuous stiffening without reaching a limit point and no specific buckling load and the structure remains stable under the applied load. However, as the included angle increases to 30°, the structure starts to buckle. As the included angle increases from 30° to 40°, the buckling load of the curved beam increases, too. This implies that larger included angles are more likely to cause geometric nonlinearity and result in instability. As the angle increases, the unstable stage of the curved beam also elongates and becomes more obvious.

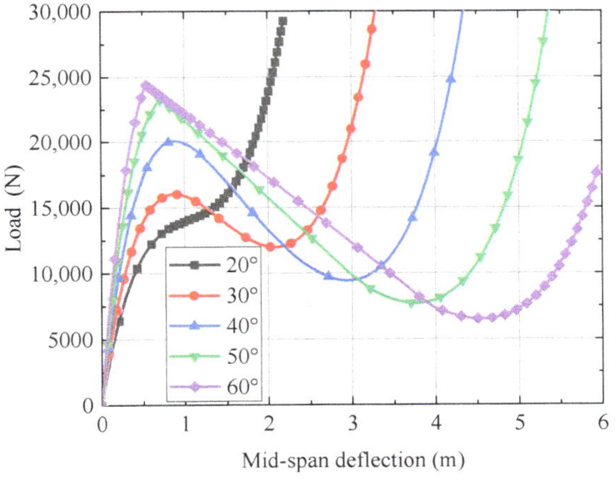

Figure 10. Mid-span load–deflection curves of the curved beam under different included angles.

5.2.2. Rise to Span Ratio and Curved Form

Example 5 investigates the effect of the rise to span ratio and curved form on the nonlinear buckling behaviors of the curved beams. A circular curved beam (Figure 11) and a parabolic curved beam (Figure 12) with rise to span ratios of xL/f = 1/7 and 1/10 are considered. The curved beams are all clamped and subjected to a concentrated load F. The Young's modulus and Poisson ratio are E = 68 Gpa and v = 0.3, respectively. The cross sections of the curved beams are all rectangular, with a height of h = 1.6 mm and a width of b = 25 mm.

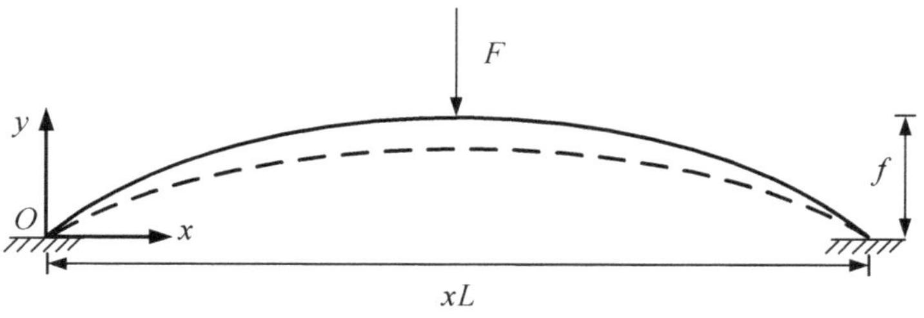

Figure 11. The clamped circular arch (curved beam) under a concentrated load.

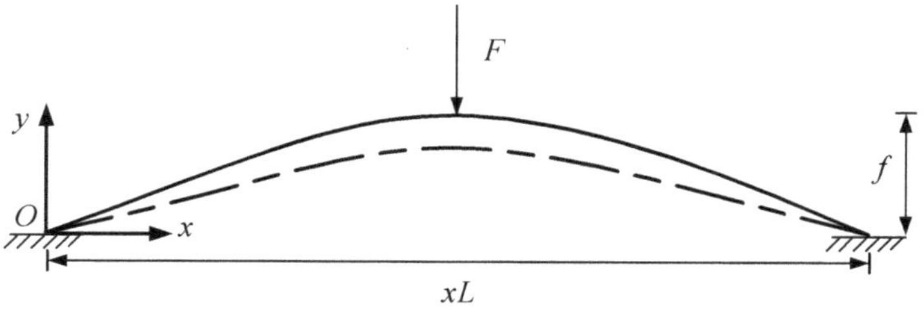

Figure 12. The clamped parabolic curved beam under a concentrated load.

In Figures 11 and 12, the solid line represents beams with a 1/7 rise to span ratio, while the dashed line represents the beams with a 1/10 rise to span ratio. The relevant parameters are shown in Table 1. To analyze the effect of the rise to span ratio, the span xL of the beams is kept constant while the height f varies.

Table 1. Parameters of circular arc curved beam.

Rise to span ratios	1/7	1/10
Arch height f (m)	0.0714	0.05
Span xL (m)	0.5	0.5
Curvature radius R(m)	0.4734	0.65
Included angle (°)	63.8	45.2

The pre- and post-buckling (mid-span load deflection) curves of the beams obtained using the proposed method are shown in Figure 13. It can be observed that the jump instability occurs for both curved beams at rise to span ratios of 1/7 and 1/10, respectively. For the same rise to span ratio, the nonlinear buckling curves of the circular and parabolic curved beams are almost the same, but the critical load of the parabolic curved beam with a rise to span ratio of 1/7 is slightly larger than that of the circular curved beam, which indicates that the effect of the curved form on the nonlinear stability of curved beams is small and negligible. It can also be found that the critical load of the curved beams increases with the increase in the rise to span ratio, which indicates that increasing the rise to span ratio can improve the stability of curved beam structures.

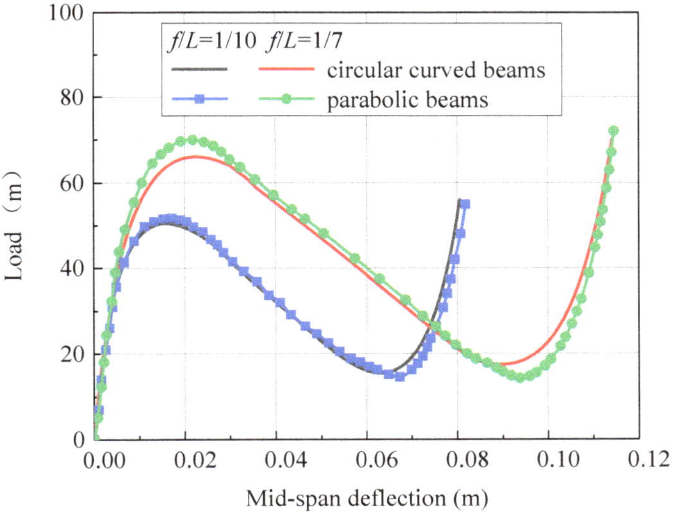

Figure 13. Effect of different rise to span ratios and curved forms on nonlinear stability of curved beams.

5.2.3. Boundary Condition and Slenderness Ratio

Example 6 investigates the effect of different boundary conditions and slenderness ratios on the nonlinear buckling behaviors of curved beams. A shallow arch with a radius curvature R of 133.1 m and a span diameter xL of 34 mm is subjected to a concentrated load F (Figure 14). The load is applied vertically to the curved beam. The Young's modulus, Poission's ratio, and shear modulus of the material of the arch are E = 10 Mpa, υ = 0.2, and $G = E/2/(1 + \upsilon)$, respectively. The cross section of the curved beam is rectangular, with a height of h = 0.1875 m and a width of b = 1 m. Different slenderness ratios ($\mu xL/i$, where μ is the length factor) of 80 and 120 and boundary conditions of clamped (clamped at both sides, μ = 0.5) and semi-clamped (clamped at one side and simply supported at the other side, μ = 0.7) supports are considered, respectively.

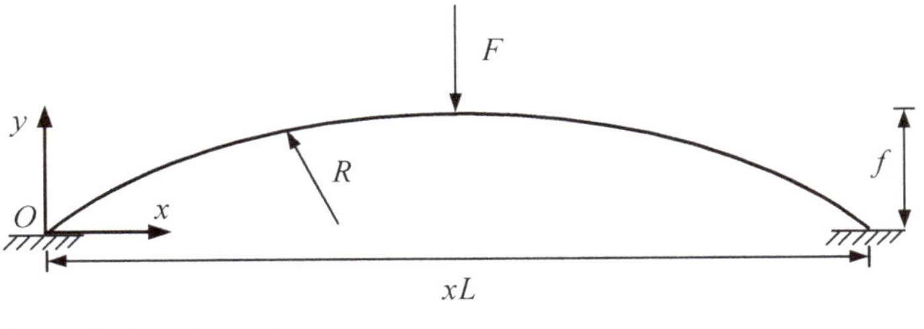

Figure 14. A shallow arch.

Figure 15 shows the mid-span load–deflection curves of the shallow arch under different boundary conditions and slenderness ratios. The results in Figure 15 suggest that both boundary condition and slenderness ratio have a significant influence on the nonlinear stability of curved beams. For the same slenderness ratio, the semi-clamped boundary condition yields a higher critical load than the clamped boundary condition. With the exception of the clamped curved beam with a slenderness ratio of 80, the other three cases undergo jumping instability, which indicates that these beams experience a sudden and significant change in mechanical behavior, leading to instability.

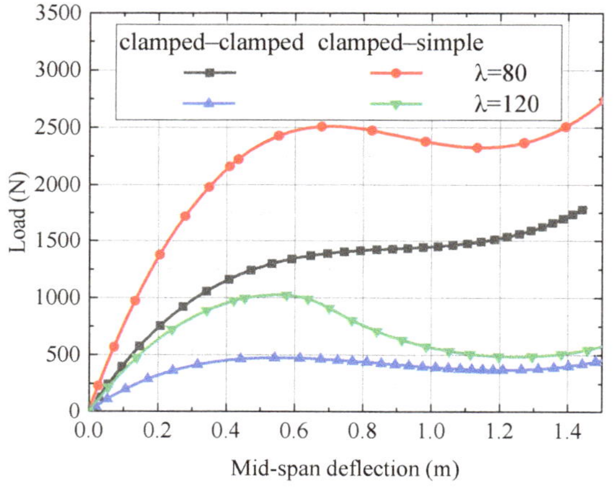

Figure 15. Mid-span load–deflection under different boundary conditions and slenderness ratios.

5.2.4. Axial to Shear Rigidity Ratio

Example 7 investigates the effect of different axial to shear rigidity ratios on the nonlinear buckling behaviors of curved beams. The circular arch in Section 5.2.2 with four different axial to shear rigidity ratios of $E/G = 3$, 50, 100, and 200 is considered, and the results are shown in Figure 16. The included angle $\varphi = 30°$. It can be observed from Figure 16 that with an increase in the axial to shear rigidity ratio, the critical load of the curved beam decreases significantly, which means that the structure is more susceptible to jump instability and, thus, sudden and violent damage. In addition, an increase in the axial to shear rigidity ratio also leads to a longer instability phase of the structure, which indicates that once buckling occurs, the structure remains unstable for a longer period of time before reaching a new equilibrium. These show that the axial to shear stiffness ratio has a significant effect on the nonlinear buckling behavior of curved beams.

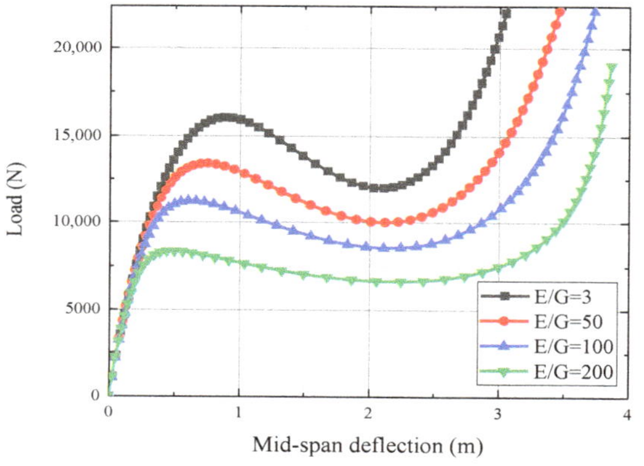

Figure 16. Mid-span load–deflection curves under different axial-to-shear rigidity ratios.

6. Conclusions

This article proposes a meshfree method for analyzing the nonlinear buckling behaviors of curved beams based on moving least-squares approximation, the first-order shear deformation theory, and the von Karman deflection theory. The accuracy of the proposed method is verified by calculating several examples of curved beams, and the effects of the angle, rise to span ratio, curved form, boundary condition, slenderness ratio, and axial shear stiffness ratio on the nonlinear buckling path of curved beams are investigated. The following conclusions are drawn:

1. The proposed method has good stability and accuracy in solving the pre- and post-buckling problems of curved beams.
2. The rise to span ratio, including the angle, boundary condition, and slenderness ratio, as well as the axial shear stiffness ratio, have important effects on the nonlinear buckling path of curved beams. The critical load of the curved beams increases with the increase in the rise to span ratio. For the included angle, larger included angles are more likely to cause geometric nonlinearity and result in instability. For the same slenderness ratio, the semi-clamped boundary condition yields a higher critical load than the clamped boundary condition. With an increase in axial to shear rigidity ratio, the critical load of the curved beam decreases significantly.
3. In practice, the selection of appropriate geometric parameters can increase the critical load of the curved beam, making it easier for the structure to reach a new equilibrium state and improving the stability of the structure. Additionally, the shape function of the proposed meshless method is only related to the adopted basis function and weight function. As long as the appropriate basis function and weight function are selected, the method can calculate all forms of curved beams.

7. Future Research

This paper proposed a mesh-free method to trace the pre-buckling as well as the post-buckling nonlinear equilibrium paths for curved beams, and the results are compared with those from the literature to verify the validity of the proposed method. Finally, the influence of geometric parameters on the nonlinear behaviors of curved beams is investigated.

However, this paper is only compared with the literature and has not been further verified by experiments. In future research, the research method should be expanded, and the validity of the proposed method should be verified by a large amount of experimental data. In order to increase accuracy, the nonlinear behavior of the curved beam should be simulated by using the refined high-order shear deformation [49–51] theory.

Author Contributions: Conceptualization, Y.L. (Yuxiao Li); Methodology, Y.L. (Yuxiao Li); Software, Y.L. (Yajing Liao); Validation, Z.X.; Resources, Z.X.; Data curation, Y.L. (Yajing Liao); Writing—original draft, Y.L. (Yuxiao Li); Supervision, L.P.; Project administration, L.P. All authors have read and agreed to the published version of the manuscript.

Funding: The work that is described in this paper has been supported by the grants awarded by the National Natural Science Foundation of China (No. 12162004), the Guangxi Natural Science Foundation (No. 2024GXNSFAA010284) and the counterpart aid project for discipline construction from Guangxi University (No. 2023N02).

Data Availability Statement: The original contributions presented in the study are included in the article, further inquiries can be directed to the corresponding author.

Conflicts of Interest: The authors declare no conflict of interest.

References

1. Yang, Y.-B. Recent researches on buckling of framed structures and curved beams. *J. Constr. Steel Res.* **1993**, *26*, 193–210. [CrossRef]
2. Ye, T.; Jin, G.; Ye, X.; Wang, X. A series solution for the vibrations of composite laminated deep curved beams with general boundaries. *Compos. Struct.* **2015**, *127*, 450–465. [CrossRef]
3. Wang, Q.; Choe, K.; Tang, J.; Shuai, C.; Wang, A. Vibration analyses of general thin and moderately thick laminated composite curved beams with variable curvatures and general boundary conditions. *Mech. Adv. Mater. Struct.* **2020**, *27*, 991–1005. [CrossRef]
4. McManus, P.F.; Nasir, G.A.; Culver, C.G. Horizontally curved girders-state of the art. *J. Struct. Div.* **1969**, *95*, 853–870. [CrossRef]
5. Timoshenko, S.P.; Gere, J.M. *Theory of Elastic Stability*; Courier Corporation: North Chelmsford, MA, USA, 2009.
6. ВПАСОВ В З. *Тонкосменные Упрутие Смержнм*, 2nd ed.; фнзматгнз: Moscow, Russia, 1959.
7. Papangelis, J.P.; Trahair, N.S. Flexural-torsional buckling of arches. *J. Struct. Eng.* **1987**, *113*, 889–906. [CrossRef]
8. Ojalvo, I.U.; Newman, M. Buckling of naturally curved and twisted beams. *J. Eng. Mech. Div.* **1968**, *94*, 1067–1087. [CrossRef]

9. Bažant, Z.P.; El Nimeiri, M. Large-deflection spatial buckling of thin-walled beams and frames. *J. Eng. Mech. Div.* **1973**, *99*, 1259–1281. [CrossRef]
10. Yang, Y.-B.; Kuo, S.-R. Buckling of frames under various torsional loadings. *J. Eng. Mech.* **1991**, *117*, 1681–1697. [CrossRef]
11. Lin, K.C.; Hsieh, C.M. The closed form general solutions of 2-D curved laminated beams of variable curvatures. *Compos. Struct.* **2007**, *79*, 606–618. [CrossRef]
12. Polit, O.; Anant, C.; Anirudh, B.; Ganapathi, M. Functionally graded graphene reinforced porous nanocomposite curved beams: Bending and elastic stability using a higher-order model with thickness stretch effect. *Compos. Part B Eng.* **2019**, *166*, 310–327. [CrossRef]
13. Wan, Z.-Q.; Li, S.-R.; Ma, H.-W. Geometrically nonlinear analysis of functionally graded Timoshenko curved beams with variable curvatures. *Adv. Mater. Sci. Eng.* **2019**, *2019*, 204145. [CrossRef]
14. Tripathy, B.; Rao, K.P. Curved composite beams—Optimum lay-up for buckling by ranking. *Comput. Struct.* **1991**, *41*, 75–82. [CrossRef]
15. Fraternali, F.; Spadea, S.; Ascione, L. Buckling behavior of curved composite beams with different elastic response in tension and compression. *Compos. Struct.* **2013**, *100*, 280–289. [CrossRef]
16. Lo, C.F.; Conway, H.D. The elastic stability of curved beams. *Int. J. Mech. Sci.* **1967**, *9*, 527–538. [CrossRef]
17. Conway, H.D.; Lo, C.F. Further studies on the elastic stability of curved beams. *Int. J. Mech. Sci.* **1967**, *9*, 707–718. [CrossRef]
18. Pi, Y.-L.; Bradford, M.A.; Uy, B. In-plane stability of arches. *Int. J. Solids Struct.* **2002**, *39*, 105–125. [CrossRef]
19. Pi, Y.-L.; Bradford, M.A.; Tin-Loi, F. Non-linear in-plane buckling of rotationally restrained shallow arches under a central concentrated load. *Int. J. Non-Linear Mech.* **2008**, *43*, 1–17. [CrossRef]
20. Pi, Y.-L.; Bradford, M.A. Lateral–torsional elastic buckling of rotationally restrained arches with a thin-walled section under a central concentrated load. *Thin-Walled Struct.* **2013**, *73*, 18–26. [CrossRef]
21. Noor, A.K.; Greene, W.H.; Hartley, S.J. Nonlinear finite element analysis of curved beams. *Comput. Methods Appl. Mech. Eng.* **1977**, *12*, 289–307. [CrossRef]
22. Akoussah, E.; Beauliieu, D.; Dhatt, G. Curved beam element via penalty/mixed formulation for nonlinear in-plane analysis. *Commun. Appl. Numer. Methods* **1986**, *2*, 617–623. [CrossRef]
23. Elias, Z.M.; Chen, K.-L. Nonlinear shallow curved-beam finite element. *J. Eng. Mech.* **1988**, *114*, 1076–1087. [CrossRef]
24. Yang, Y.-B.; Kuo, S.-R.; Yau, J.-D. Use of straight-beam approach to study buckling of curved beams. *J. Struct. Eng.* **1991**, *117*, 1963–1978. [CrossRef]
25. Kuo, S.-R.; Yang, Y.-B. New theory on buckling of curved beams. *J. Eng. Mech.* **1991**, *117*, 1698–1717. [CrossRef]
26. Litewka, P.; Rakowski, J. An efficient curved beam finite element. *Int. J. Numer. Methods Eng.* **1997**, *40*, 2629–2652. [CrossRef]
27. Hashemian, A.; Hosseini, S.F. Nonlinear bifurcation analysis of statically loaded free-form curved beams using isogeometric framework and pseudo-arclength continuation. *Int. J. Non-Linear Mech.* **2019**, *113*, 1–16. [CrossRef]
28. Lucy, L.B. A numerical approach to the testing of the fission hypothesis. *Astron. J.* **1977**, *82*, 1013–1024. [CrossRef]
29. Belytschko, T.; Lu, Y.Y.; Gu, L. Element-free Galerkin methods. *Int. J. Numer. Methods Eng.* **1994**, *37*, 229–256. [CrossRef]
30. Lu, Y.Y.; Belytschko, T.; Gu, L. A new implementation of the element free Galerkin method. *Comput. Methods Appl. Mech. Eng.* **1994**, *113*, 397–414. [CrossRef]
31. Liu, W.K.; Jun, S.; Zhang, Y.F. Reproducing kernel particle methods. *Int. J. Numer. Methods Fluids* **1995**, *20*, 1081–1106. [CrossRef]
32. Yang, Y.; Lam, C.C.; Kou, K.P. Forced vibration analysis of functionally graded beams by the meshfree boundary-domain integral equation method. *Eng. Anal. Bound. Elem.* **2016**, *72*, 100–110. [CrossRef]
33. Kou, K.P.; Yang, Y. A meshfree boundary-domain integral equation method for free vibration analysis of the functionally graded beams with open edged cracks. *Compos. Part B Eng.* **2019**, *156*, 303–309. [CrossRef]
34. Dodagoudar, G.R.; Rao, B.N.; Sunitha, N.V. A meshfree method for beams on elastic foundation. *Int. J. Geotech. Eng.* **2015**, *9*, 298–306. [CrossRef]
35. Chinh, T.H.; Tu, T.M.; Duc, D.M.; Hung, T.Q. Static flexural analysis of sandwich beam with functionally graded face sheets and porous core via point interpolation meshfree method based on polynomial basic function. *Arch. Appl. Mech.* **2021**, *91*, 933–947. [CrossRef]
36. He, C.; Wu, X.; Wang, T.; He, H. Geometrically nonlinear analysis for elastic beam using point interpolation meshless method. *Shock. Vib.* **2019**, *2019*, 9065365. [CrossRef]
37. Wang, D.; Chen, J.-S. A locking-free meshfree curved beam formulation with the stabilized conforming nodal integration. *Comput. Mech.* **2006**, *39*, 83–90. [CrossRef]
38. Yeh, W.-C.; Wang, Y.-M. Meshfree method for geometrical nonlinear analysis of curved and twisted beams using a three-dimensional finite deformation theory. *Int. J. Struct. Stab. Dyn.* **2019**, *19*, 1950116. [CrossRef]
39. Agarwal, S.; Chakraborty, A.; Gopalakrishnan, S. Large deformation analysis for anisotropic and inhomogeneous beams using exact linear static solutions. *Compos. Struct.* **2006**, *72*, 91–104. [CrossRef]
40. Chen, D.; Huang, W.; Sloan, S.W. An alternative updated Lagrangian formulation for finite particle method. *Comput. Methods Appl. Mech. Eng.* **2019**, *343*, 490–505. [CrossRef]
41. Lancaster, P.; Salkauskas, K. Surfaces generated by moving least squares methods. *Math. Comput.* **1981**, *37*, 141–158. [CrossRef]
42. Yin, Y. *Fundamentals of Nonlinear Finite Element Method*; Peking University Press: Beijing, China, 2007.

43. Chen, J.-S.; Pan, C.; Wu, C.-T.; Liu, W.K. Reproducing kernel particle methods for large deformation analysis of non-linear structures. *Comput. Methods Appl. Mech. Eng.* **1996**, *139*, 195–227. [CrossRef]
44. Riks, E. An incremental approach to the solution of snapping and buckling problems. *Int. J. Solids Struct.* **1979**, *15*, 529–551. [CrossRef]
45. Yang, X.; Zhou, Y.; Wang, X. The arc-length method for solving the coupled equation of mechanics-electricity. *Chin. J. Comput. Mech.* **2009**, *26*, 142–145.
46. Lam, W.F.; Morley, C.T. Arc-length method for passing limit points in structural calculation. *J. Struct. Eng.* **1992**, *118*, 169–185. [CrossRef]
47. Memon, B.-A.; Su, X.-Z. Arc-length technique for nonlinear finite element analysis. *J. Zhejiang Univ. Sci. A* **2004**, *5*, 618–628. [CrossRef] [PubMed]
48. Luu, A.-T.; Lee, J. Non-lineOar buckling of elliptical curved beams. *Int. J. Non-Linear Mech.* **2016**, *82*, 132–143. [CrossRef]
49. Sayyad, A.S.; Avhad, P.V. A new higher order shear and normal deformation theory for the free vibration analysis of sandwich curved beams. *Compos. Struct.* **2022**, *280*, 114948. [CrossRef]
50. Lezgy-Nazargah, M.; Etemadi, E.; Hosseinian, S. Assessment of four-variable refined shear deformation theory for low-velocity impact analysis of curved sandwich beams. *Eur. J. Mech. A Solids* **2022**, *94*, 104604. [CrossRef]
51. Lezgy-Nazargah, M. A four-variable global–local shear deformation theory for the analysis of deep curved laminated composite beams. *Acta Mech.* **2020**, *231*, 1403–1434. [CrossRef]

buildings

MDPI

Article

Numerical Study on the Sensitive Wavelength and Amplitude Management of Track Irregularities in Monorail Systems

Hong Zhang [1], Pengjiao Wang [2,*], Chen Chen [3], Yanqiang Ji [2], Shiqi Wei [2], Boyou Guan [2], Zhiqiang Liu [3] and Jie Zhang [3]

[1] College of Mechanical and Intelligent Manufacturing, Central South University of Forestry and Technology, Changsha 410075, China; t20040415@csuft.edu.cn
[2] School of Civil Engineering, Central South University, Changsha 410075, China; 15138051614@163.com (Y.J.); 224811107@csu.edu.cn (S.W.); guanboyou@csu.edu.cn (B.G.)
[3] 3rd Construction Co., Ltd., of China Construction 5th Engineering Bureau, Changsha 410004, China; cc57138961@163.com (C.C.); liuzhqiang0927@163.com (Z.L.); zj3280188@163.com (J.Z.)
* Correspondence: w2592570@gmail.com

Abstract: Track irregularity is one of the principal excitations that induces coupled vibrations in vehicle–bridge systems. Understanding the sensitive wavelength of track irregularities is critical for the evaluation and management of track conditions. Notably, existing studies generally focus on railway systems, but the characteristics of sensitive wavelengths in monorail systems are insufficiently understood. This study aims to investigate the influence of longitudinal level irregularity (LLI) on the dynamic response of the monorail tour transit system (MTTS), as well as the sensitive wavelength of LLI. First, a joint model was developed by integrating multi-body dynamics with the finite element method. The LLI utilized in the numerical mode was simulated by trigonometric functions with various frequencies (i.e., wavelengths) and amplitudes. The dynamic responses of monorail vehicles, including the wheel load reduction rate and vertical acceleration, were obtained and then used to evaluate track conditions. Results indicate that the dynamic responses of MTTS are mainly affected by the LLI with a wavelength of less than 5 m. In addition, it was found that, in the studied ranges, the sensitivity wavelength grows as the vehicle speed increases. Finally, the recommended value of LLI control under various track conditions was evaluated.

Keywords: monorail tour transit system; track irregularity; dynamic response; sensitive wavelength; multi-body dynamics; finite element method

1. Introduction

The transportation system plays an important role in the economic development of a region. For tourist attractions, an efficient passenger transportation system generally indicates good management and economic benefits. Nowadays, with the increasing awareness of natural landscape protection, the question of how to protect the original landscape while adapting to the complex and changing terrain conditions in order to develop the tourism industry has gradually become the focus of attention. In this case, monorail tour transit systems (MTTSs) have been rapidly developed in various regions [1]. In fact, as early as the 20th century, countries, including the United States and Japan, began to develop monorail systems for urban and tourist transportation. The main structural features of these monorails are the use of straddle-type monorail and concrete or steel–concrete composite beams. It should be noted that MTTSs generally adopt steel track beams because of the convenience of their prefabrication and rapid installation, which represents a significant structural diversity when compared with conventional monorail systems. In addition, there are some technical parameters and differences in dynamic characteristics [2].

Similarly to many transportation systems, such as railways [3,4] and highways [5,6], engineers and researchers are concerned about the service performance of both the vehicle subsystem and the supporting structure (i.e., track beam bridge for monorails) subsystem of an MTTS. To accurately capture the patterns of the dynamic responses of the system, coupling methods should be used to consider the interaction between these two subsystems. At the same time, to achieve the simulation of vehicle–bridge interactions, it is undoubtedly necessary to consider the existence of track irregularities. For instance, Naseri et al. developed a multi-point contact algorithm to simulate the wheel–rail interface and implemented it so as to properly capture rail surface irregularities [7]. Melo et al. proposed a specific finite element to model the track–deck interface by assembling a friction contact element and a non-linear spring [8]. With these efforts, the non-linear behavior of the coupled system can be better investigated. Specifically, the irregularity-induced vibrations of supporting structures and vehicles can be accurately evaluated. As a result, this will facilitate the subsequent assessment of the service performance of these transportation systems. Research on track irregularities can be conducted from various perspectives, including measuring the actual profile of engineering structures, integrating track irregularities into vehicle–bridge coupling models, and investigating the characteristics of track irregularities in both the time and frequency domains [9–11]. Among these perspectives, the sensitive wavelength of track irregularity has continued to draw the attention of the research community.

Generally, actual track irregularity on a structure surface is recognized as a random process that changes with rail (or road, track, etc.) mileage. In engineering practice, a track irregularity signal can be regarded as a superposition of harmonics with different wavelengths, different amplitudes, and different phases. At a certain operating speed, different wavelengths of irregularity have varying effects on the disturbance frequency of vehicles, with certain wavelengths, known as sensitive wavelengths, inducing more significant vibrations in the vehicles due to the dynamic amplification effect [12,13]. The range of wavelengths for track irregularities is broad and can be categorized into short-wave, medium-wave, and long-wave track irregularities [14]. It should be noted that, for different transportation systems (vehicle speeds), the critical value for distinguishing wavelength categories can be diverse.

1. Short-wave wavelengths (less than 1 m) have a high excitation frequency, significantly impacting wheel–rail interactions, wheel weight reduction [15], and wheel–rail noise [16], though they have minimal effects on train body vibration and ride comfort.
2. Medium-wave wavelengths (between 1 to 30 m) primarily affect the smoothness and comfort of train operation, with a smaller impact on vehicle safety and wheel–rail dynamics compared with short-wave roughness.
3. Long-wave wavelengths (between 30 to 200 m) play a critical role in the riding comfort of high-speed train operations [17,18].

It is worth noting that existing studies generally focus on railway systems [19–22], whereas very limited attention has been paid to monorail systems. Monorail systems are primarily utilized for transportation in urban and scenic areas, rather than the long-distance travel that railways are utilized for. Although their speeds are relatively low, managing track irregularities is essential due to the lower stiffness of the steel rail girders. As pointed out in [23], with an increase in the track stiffness irregularity of the track–bridge system, dynamic irregularity also increases markedly. Moreover, with a decrease of the supporting stiffness of the track slabs, the probability distribution of the dynamic irregularity extrema is shifted to a higher-level area. In their studies, the impact of stiffness refers to various structural components, such as floating slabs and substructures, in railways. For monorails, however, only the stiffness of track beams is related. However, the length of the track span affects the geometry of local track irregularity, which is related to the occurrence of different train motions [17]. In addition, given that monorail systems operate at relatively low speeds, short-wave roughness has become a key factor influencing operational safety, smoothness, and ride comfort. Specific uneven wavelengths can cause significant excitation in the train body, adversely affecting comfort and safety. Therefore, it is essential to enhance the control and management of medium- and short-wave irregularities in monorail track systems. To sum up, structural diversity between monorails and railways causes a distinct difference regarding the analysis of sensitive wavelength, hence it is of great significance to investigate the features of sensitive wavelengths for monorail systems, particularly in MTTSs, under the accurate simulation of structural configurations. For the purposes of this analysis, long-wave irregularities will be temporarily excluded, focusing instead on the effects of medium- and short-wave roughness. With these efforts, we expect to provide a necessary database and an important theoretical basis for the management and control of track irregularities on monorail track beams.

On a related note, studies on sensitive wavelengths have been predominantly conducted via numerical simulation, which is determined by the nature of the problem. This is because natural signals of track irregularity typically contain continuous frequency information, making it very difficult to isolate the effects of an individual frequency (i.e., wavelength) within the dynamic responses of complex coupled systems. So far, research regarding the sensitive length of track irregularity and irregularity-induced dynamic responses in MTTSs or general monorail systems are very limited. The first available field-measured track irregularity data and the associated spectra were obtained in Japan in 2006 [24]. Recently, the database of longitudinal level irregularity (LLI), which affects the system most among the various kinds of irregularities, has been significantly enlarged with the implementation of data measured from MTTSs [2,25]. By using these measured signals, the dynamic responses of monorail systems, including the bridge and vehicle subsystems were accurately evaluated through field tests and numerical simulation [26–29]. In addition, concepts from railway systems, such as the contact force and load reduction of wheels, were referred to and utilized in order to comprehensively evaluate monorail systems [30,31]. Nevertheless, developing more specialized evaluation indicators has always been an essential topic in the research field. Moreover, reviews on existing studies have utilized the optimal choice among the numerical simulation techniques, which is a combination of the finite element method (FEM) and multi-body dynamics (MBDs). For example, Huang et al. utilized a co-simulation method to analyze the mechanical properties of bridges under wind–vehicle–bridge coupling actions [32]. Chen et al. [33] and Wang et al. [34] studied the aerodynamic performance of a vehicle–bridge system under the action of a crosswind. In this work, however, evaluation under complex or extreme service conditions, such as strong winds, are not considered. This is because external loadings do not alter the natural characteristics of trains and structures related to sensitive wavelengths. Therefore, the investigation of wind-induced effects, as well as active control of structural or vehicle motions [35,36], are not included.

To sum up, existing studies on the sensitive wavelength of track irregularities in monorail systems, especially in MTTSs, are relatively rare. Nevertheless, the usage of FEM and MBDs on monorail systems have been verified, providing an effective method by which to investigate the effects related to sensitive wavelengths. To the best of our knowledge, no existing studies have focused on the determination of the sensitive wavelengths of MTTSs. Based on a comprehensive analysis of the existing research, this work employs FEM (ABAQUS 2013) and MBD (SIMPACK 2021x) software to establish a vehicle–track coupling model, which is then used to investigate the sensitive wavelength and the amplitude limit of LLI, as this contributes more significantly than other types of track irregularities. The results of the study can provide a meaningful reference for subsequent studies regarding track irregularities, sensitive wavelengths and related engineering management.

2. Methodologies

2.1. Modelling Track Beams

For easy demonstration, a 15 m-long track beam in an MTTS test-line located in Zhuzhou city, China is taken as an example. According to the drawings provided by the design institute, and using solid element, a finite element analysis model of the track beam was established based on Abaqus [37]. The track beam is made of S355 steel material with a density of 7850 kg/m, an elastic modulus of 210 GPa, and a Poisson's ration of

0.3. Boundary conditions of the structure were simplified during modelling, with a fixed constraint at one end to release the bending moment and the longitudinal degree of freedom released at the other end. The geometries of the track beam and related FEM model are shown in Figure 1. Furthermore, a uniform load was applied on the top surface, then a static analysis was conducted to observe the deformation form and stress distribution and, more importantly, to roughly examine the modelling process.

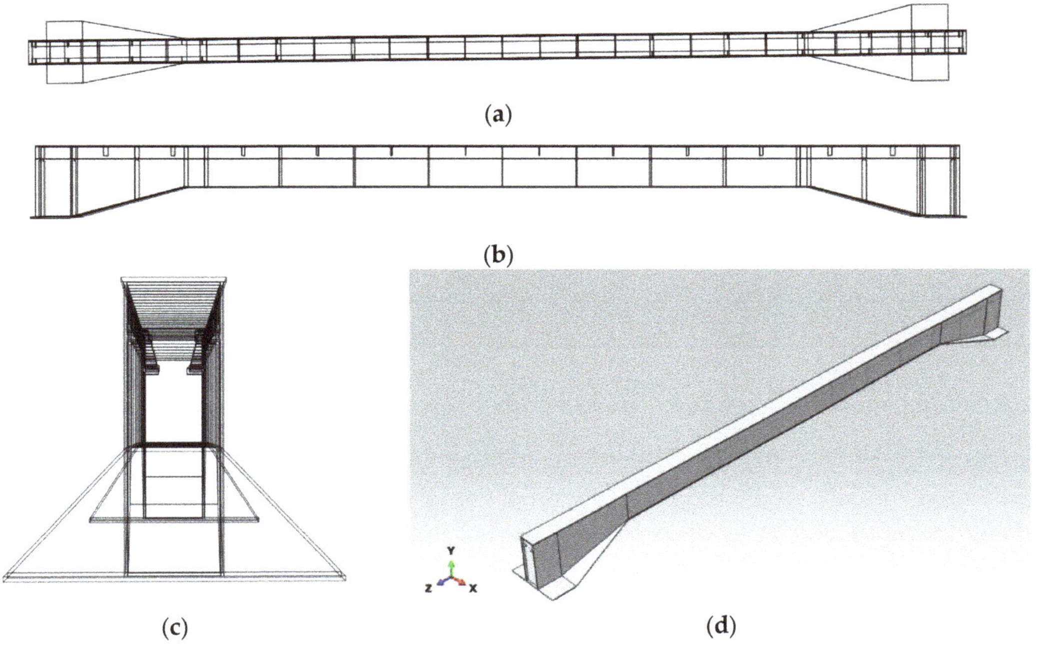

Figure 1. Representative design diagrams of monorail substructures. (**a**) Floor plan; (**b**) elevation view; (**c**) cross section view; (**d**) axonometric view of the model.

2.2. Modelling Monorail Trains

Owing to the complexity of vehicle systems, necessary simplifications should be performed to efficiently simulate their mechanical characteristics, e.g., by using MBD [29]. In this work, the modelling of monorail trains adopts the substructure technique in SIMPACK [38]. In general, the mass and the geometric and dynamic characteristics of each component of the monorail train were considered. To improve the model's precision, hourglass springs, shock absorbing equipment, etc. are simulated by force elements. Note, that, when compared with a conventional locomotive, a monorail bogie does not contain primary suspension, and the secondary suspension consists of hourglass springs. Parameters of the vehicle were provided by the manufacturer (CRRC Special Equipment Technology Co., Ltd., Zhuzhou city, China). The developed model of an MTTS train is shown in Figure 2. Detailed parameters for the train model are given in the Supplementary materials.

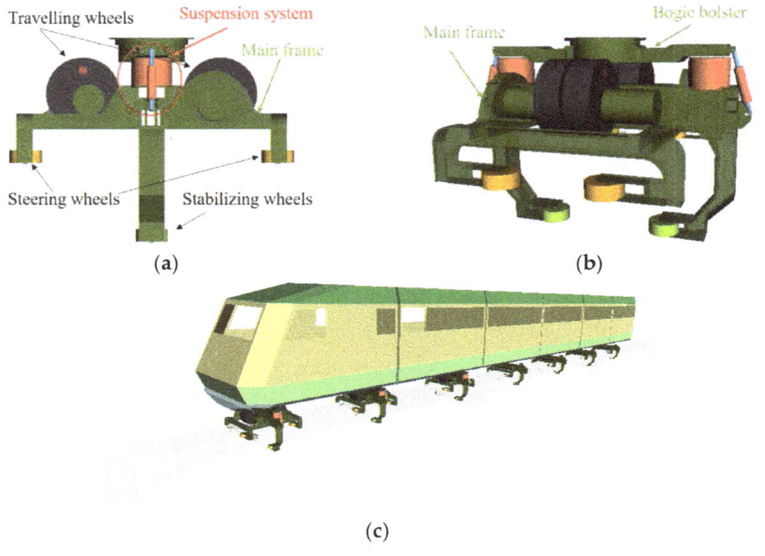

Figure 2. Modelling of an MTTS train: (**a**) Elevation view of a bogie; (**b**) axonometric view of the bogie; (**c**) the entire train model.

2.3. Coupling Model and Its Verification

Existing studies have presented approaches by which to integrate the FEM and MBD model. In this work, this process was completed using an interface file, as follows: firstly conduct a substructure analysis in Abaqus and transform the result file into the so called flexible body interface file, which can be processed as a flexible object and applied with various force elements in subsequent simulations [25,26]. The relationship between the monorail train and track beams is established by utilizing force elements, and time-dependent markers and integrating the track irregularity profile, which can be input directly or calculated from the spectrum, as in Figure 3. The latter approach allows for the analysis of sensitive wavelengths, as the simulated spectrum contains only a few isolated frequencies.

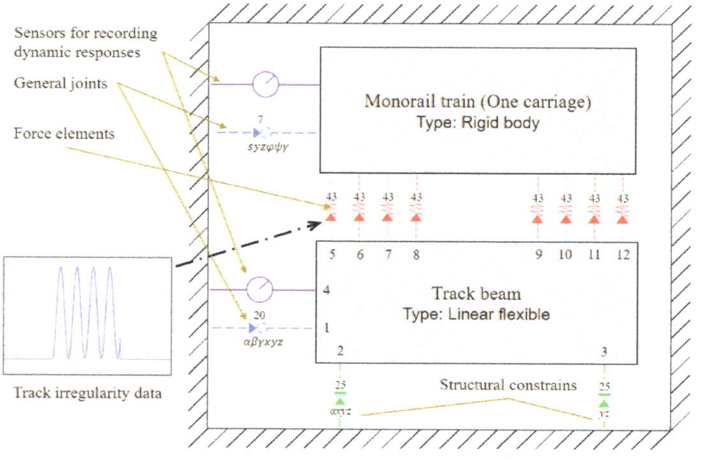

Figure 3. Topology graph of MTTS with track irregularity integrated.

To demonstrate the effectiveness of the developed coupling model, a series of field tests on the dynamic responses of MTTSs were conducted in the test line. Equipment such as linear variable differential transform and vibration pickups were used to capture the dynamic signals. For the model, parameters of the monorail vehicle were set so as to be consistent with the actual conditions. At the beginning of the simulation, the monorail train started from 30 m away from the end of the track beam to eliminate the initial vibration effects, and then passed through the track beam at a constant speed of 30 km/h. The vertical displacement and the vertical and lateral acceleration at the middle span of the track beam were monitored. Results of the joint simulation under various speeds are shown in Figure 4. These results show that the simulated responses of acceleration and displacement of the track beam are highly consistent. It is observed that the simulated vertical displacement at the midspan of the track beam was slightly larger than the tested signal, which is believed to be due to the model not considering the influence of steel stiffeners inside the track beam. Nevertheless, the result reveals that the joint model can be applied to subsequent analysis.

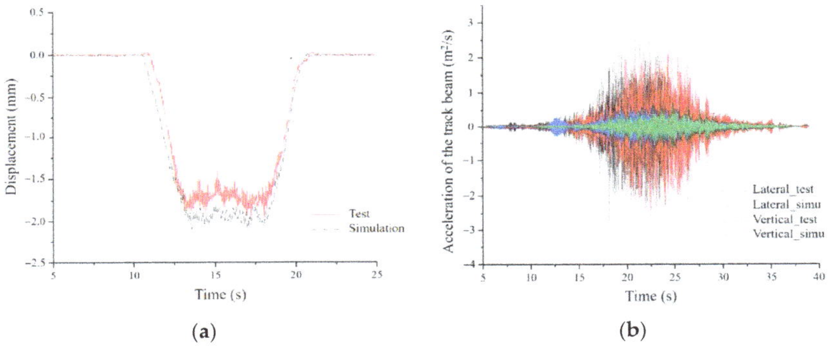

Figure 4. Comparison of results at the mid-span of a track beam from field tests and simulation. (**a**) Vertical displacement; (**b**) vertical and lateral acceleration.

2.4. Sensitive Wavelength

Due to the inclusion of numerous wavelength components, it is difficult to represent the properties of track irregularity with a single function. More importantly, not all wavelengths have the same impact on the coupled system. In fact, certain wavelengths of irregularity significantly affect the quality of vehicle operation and are classified as sensitive wavelengths. The identification of sensitive wavelengths is generally determined by the dynamic responses, such as the vibration acceleration of the vehicle body or structure, and it depends on the specific issue being studied. Generally, when induced frequency of track irregularity is close to the resonance frequency of the train, the riding comfort will be greatly reduced. Related research has been undertaken on the effects of track irregularity on the dynamic performance of railway trains, using the acceleration at the car body floor. Furthermore, some data acquisition systems have been proposed to monitor the safety in railways [39]. Nevertheless, limited studies on monorails have been reported.

In this work, a unilateral cosine wave model, as shown in Equation (1), is used to simulate the LLI. According to the input mode shown in Figure 5, various harmonic irregularities with different directions and phase differences were applied to the top plate of track beams, which are consistent to the running position of travelling wheels. To avoid the influence of the uneven state of the web on the response, it was assumed that the web is in an ideally smooth state during the analysis.

$$Z(t) = \frac{A}{2}(1 - \cos \omega t)\left(0 \leq t \leq N\frac{L}{v}\right) \quad (1)$$

where Z is the LLI at a certain point, t is the total simulation time, A is the amplitude of the LLI, L is the wavelength, v is the running speed, N is the number of cycles of the continuous harmonic, and w is the angular frequency.

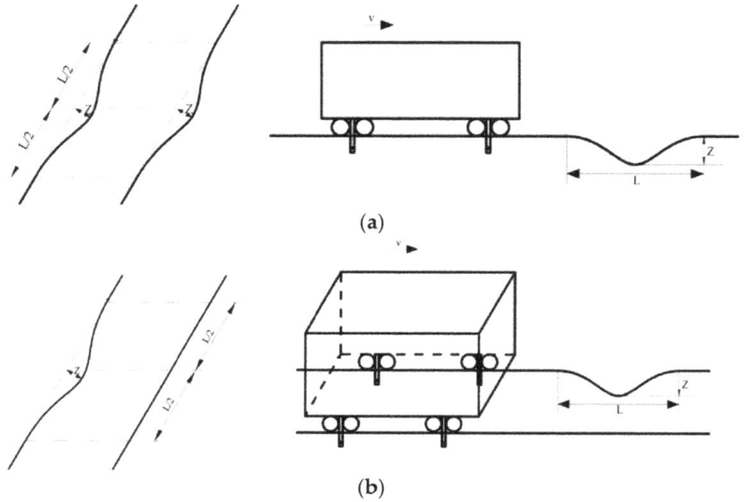

Figure 5. Track geometry irregularities: (**a**) Longitudinal level (top); (**b**) horizontal level.

3. Results and Analysis

3.1. Distribution of Sensitive Wavelengths

To investigate the distribution of sensitive wavelengths associated with LLI under varying speed conditions, simulation calculations based on the aforementioned model were performed. This study focuses primarily on medium- and short-wave irregularities, with the wavelength of track irregularities varying from 0.1 to 30 m. Hence, we considered each meter (0.1, 1, 2, etc.) as a distinct working condition, resulting in a total of 31 conditions within this wavelength range. To examine the influence of vehicle speed on the sensitive wavelengths of the coupling system, three indicators, including the wheel–rail contact force (WRCF), the wheel load reduction rate (WLRR) of travelling wheels, and the vertical acceleration of the monorail train were utilized. In a practical MTTS project, potential differences between operation conditions include the changes of train speed and train load. To investigate the characteristics of sensitive wavelength in detail, both light (empty train) and heavy (with full passengers' weight) monorails were simulated. In addition, the amplitudes of LLI remain constant at 3 mm to reflect a medium level condition; the simulation calculations were conducted at speeds of 10 km/h, 20 km/h, 30 km/h, 40 km/h, and 50 km/h.

3.1.1. Wavelengths for Light Trains

By utilizing the established joint model, dynamic responses were extracted separately for different speed conditions, as shown in Figure 6. It is observed that indicators related to wheel–rail interactions (WRIs), i.e., WRCF and WLRR, generally vary simultaneously with one other. From Figure 6a to Figure 6b, it is evident that, when the wavelength is less than 4 m, WRLs experience a rapid increase as the wavelength decreases. This trend indicates that shorter wavelengths result in significant wheel–rail dynamic interactions, leading to significant fluctuations in the WRI indicators. Moreover, higher vehicle speeds correlate with larger WRIs. In contrast, when the wavelength exceeds 4 m, WRIs tend to stabilize, with only slight increases observed within the range of 4 to 25 m. This suggests that, for the light weight condition, this kind of interaction is primarily influenced by the LLI associated with wavelengths shorter than 4 m.

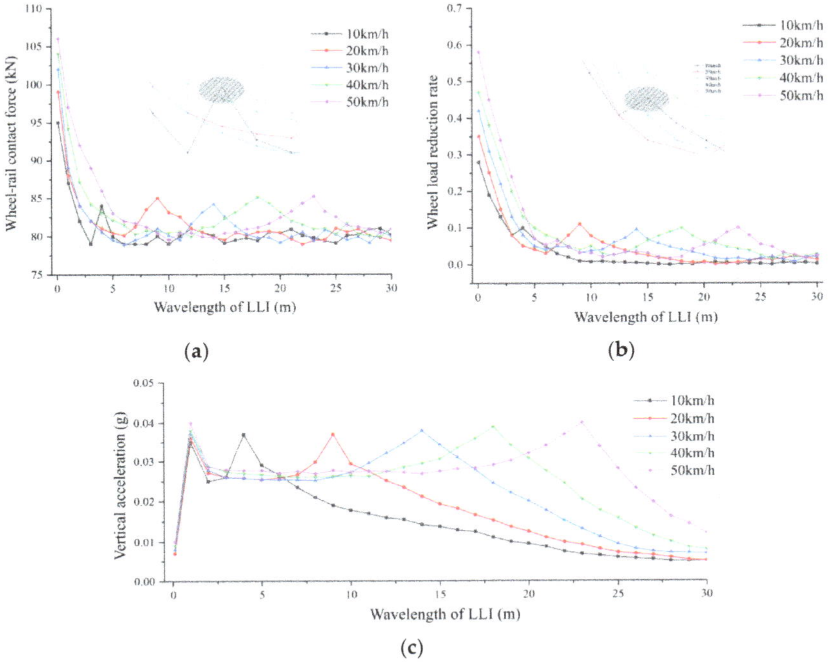

Figure 6. Effects of wavelengths on the dynamic performance. (**a**) WRCF: wheel–rail contact force; (**b**) WLRR: wheel load reduction rate; (**c**) vertical acceleration at the car body floor.

Analysis of the vertical acceleration of car bodies reveals two significant peaks as the wavelength increases, as illustrated in Figure 6c. The first peak, which occurs within the wavelength range of 0.1 to 4 m, may be attributed to the heightened WRIs that result from the increased vertical vibrations of the car body, as it is irrelevant to the train speed. The location of the second peak, whose wavelength shifts with increasing speed, is identified as that of the sensitive wavelength. In this context, wavelengths of 4, 9, 14, 19, and 23 m, corresponding with vehicle speeds of 10, 20, 30, 40, and 50 km/h, respectively, are identified. Moreover, in Figure 6a,b, WRIs also exhibit slight increases at these specific wavelengths. These increases can be attributed to the enhanced coupled vibrations of the vehicle body.

3.1.2. Wavelengths for Heavy Trains

The variations of WRI and vertical accelerations of a heavy train at different speeds are illustrated in Figure 7. Comparison between Figures 6 and 7 yields similar findings. For example, it is observed that, when the wavelength is less than 5 m, WRIs of the heavy train increase rapidly as the wavelength decreases. In addition, as the vehicle speed increases, the variation trends of WRIs for heavy trains exhibit similar patterns compared with light trains. Furthermore, when the wavelength exceeds 5 m, WRIs of the heavy train tend to stabilize, with slight increases observed within the range of 5 to 28 m.

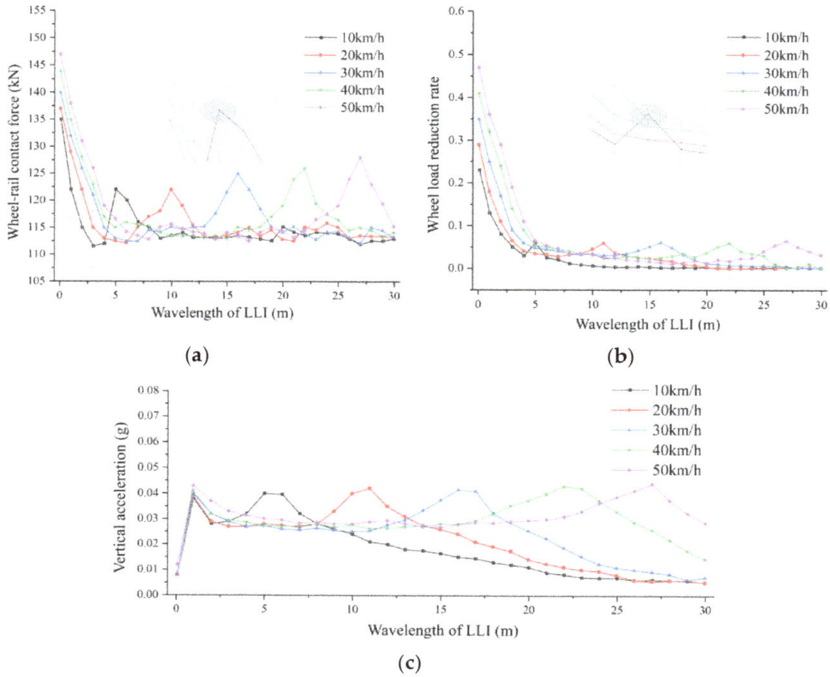

Figure 7. Effects of wavelengths on the dynamic performance of a heavy monorail train. (**a**) Wheel–rail contact force; (**b**) wheel load reduction rate; (**c**) vertical acceleration at the car body floor.

Analysis of the vertical acceleration responses led to the determination of the sensitive wavelengths of heavy trains, as shown in Figure 7c. The first peak corresponds with the wavelength range of 0.1 to 5 m. The second peak's wavelength increases with vehicle speed, with wavelengths of 5, 11, 16, 22, and 27 m corresponding with speeds of 10, 20, 30, 40, and 50 km/h, respectively. It should be noted that the curves in Figure 7c, especially those for 40 km/h and 50 km/h, illustrate subtle differences in terms of the long tail area. In fact, this is mainly due to the limited coordinate range of the x axis. The reason behind this issue is that, as the speed and train load increase to a certain level (in this case, 40 km/h), they induce more significant dynamic responses to the coupled system. Hence, although at the same speed, a heavier train load may induce stronger responses which cannot be reduced as quickly as in the cases of trains travelling at 10, 20, or 30 km/h. In other words, this issue partially reflects the impact of train load on dynamic responses.

3.1.3. Distribution Patterns of Sensitive Wavelengths

Previous results demonstrate that, at the same speed, different vehicle weights similarly affect the distribution of sensitive wavelengths. This section provides a qualitative analysis of how sensitive wavelengths vary with vehicle weight, using selected speed conditions as examples. The wheel–rail vertical force, wheel load reduction rate, and vertical acceleration of the vehicle for various weights at a speed of 30 km/h are illustrated in Figure 8. Analysis indicates that the wheel–rail vertical force generally increases with the increase in vehicle weight. Additionally, both the wheel load reduction rate and the peak vertical acceleration also rise as vehicle weight increases.

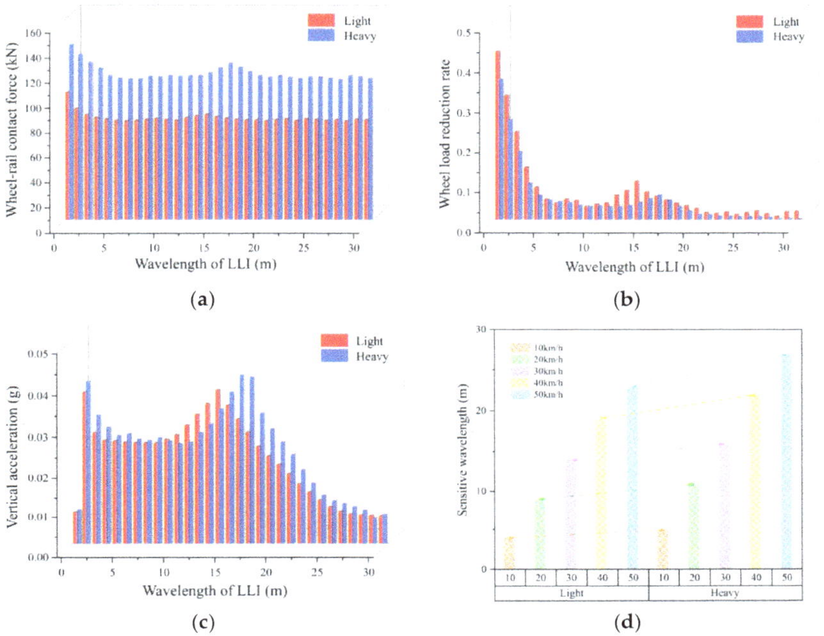

Figure 8. Dynamic performance of the monorail train at 30 km/h. (**a**) Wheel–rail contact force; (**b**) wheel load reduction rate; (**c**) vertical acceleration at the car body floor; (**d**) Variation of sensitive wavelengths.

Figure 8d illustrates the relationship between sensitive wavelengths due to unevenness and train weight under different speed conditions. Analysis reveals that, at the same speed, sensitive wavelengths increase with the increase in vehicle weight. Furthermore, as speed rises, the variation in sensitive wavelengths with respect to vehicle weight becomes more pronounced. For a given vehicle weight, sensitive wavelengths also increase with increasing speed and tend to exhibit a linear growth pattern.

In summary, the wheel–rail vertical force and wheel load reduction rate of light vehicles are particularly sensitive to wavelengths shorter than 4 m, with response peak wavelengths of 4, 9, 14, 19, and 23 m, corresponding with speeds of 10, 20, 30, 40, and 50 km/h, respectively. For heavy vehicles, the wheel–rail vertical force and wheel load reduction rate are sensitive to wavelengths shorter than 5 m, with response peak wavelengths of 5, 11, 16, 22, and 27 m at the same speeds. These results are summarized in Table 1. Additionally, the most unfavorable wavelength increases with vehicle weight, and this effect becomes more pronounced at higher speeds. For a given vehicle weight, the most unfavorable wavelength increases linearly with increasing speed.

Table 1. Sensitive wavelengths of LLI under different conditions.

Train Speed (km/h)	Light	Heavy
10	3, 4, 5	4, 5, 6
20	8, 9, 10	9, 10, 11
30	13, 14, 15	15, 16, 17
40	17, 18, 19	21, 22, 23
50	22, 23, 24	26, 27, 28

3.2. Amplitude Control of LLI

In the engineering design of track beams, design standards about the track irregularity of monorail systems remain unclear, and the range of some key parameters is uncertain. This ambiguity primarily arises from a lack of comprehensive analysis. During the production of track beams for monorail systems, issues such as poor fabrication can lead to significant track irregularity. Additionally, improper maintenance during operation can exacerbate track irregularity due to, for example, corrosion of the steel beams.

Currently, the track beam of an MTTS is designed in accordance with the "Large-scale amusement device safety code" (GB 8408-2018) and the "Specifications of amusement rides fairy train category" (GB/T 18165-2008). Relevant guidelines include the design specification for urban rail transit bridges (GB/T 51234-2017) and monorail systems (GB 50458-2022). These specifications encompass a broad range of topics, including structural design and traffic organization, yet do not specifically address the amplitude of unevenness in track beam structures. As a critical source of excitation for vibrations in the monorail system, track unevenness directly affects the smooth operation of a tour bus. In track management, both the wavelength and amplitude of track unevenness serve as key indicators for assessing track quality. Therefore, further research in this area is essential.

3.2.1. Evaluation Metrics

The management of track irregularities is based on the principle of ensuring safe and smooth operation while also considering maintenance capabilities and the economic efficiency of normal operations. Existing standards for managing track irregularities have been formulated with a focus on the induced dynamic response of trains. In developing such standards, the primary objective is to ensure operation safety. In railway systems, key safety evaluation indicators typically include the derailment coefficient, which is not the case for monorails, and wheel load reduction rate. Additionally, the vertical acceleration of the vehicle body serves as another important evaluation metric, primarily used to assess the smoothness and comfort of the vehicle's operation.

Based on related design specifications and maintenance rules in railway systems, such as GB5599-2019 (Specification for dynamic performance assessment and testing verification of rolling stock), the wheel load reduction rate and acceleration of car body are proposed as the indicators for controlling the amplitude of track irregularity in MTTS, summarized in Table 2.

Table 2. Indicators for vehicle operating comfort and safety and their management values.

Objectives	Indicators	Management Levels			
		Level I	Level II	Level III	Level IV
Safety	Wheel load reduction rate	$\leq$0.65			
Comfort	Vertical acc.	0.1 g	0.15 g	0.20 g	0.25 g
	Horizontal acc.	0.06 g	0.09 g	0.15 g	0.20 g

Note: Levels I to IV represent regular maintenance, comfort level, temporary repairs needed, and speed limit needed, respectively; acc. denotes acceleration response.

3.2.2. Simulation of Track Irregularity

In this section, the geometric track irregularities are described using the one-sided cosine wave model as outlined in the previous section. Taking a wavelength of 10 m as an example, the one-sided cosine wave model is illustrated in Figure 9. This model is utilized to study the amplitude limits in this chapter by adjusting the amplitude values. Subsequently, this model will be input into the joint dynamic analysis model, where it will be converted into track irregularities based on the embedded algorithms. This conversion will facilitate the dynamic analysis of the system.

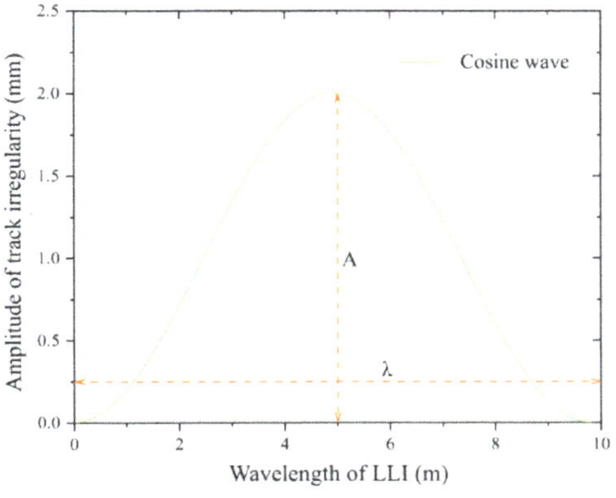

Figure 9. Simulation of LLI using cosine function.

To analyze the critical amplitude of wavelength sensitivity to irregularities, the typical wavelength was the one from which the dynamic index peaks was selected, based on the analysis results within the sensitive wavelength range. This study investigates the variation in vehicle dynamic response with respect to different amplitudes of LLI under various speed and load conditions. The typical sensitive wavelengths for different speeds and vehicle weights have been summarized in Table 1.

3.2.3. Light Train Condition

From the previous analysis, it can be concluded that the vertical acceleration of the vehicle body and the load reduction rate on the wheels are most sensitive to LLI. The impact of varying amplitudes of LLI on the vehicle body response at different speeds is illustrated in Figure 10. For simple demonstration, only the cases of 40 km/h and 50 km/h are shown. It can be observed that, under typical wavelength conditions, the dynamic indicators tend to increase linearly with the increase in the amplitudes of high and low irregularities. Taking the speed condition of 40 km/h as an example, the dynamic response corresponding with a wavelength of 18 m is greater than that of the other two wavelengths. Next, the amplitude limits of LLI are determined by the cut-off points in Figure 10 according to the limited responses in Table 2. Therefore, for the case of 40 km/h, taking the vertical acceleration as the indicator, the amplitude limits of levels I to IV are determined as 4, 6, 8, and 10 m, respectively. The amplitude limits of LLI for light trains under various speed conditions are summarized in Table 3.

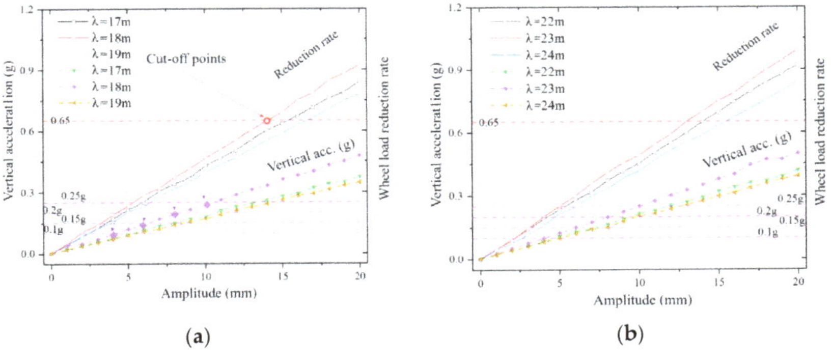

Figure 10. Effects of amplitude changes on the dynamics performance of light vehicles: (**a**) 40 km/h; (**b**) 50 km/h. Note: λ denotes the wavelength.

Table 3. Sensitive wavelengths of LLI under different conditions.

Train Speed (km/h)	Indicator	Limit Value (mm)				Indicator	Limit Value (mm)
		Level I	Level II	Level III	Level IV		Level I to Level IV
10	Vertical acc.	5	8	11	14	WLRR	15
20		5	8	10	13		14
30		5	7	9	12		14
40		4	6	8	10		14
50		4	6	8	10		13

Note: WLRR and acc. denote wheel load reduction rate and acceleration, respectively.

3.2.4. Heavy Train Condition

Analogous to light trains, under typical wavelength conditions, the wheel load reduction rate and vertical acceleration tend to increase linearly with the increase in the amplitude of vertical and lateral irregularities. The impact of changes in the amplitude of LLI for heavy vehicles is illustrated in Figure 11. Because the higher the speed, the lower the amplitude limits, Figure 11 only presents responses under the conditions of 40 km/h and 50 km/h. Similarly, taking the speed condition of 50 km/h as an example, the dynamic response corresponding with a wavelength of 27 m is greater than that of the other two wavelengths. According to the analysis of Table 2 and Figure 11, the amplitude limits for level I to IV are determined as 3, 6, 8, and 10 m, respectively, in terms of vertical acceleration; from the perspective of wheel load reduction rate, the limit is 13 m. The amplitude limits of LLI for heavy vehicles under various speed conditions are presented in Table 4.

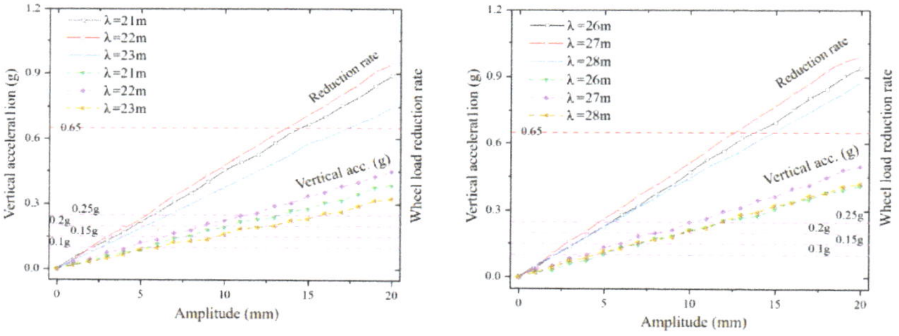

(a) (b)

Figure 11. Effects of amplitude changes on the dynamics performance of heavy trains: (**a**) 40 km/h; (**b**) 50 km/h.

Table 4. Sensitive wavelengths of LLI under different conditions (for heavy trains).

Train Speed (km/h)	Indicator	Limit Value (mm)				Indicator	Limit Value (mm)
		Level I	Level II	Level III	Level IV		Level I to Level IV
10	VA	4	7	9	12	WLRR	15
20		4	7	10	11		14
30		4	6	9	11		13
40		4	6	8	10		13
50		3	6	8	10		12

Note: WLRR and VA denote wheel load reduction rate and vertical acceleration, respectively.

Analysis of Tables 3 and 4 reveals that the amplitude limits defined by the acceleration criterion are more stringent than those defined by the wheel load reduction rate. Therefore, it is suggested to utilize the acceleration criterion to set the limits on the amplitude of vertical and lateral irregularities for track beams.

4. Discussion

4.1. Parametric Study of the Developed Joint Model

To illustrate a system with multiple degrees of freedom, the equation of motion under an externally applied time-dependent force is represented by Equation (2). In Section 2, a joint model utilizing FEM and MBD is developed. To examine the robustness of the model, a parametric study is presented. To achieve this, the damping of the track beam is chosen as the parameter. A common method of modelling damping is so-called Rayleigh damping, which approximates the damping coefficient as a linear combination of mass and stiffness, as in the following Equation (3).

$$[M]\{\ddot{X}\} + [C]\{\dot{X}\} + [K]\{X\} = \{P_t\} \tag{2}$$

$$[C] = \alpha[M] + \beta[K] \tag{3}$$

where the force vector $\{P_t\}$ is a function of time; $[M]$, $[C]$, and $[K]$ denote the mass, damping, and stiffness matrix; and α and β are two coefficients to determine the Rayleigh damping.

In this section, coefficient of the stiffness matrix, β, is selected as the key parameter, ranging from 2×10^{-4} to 2×10^{-3}. Then, the vertical displacement of track beam under trains passing through at speeds of 30 km/h and 50 km/h are simulated. The results are shown in Figure 12. For comparison, the dynamic response of MTTS with an overall structure damping of 0.02 is obtained and serves as the normalization baseline. As shown, as the damping coefficient increases, the induced responses generally reduce. The higher the train speed, or the more pronounced the train-induced dynamic effects, the greater the impact of the damping coefficient, as illustrated in Figure 12b. Nevertheless, within the simulated neighboring area (colored) of selected damping coefficient (0.02), the maximum responses do not change significantly, which indicates the robustness of the developed model.

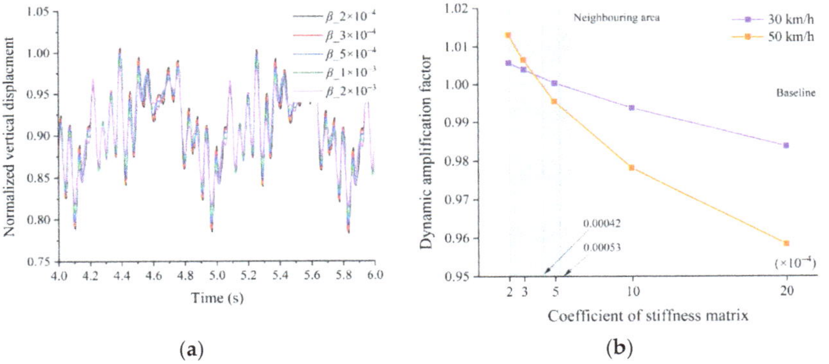

Figure 12. Simulation results of the dynamic performance under various β values: (**a**) Time history curves of normalized vertical displacement (train-induced part); (**b**) amplification factor of the maximum displacement.

4.2. Sensitive Wavelength and Amplitude Control of MTTS

Results of sensitive wavelengths obtained in the previous section are of great significance to the control or management of the sensitive wavelength of a monorail track beam. Studies on railways generally show that long-wavelength track irregularities significantly influence the ride comfort and running stability of trains. However, according to the variation of vertical acceleration responses, it is observed that relatively shorter wavelengths can introduce significant dynamic responses in MTTS, which highlights the priority of managing short wavelengths in monorails. In addition, this finding corresponds with research regarding suspended monorails. For instance, the results in [40] indicate that the sensitive wavelengths that amplify the vertical and lateral accelerations of suspended monorail trains mainly fall within the range of 1.3 to 2.3 m. Nevertheless, there are also distinct differences in structural configurations between MTTS (straddle-type) and suspended monorails, hence, similar results may not arise from similar reasons.

Moreover, results from simulated LLI with various amplitudes demonstrate that, for monorails, the tolerance level for track irregularities is close to that in railways, where simulations [41] and rules of thumb suggest a tolerance of 5 to 6 mm for high-level riding comfort. Nevertheless, straddle-type monorails can be safely operated even on track beams with relatively high-level irregularities thanks to their unique structural configuration and running mode.

4.3. Limitations

In this work, the sensitive wavelength of LLI and its effects on the dynamic responses of MTTS were evaluated. Based on the calculated results under various conditions of amplitude and frequency of LLI, limit values of irregularity amplitude were obtained. During the determination of sensitive wavelengths, the selection of wavelength intervals considers both the simulation efficiency and convenience in practical management. As distinct peaks are observed in the contribution curves, a more detailed division on wavelengths was not conducted. It should also be noted that, during the analysis, horizontal level irregularity and the irregularities on the web plates are not included. This is mainly because of the limited amount of field-measured data regarding these types of irregularities. In addition, MTTS trains adopt secondary suspensions with excellent dynamic performance of vibration control. In this context, the position changes caused by the web plates, just like the gauge level irregularity in railways, will not be directly reflected on the actual position of the bogie. Moreover, an accurate simulation of this effect involves an enhanced modelling on the dynamic characteristics of the suspension systems.

Nevertheless, as the monorail track beam has relatively weak stiffness, this may be a potential effect that can affect the results of irregularity-induced responses. In addition, the analysis in this work is conducted based on numerical simulation. If a comprehensive database of measured track irregularities is available, further analysis can be conducted based on the excitations of actual track irregularity signal with distinct frequency features.

5. Conclusions

To address the limited knowledge on the sensitive wavelength of track irregularities in MTTS and general monorail systems, this work presents a comprehensive simulation analysis based on a validated joint model. The main conclusions are as follows:

1. By combining MBD and FEM, a joint simulation model for monorail systems can be efficiently established—simply by extracting the results of the modal analysis from FEM and transforming them into flexible objects in the MBD. Moreover, the developed model was verified by the field test results, thereby assuring the effectiveness of the subsequent analysis.
2. To evaluate the sensitive wavelength of track irregularity, cosine wave functions with various amplitudes and frequencies were utilized. By integrating the simulated signal, the vertical acceleration and wheel load reduction rate of MTTS were obtained and identified as the indicators of sensitive wavelengths.
3. The results of the simulation demonstrate that, within the tested speed range, the sensitive wavelength arises gently with the increase in train load. Under the same train load, the sensitive wavelength increases linearly with increasing speed.
4. Analysis of the limit of the irregularity amplitude was performed: both the wheel load reduction rate and the vertical acceleration increase linearly with the rise in the amplitude of track irregularities. Limits for the amplitude of LLI in MTTS are given.

5. Comparison between various transportation systems indicates that more attention should be paid to track irregularities with short wavelengths. To maintain a high level of ride comfort, the amplitude of track irregularity should be less than approximately 4 to 5 mm under most operating conditions.

To sum up, this work presents meaningful results of sensitive wavelengths in MTTS. Obtained results can form a good reference for future studies and enlarge the research borders associated with the dynamic responses of monorail systems.

Supplementary Materials: The following supporting information can be downloaded at: https://www.mdpi.com/article/10.3390/buildings14124041/s1, Table S1: Major parameters for the MBD model of the train in the test line.

Author Contributions: Methodology, P.W.; software and formal analysis, H.Z. and Y.J.; data curation, C.C., S.W., B.G., Z.L. and J.Z.; writing—original draft preparation, Y.J. and P.W.; writing—review and editing, H.Z., C.C., S.W., B.G., Z.L. and J.Z.; supervision and project administration, H.Z. All authors have read and agreed to the published version of the manuscript.

Funding: This research received no external funding.

Data Availability Statement: The data presented in this study may be available on reasonable request.

Conflicts of Interest: Authors Chen Chen, Zhiqiang Liu, and Jie Zhang were employed by the 3rd Construction Co., Ltd. of China Construction 5th Engineering Bureau. The remaining authors declare that the research was conducted in the absence of any commercial or financial relationships that could be construed as potential conflicts of interest.

References

1. Guo, F.; Chen, K.; Gu, F.; Wang, H.; Wen, T. Reviews on current situation and development of straddle-type monorail tour transit system in China. *J. Cent. South Univ. Sci. Technol* **2021**, *52*, 4540–4551.
2. Guo, F.; Wang, P. Measurement and analysis of the longitudinal level irregularity of the track beam in monorail tour-transit systems. *Sci. Rep.* **2022**, *12*, 19219. [CrossRef] [PubMed]
3. Wiberg, J.; Karoumi, R. Monitoring dynamic behaviour of a long-span railway bridge. *Struct. Infrastruct. Eng.* **2009**, *5*, 419–433. [CrossRef]
4. Galvín, P.; Romero, A.; Moliner, E.; De Roeck, G.; Martínez-Rodrigo, M.D. On the dynamic characterisation of railway bridges through experimental testing. *Eng. Struct.* **2021**, *226*, 111261. [CrossRef]
5. Li, H.; Wekezer, J.; Kwasniewski, L. Dynamic response of a highway bridge subjected to moving vehicles. *J. Bridge Eng.* **2008**, *13*, 439–448. [CrossRef]
6. Erdogan, Y.S.; Catbas, N.F. Seismic response of a highway bridge in case of vehicle-bridge dynamic interaction. *Earthq. Struct.* **2020**, *18*, 1–14.
7. Naseri, R.; Mohammadzadeh, S.; Rizos, D.C. Rail surface spot irregularity effects in vehicle-track interaction simulations of train-track-bridge interaction. *J. Vib. Control* **2024**, 10775463241232024. [CrossRef]
8. Melo, L.T.; Malveiro, J.; Ribeiro, D.; Calçada, R.; Bittencourt, T. Dynamic analysis of the train-bridge system considering the non-linear behaviour of the track-deck interface. *Eng. Struct.* **2020**, *220*, 110980.
9. Matsuoka, K.; Tanaka, H.; Kawasaki, K.; Somaschini, C.; Collina, A. Drive-by methodology to identify resonant bridges using track irregularity measured by high-speed trains. *Mech. Syst. Signal Process.* **2021**, *158*, 107667. [CrossRef]
10. Muñoz, S.; Ros, J.; Urda, P.; Escalona, J.L. Estimation of lateral track irregularity using a Kalman filter. Experimental validation. *J. Sound Vib.* **2021**, *504*, 116122. [CrossRef]
11. Bosso, N.; Bracciali, A.; Megna, G.; Zampieri, N. Effects of geometric track irregularities on vehicle dynamic behaviour when running through a turnout. *Veh. Syst. Dyn.* **2023**, *61*, 782–798. [CrossRef]
12. Wang, K.; Liu, P.; Zhai, W.; Huang, C.; Chen, Z.; Gao, J. Wheel/rail dynamic interaction due to excitation of rail corrugation in high-speed railway. *Sci. China Technol. Sci.* **2015**, *58*, 226–235. [CrossRef]
13. Karis, T.; Berg, M.; Stichel, S.; Li, M.; Thomas, D.; Dirks, B. Correlation of track irregularities and vehicle responses based on measured data. *Veh. Syst. Dyn.* **2018**, *56*, 967–981. [CrossRef]
14. Gao, J.; Zhai, W.; Wang, K. Study on Sensitive Wavelengths of Track Irregularities in High-speed Operation. *J. China Railw. Soc.* **2012**, *34*, 83–88. [CrossRef]
15. Croft, B.; Jones, C.; Thompson, D. Modelling the effect of rail dampers on wheel–rail interaction forces and rail roughness growth rates. *J. Sound Vib.* **2009**, *323*, 17–32. [CrossRef]

16. Nielsen, J.C.; Pieringer, A.; Thompson, D.J.; Torstensson, P.T. Wheel–rail impact loads, noise and vibration: A review of excitation mechanisms, prediction methods and mitigation measures. In *Noise and Vibration Mitigation for Rail Transportation Systems: Proceedings of the 13th International Workshop on Railway Noise, Ghent, Belgium, 16–20 September 2019*; Springer International Publishing: Berlin/Heidelberg, Germany, 2021; pp. 3–40.
17. Hung, C.; Hsu, W. Influence of long-wavelength track irregularities on the motion of a high-speed train. *Veh. Syst. Dyn.* **2018**, *56*, 95–112. [CrossRef]
18. Jiang, L.; Li, Y.; Zhao, Y.; Cen, M. The Characteristics of Long-Wave Irregularities in High-Speed Railway Vertical Curves and Method for Mitigation. *Sensors* **2024**, *24*, 4403. [CrossRef]
19. Li, M.; Berggren, E.; Berg, M.; Persson, I. Assessing track geometry quality based on wavelength spectra and track–vehicle dynamic interaction. *Veh. Syst. Dyn.* **2008**, *46*, 261–276. [CrossRef]
20. Giannakos, K. Modeling the influence of short wavelength defects in a railway track on the dynamic behavior of the Non-Suspended Masses. *Mech. Syst. Signal Process.* **2016**, *68*, 68–83. [CrossRef]
21. Xin, T.; Wang, P.; Ding, Y. Effect of Long-Wavelength Track Irregularities on Vehicle Dynamic Responses. *Shock Vib.* **2019**, *2019*, 4178065. [CrossRef]
22. Lu, T.; Chen, J.; Sun, X.; Wei, Z.; Yang, F.; Chen, R.; Wang, P. The stochastic characteristics of wideband-wavelength track irregularity on Chinese HSRs and its application to dynamic wheel–rail interaction. *Veh. Syst. Dyn.* **2024**, 1–24. [CrossRef]
23. Ding, W.; Wei, K.; Zhao, Z.; Cheng, F.; Wang, P. A reliability analysis method of dynamic irregularity for track–viaduct system with low stiffness. *Mech. Syst. Signal Process.* **2022**, *167*, 108518. [CrossRef]
24. Lee, C.H.; Kawatani, M.; Kim, C.W.; Nishimura, N.; Kobayashi, Y. Dynamic response of a monorail steel bridge under a moving train. *J. Sound Vib.* **2006**, *294*, 562–579. [CrossRef]
25. Wang, P.; Guo, F.; Zhang, H.; Jin, J.; Liao, Q.; Yan, Y. Acquiring the High-Precision Spectrum of Track Irregularity by Integrating Inclination in Chord Methods: Mathematics, Simulation, and a Case Study. *Mathematics (2227–7390)* **2024**, *12*, 2197. [CrossRef]
26. Naeimi, M.; Tatari, M.; Esmaeilzadeh, A.; Mehrali, M. Dynamic interaction of the monorail–bridge system using a combined finite element multibody-based model. *Proc. Inst. Mech. Eng. Part K J. Multi-Body Dyn.* **2015**, *229*, 132–151. [CrossRef]
27. Wang, H.; Zhu, E. Dynamic response analysis of monorail steel-concrete composite beam-train interaction system considering slip effect. *Eng. Struct.* **2018**, *160*, 257–269. [CrossRef]
28. Zhou, J.; Du, Z.; Yang, Z.; Xu, Z. Dynamics study of straddle-type monorail vehicle with single-axle bogies-based full-scale rigid-flexible coupling dynamic model. *IEEE Access* **2019**, *7*, 110249–110257. [CrossRef]
29. Zhang, H.; Wang, P.; Li, Q.; Jin, J.; Wei, S.; Guo, F.; Feng, C.; Deng, Q. An Experimental and Numerical Study on the Lateral Stiffness Limits of Straddle-Type Monorail Tour-Transit Systems. *Buildings* **2024**, *14*, 3111. [CrossRef]
30. Zou, Y.; Liu, Z.; Shi, K.; He, X.; Zhou, S.; Ma, W. Analysis of effects of aerodynamic interference on dynamic response of suspended monorail wind–vehicle–bridge system using joint simulation approach. In *Structures*; Elsevier: Amsterdam, The Netherlands, 2022; pp. 179–198.
31. Li, Y.; Xu, X.; Zhou, Y.; Cai, C.; Qin, J. An interactive method for the analysis of the simulation of vehicle–bridge coupling vibration using ANSYS and SIMPACK. *Proc. Inst. Mech. Eng. Part F J. Rail Rapid Transit* **2018**, *232*, 663–679. [CrossRef]
32. Huang, G.; Ren, N.; Zheng, H. Structural Dynamics of Bridges Under the Coupling Effect of Windmills and Bridges. *Eur. J. Comput. Mech.* **2022**, *31*, 601–620. [CrossRef]
33. Chen, J.; Li, R.; Xu, J.; Li, P.; Zhao, W.; Peijie, Z.; Jin, K. Study on the aerodynamic performances and the operational safety of the vehicle-bridge system under the crosswind. *Mech. Based Des. Struct. Mach.* **2024**, *52*, 7962–7984. [CrossRef]
34. Wang, S.; Wan, X.; Guo, M.; Qiao, H.; Zhang, N.; Ye, Q. Nonlinear dynamic analysis of the wind–train–bridge system of a long-span railway suspension truss bridge. *Buildings* **2023**, *13*, 277. [CrossRef]
35. Zhang, C. The active rotary inertia driver system for flutter vibration control of bridges and various promising applications. *Sci. China Technol. Sci.* **2023**, *66*, 390–405. [CrossRef]
36. Du, Z.; Xin, L.; Yang, Z.; Xu, Z. Improving ride comfort of straddle-type monorail vehicle with single-axle bogie by active suspension. *Noise Vib. Worldw.* **2021**, *52*, 334–344. [CrossRef]
37. Dassault_Systemes ABAQUS, ABAQUS User's Manual. 2016. Available online: http://62.108.178.35:2080/v2016/index.html (accessed on 16 December 2024).
38. Dassault_Systemes SIMPACK. Version 2021x. 2021. Available online: https://blog.3ds.com/brands/simulia/simpack-r2021x-multibody-system-simulation-updates/ (accessed on 16 December 2024).
39. Cano-Moreno, J.; de Pedro, J.; Villagrá, M.R.; González, R.G. Design and development of a data acquisition system to monitor comfort and safety in railways. *Insight-Non-Destr. Test. Cond. Monit.* **2021**, *63*, 47–54. [CrossRef]
40. He, Q.; Cai, C.; Zhu, S.; Zhang, J.; Wang, K.; Zhai, W. Field measurement of the dynamic responses of a suspended monorail train–bridge system. *Proc. Inst. Mech. Eng. Part F J. Rail Rapid Transit* **2020**, *234*, 1093–1108. [CrossRef]
41. Xu, L.; Zhai, W. A novel model for determining the amplitude-wavelength limits of track irregularities accompanied by a reliability assessment in railway vehicle-track dynamics. *Mech. Syst. Signal Process.* **2017**, *86*, 260–277. [CrossRef]

buildings

MDPI

Article

Investigation of the Hysteresis Performance of Multi-Story Y-Shaped Eccentrically Bare Braced Steel Frame with Block Slit Damper (BSD)

Chen Zhou [1], Haibo Wen [1,*], Jing Zhao [2], Lizhong Jiang [1], Xiaodong Xu [3], Hong Zheng [3], Yuxiao Tian [4], Muhua Liang [4], Xiyao Wang [2] and Hai Zhang [4]

[1] School of Civil Engineering, Central South University, Changsha 410075, China; 13578623580@126.com (C.Z.); lzhjiang@csu.edu.cn (L.J.)
[2] 3rd Construction Co., Ltd., China Construction 5th Engineering Bureau, Changsha 410004, China; yiyun21709@163.com (J.Z.); cscec53hnhw@163.com (X.W.)
[3] School of Civil Engineering, Chang'an University, Xi'an 710061, China; 21020201085@mail.hnust.edu.cn (X.X.); cehzheng@chd.edu.cn (H.Z.)
[4] China Railway Urban Construction Group Co., Ltd., Changsha 410036, China; zjq54610@163.com (Y.T.); 13786282768@139.com (M.L.); fusu0116@126.com (H.Z.)
* Correspondence: haibowen_2000@163.com

Abstract: This paper investigated the hysteresis performance of multi-story Y-shaped eccentrically bare braced steel frames with block slit dampers (BSDs). After validating existing models through finite element (FE) analysis, an FE model of Y-shaped eccentrically bare braced steel frames (YEBFs) with BSD (BSD-YEBFs) was established. The influences of different types of BSDs on the hysteresis performance of YEBFs were subsequently analysed, as were the seismic performances of YEBFs with BSD (BSDF) and traditional Y-shaped eccentrically bare braced steel frames (TEBF) under cyclic loading, including the stress failure mode, hysteresis curve, stiffness, strength, and energy dissipation capacity. The results showed that: (1) compared with those of BASE-Y, the ductility coefficients of YEBFs with one level BSD (BSD-1) and two level BSD (BSD-2) increased by 59.57% and 39.47%, respectively, and the lateral bearing capacity decreased by 3.26% and 2.78%, respectively. (2) Compared with those of TEBF, the yield bearing capacity and ultimate bearing capacity of BSDF increased by 3.13% and 11.12%, respectively, and the ductility coefficient increased by 56.7%. Moreover, BSDF possesses higher initial stiffness and more sustained energy dissipation capabilities.

Keywords: Y-shaped eccentrically bare braced steel frames; block slit dampers (BSDs); hysteresis performance; finite element modeling

1. Introduction

Earthquakes are highly hazardous natural disasters characterized by uncertainty and contingency. Each outbreak poses a threat to human life and property safety. Steel structures have been widely used in buildings because of their excellent seismic performance [1,2].

In many earthquake-resistant areas of China, multi-story steel structure buildings usually adopt a bending steel frame structure composed of frame beams and frame columns with good energy dissipation and ductility performance. However, the lateral stiffness of steel frames is relatively low, and setting eccentric braces in the steel frame can effectively improve this situation. Eccentrically braced steel frame (EBF) structures represent effective seismic-resistant systems. During seismic events, these structures utilize the plastic deformation of energy-dissipating beam sections to dissipate seismic energy, limiting brace buckling and maintaining overall structural stability [3,4]. Currently, many works have tested different types of EBFs, and the results show that EBFs can significantly improve the seismic resistance of steel frame structures [5–10]. Eccentric braces can be categorized into various types, such as D-type, K-type, V-type and Y-type braces, on the basis of their geometric configuration. Compared with other braces, the Y-shaped eccentric braces (Figure 1a) have relatively high independence because the energy-dissipating beam section is set separately below the frame beam. Xiao et al. [6] conducted a study on the seismic performance of Y-type EBF and K-type EBF under seismic action, and the results indicated that the Y-type EBF has superior energy dissipation capabilities compared with the K-type EBF. Wang et al. [10] evaluated the seismic performance of four types (D, K, Y, and V) of EBF structures via pushover analysis and nonlinear time-history dynamic analysis, and the findings demonstrated that the Y-shaped EBF exhibited better ductility.

(**a**) Traditional YEBFs

(**b**) Block slit damper

(**c**) YEBFs with BSD

Figure 1. Schematic diagram of steel frames and dampers.

However, under strong seismic events, the energy-dissipating beam section of EBF structures may reach their ultimate deformation, potentially leading to severe structural damage. Concurrently, significant residual deformation of a structure can result in a complex and costly post-earthquake repair process [11,12]. It is necessary to further enhance the ductility and energy dissipation capacity of the energy-dissipating beam section in EBF structures to reduce the damage and residual deformation in these sections, thus adapting to the higher seismic fortification standards currently proposed. Many studies have considered the use of dampers, including viscous dampers [13], friction dampers [14], and metal dampers [15], to enhance the seismic performance of steel frames. Zhao et al. [16] utilized a novel bent shear panel damper to improve the seismic performance of YEBFs, thereby reducing damage to concrete slabs under seismic loading. Li et al. [17] explored the application of a demountable-metallic-corrugated-shear-panel damper (DCSPD) in YEBFs and evaluated the dynamic seismic performance of DCSPDs through dynamic time history analysis and seismic fragility analysis. Mirzai et al. [18] investigated the application of self-centring dampers in YEBFs, which consist of a combination of shape memory alloy (SMA) and lead rubber damper (LRD). This device utilizes the self-centring characteristics of SMA and the energy dissipation capabilities of LRD to enhance the seismic performance of structures. Meanwhile, several studies [1,19] have considered employing easily replaceable energy dissipation devices within the EBF structures. In addition, there has been considerable interest in the application of negative Poisson's ratio (NPR) materials in dampers [20]; NPR structures are capable of significantly enhancing energy dissipation by undergoing substantial plastic deformations under both tensile and compressive stresses.

As is well known, metal dampers dissipate energy through the plastic deformation of their constitutive materials, offering advantages such as simple construction, stable hysteresis performance, and low-temperature sensitivity [21,22]. Recently, Amiri et al. [21] introduced a novel type of metal damper known as the block slit damper (BSD), which consists of a steel block with several slits and two steel plates (Figure 1b). This assembly provides high bearing capacity while maintaining the ability to dissipate energy continuously under large displacements. Their findings also revealed that as the aspect ratio (h_0/b_1) decreased, the shear and energy dissipation capabilities were augmented, whereas the displacement capacity was diminished. Subsequent research [23] utilized the endurance time (ET) dynamic analysis method to evaluate the horizontal seismic performance of low-rise steel frame structures equipped with BSDs. Mohseni et al. [24] employed numerical simulation methods to investigate the impact of BSD geometric parameters on seismic energy absorption.

Given the aforementioned information, this study proposed a YEBF with a block slit damper, as shown in Figure 1c. The hysteresis performance of YEBFs with different types of BSDs was investigated via the FE software ABAQUS 6.14. Furthermore, a detailed comparative analysis was conducted between YEBF with BSD (BSDF) and traditional YEBF (TEBF) to evaluate their hysteresis performance under cyclic loading. This work aimed to provide insights into the use of BSDs in YEBF structures.

2. FE Model Validation

2.1. Establishment of the FE Model

To validate the accuracy of the EBF model, this study established an FE model of specimen Y-1 from the 1/3 scale single-span, two-story frame test in reference [25] according to the relevant provisions of the Chinese standard (GB 50011-2010; JGJ 99-2015; GB 50017-2017) [26–28]. A schematic diagram of the YEBF is shown in Figure 2, and the detailed dimensions of the relevant components are presented in Table 1.

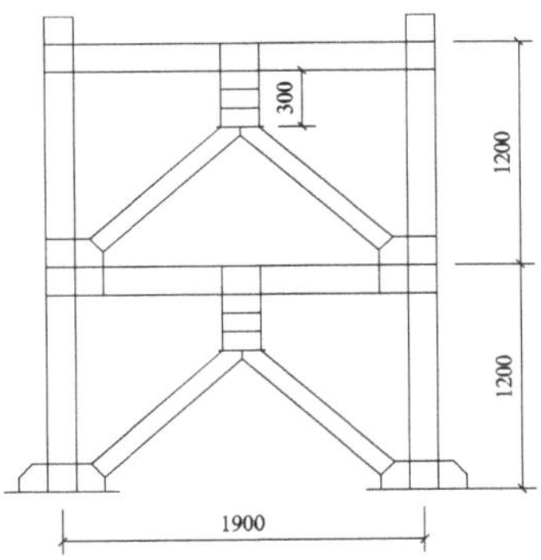

Figure 2. Dimensions of YEBF in reference [22] (unit: mm).

Table 1. Section size of each component of specimen Y-1 (unit: mm).

Element	Section ($h \times b_f \times t_w \times t_f$)
Column	H150 × 150 × 7 × 10
Beam	H150 × 150 × 7 × 10
Brace	H100 × 100 × 7 × 10
Energy Dissipation Beam	H200 × 100 × 5.5 × 8

In the experiment, all of the components were made of Q235 steel, with an elastic modulus of 206,000 MPa and a Poisson's ratio of 0.3. Detailed information regarding the material properties is shown in Table 2.

Table 2. Properties of the steel material.

Steel	σ_y/MPa	ε_y	σ_u/MPa	ε_u	σ_{st}/MPa	ε_{st}
Q235	235	0.001	420	0.15	330	0.22
Q355	355	0.002	550	0.116	450	0.17

The following assumptions are made regarding the properties of steel materials: (1) the steel material is homogeneous and isotropic; (2) initial defects in the steel are not considered; and (3) the yield criterion adopts the von Mises criterion. To accurately simulate the stress situation of steel under low-cycle reversed loading and large deformations [29] while taking into account the influence of the Bauschinger effect on steel properties, a multilinear mixed strengthening criterion and a tri-linear model with a descending segment were adopted, as shown in Figure 3.

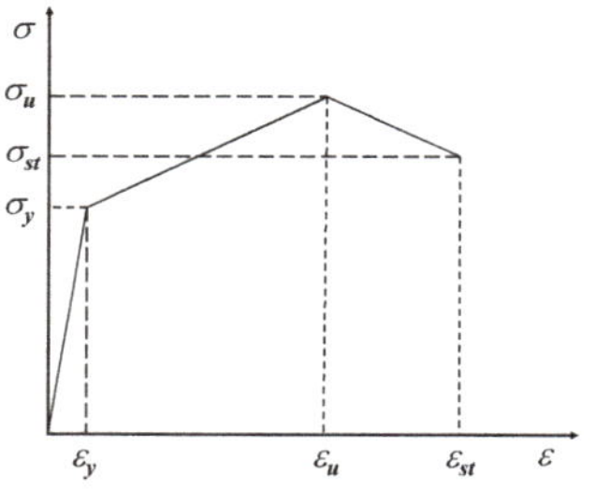

Figure 3. Stress–strain relationship of steel.

A three-dimensional solid hexahedral nonconforming element model (C3D8I) was employed to establish the YEBF model. The sharp parts of the bracing members were transitioned using a limited number of six-node mother prismatic elements (C3D6). The boundary conditions of the FE model are defined as follows: the frame columns and gusset plates are rigidly connected to the ground (U1 = U2 = U3 = UR1 = UR2 = UR3 = 0); and lateral constraints in the X direction of the frame beams (U1 = 0) are applied to restrict out-of-plane instability of the sample, as shown in Figure 4.

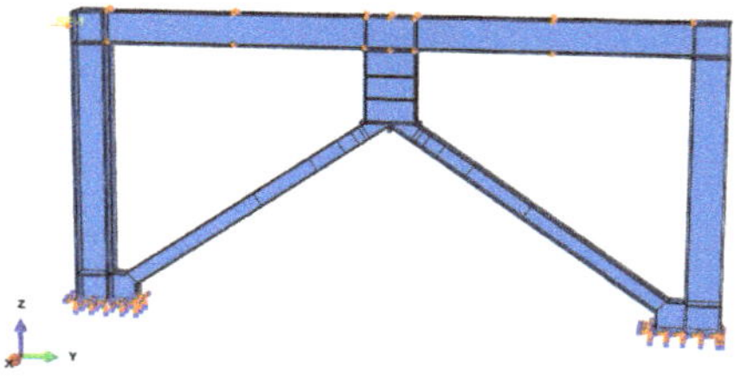

Figure 4. Boundary conditions.

The FE loading protocol adheres to the complete loading system specified in the standard [30]: 0.25 Δ_y, 0.5 Δ_y, 0.75 Δ_y, 1 Δ_y, 2 Δ_y, 3 Δ_y, 4 Δ_y, 5 Δ_y. . ., where Δ_y represents the yield displacement of the sample. Each loading increment is cycled twice, as shown in Figure 5.

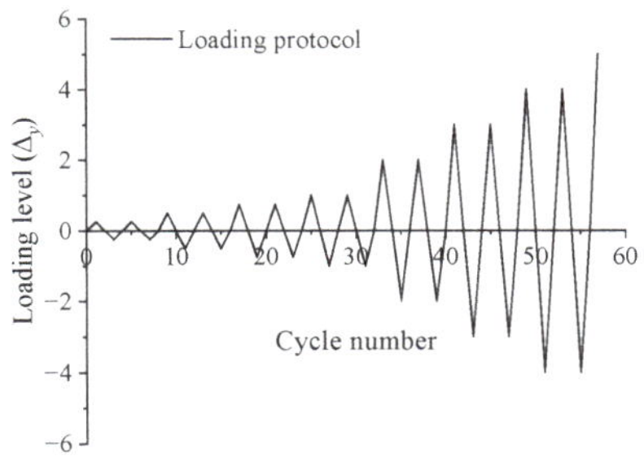

Figure 5. Loading protocol.

According to the calculation results, the grid size was adjusted as follows: the element size for the frame beams and frame columns was set to 60 mm, whereas the element size for the eccentric bracing members was 30 mm. To analyse the stress–strain conditions of the energy-dissipating beam section precisely, the grid in these areas was refined, with each plate component having an element size of 20 mm, as shown in Figure 6.

(**a**) Grid division of the steel frame. (**b**) Energy dissipation beam

Figure 6. Grid division of specimen Y-1.

2.2. *Comparison of the FEM and Test Results*

Figure 7 shows the failure modes of specimen Y-1 from both the experimental and FE analysis results. It is evident that the specimen experienced tearing and failure of the weld seam at the connection between the web plate and end plate of the energy dissipation beam section, as well as at the connection between the energy dissipation beam section and the flanges. The FE simulation results show that the failure locations are consistent with the experimental results and indicate that the stress values at these locations are significantly higher than the ultimate stress of the steel.

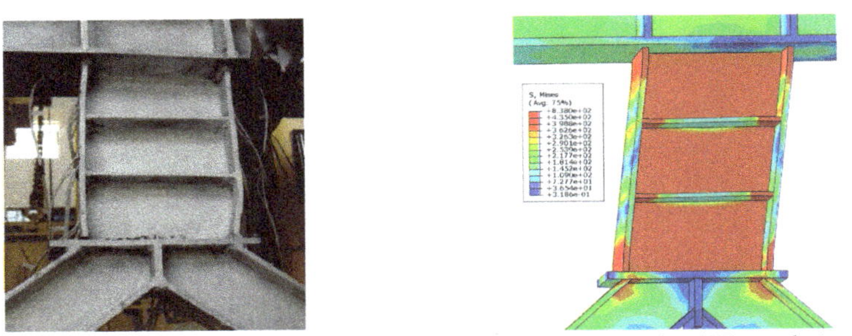

(**a**) Failure mode of Test (**b**) Failure mode of FE simulation

Figure 7. Failure mode of specimen Y-1.

The hysteresis and skeleton curves of specimen Y-1 from both the FE analysis and experimental tests are presented in Figure 8. The hysteresis curves of the two results are symmetrical and full, with a shape similar to a shuttle, and the enveloping areas of the hysteresis loops at each loading stage are roughly equal. The detailed parameters are shown in Table 3. Compared with the experimental results, the buckling load in the FE analysis results decreased by 4.89% (increased by 0.53% in the negative direction), and the ultimate load decreased by 3.96% (increased by 2.35% in the negative direction). The reasons for this error may include the following two points: (1) The FE simulation is more idealized than the experimental simulation is, without considering defects such as steel properties and weld stress, and the constraint conditions for complete rigidity at the bottom of the column cannot be guaranteed during the experiment, resulting in slightly greater initial stiffness in the FE simulation than in the experiment; (2) the material constitutive relationship in the finite element simulation is a tri-linear model, but during the experimental loading, as the steel enters the plastic development stage, its stress–strain relationship is not linear.

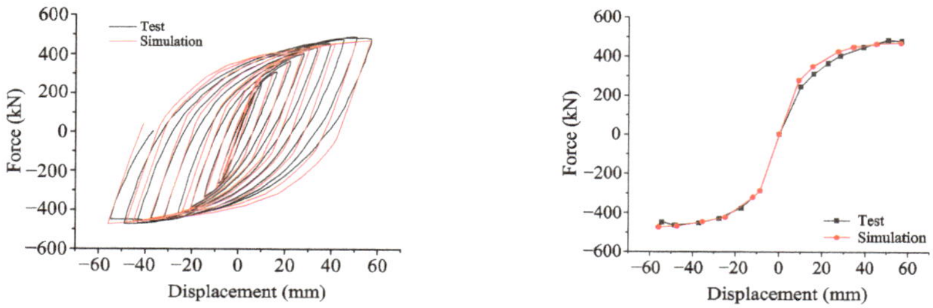

(a) Hysteresis curves (b) Backbone curves

Figure 8. Comparison of hysteresis curves and backbone curves of specimen Y-1.

Table 3. Comparison of the analytical and experimental results.

Parameter	Loading Direction	Test	FEM	Error
P_y/kN	Positive	401.27	381.66	4.89%
	Negative	−386.24	−388.26	0.53%
P_{max}/kN	Positive	488.45	469.09	3.96%
	Negative	−463.64	−474.54	2.35%

In summary, the force-displacement curves obtained from the FE simulation and experiments exhibit similar shapes, with a high degree of agreement. In terms of the mechanical performance indicators, the errors between the two are within a reasonable range. The FE modelling method used in this paper can accurately simulate the stress characteristics of the YEBFs and provide braces for subsequent research.

3. Influence of BSD on the Hysteresis Performance of YEBFs

3.1. BSD-YEBFs Model

Based on the geometric dimensions of the specimen BASE-Y in reference [31], this section established a traditional YEBFs model and a YEBFs model equipped with BSD-1 and BSD-2. The thickness of the stiffeners is 20 mm, and the cross-sectional dimensions of the frame are presented in Table 4. A schematic diagram of the BSD-YEBF is shown in Figures 9 and 10, with the detailed dimensions of BSD-1 and BSD-2 provided in Table 5.

Table 4. Section size of each component of YEBF (unit: mm).

Element	Section ($h \times b_f \times t_w \times t_f$)
Column	H400 × 400 × 16 × 20
Beam	H400 × 400 × 16 × 20
Brace	H200 × 200 × 12 × 16
Joist	H200 × 100 × 12 × 16

Table 5. Dimensions of BSD-1 and BSD-2 (unit: mm).

Specimen	H	h_0/b_1	b_1	b_2	t	h_u	h_d	r_1	r_2
BSD-1	330	240	45	30	60	30	30	35	15
BSD-2	200	150	30	15	60	30	15	20	10

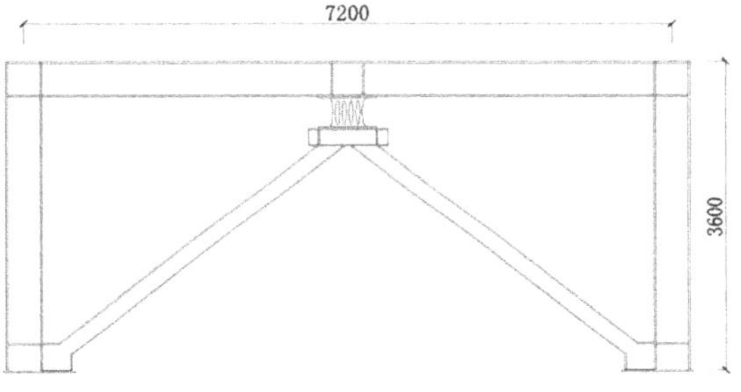

Figure 9. Schematic diagram of BSD-YEBF (unit: mm).

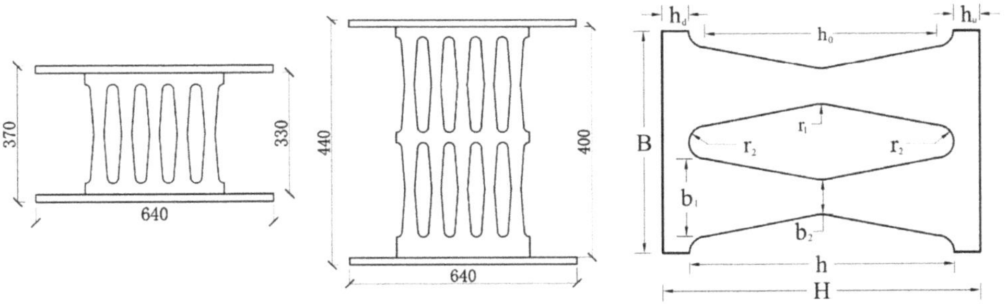

(a) BSD-1 (b) BSD-2 (c) Steel strip

Figure 10. Schematic diagram of BSD-1 and BSD-2 (unit: mm).

In this model, the frame beams, frame columns, and stiffeners of the joist are fabricated from Q355 steel, whereas all other components are made from Q235 steel. The steel has an elastic modulus of 206,000 MPa and a Poisson's ratio of 0.3; additional properties are presented in Table 2.

In addition, the columns and beams, BSD devices and their connected frame beams, as well as the joists, are all constrained via "Tie" constraints to bind them together; The bracing members and bracing feet, columns and their stiffeners, beams and their stiffeners, and joists and their stiffeners are merged into a whole via "Merge". The boundary conditions and loading method are consistent with those in the previous section; notably, each loading level cycle is set to three times.

Figure 11 shows the grid division of the BSD-YEBF. The energy dissipation steel strips of the BSD are divided into C3D6 elements. To analyse the stress–strain variation of the

BSD device's energy dissipation steel strips accurately, the grid in this region is refined with an element size of 10 mm. The grid division and element sizes for the remaining components are consistent with those described in the previous section.

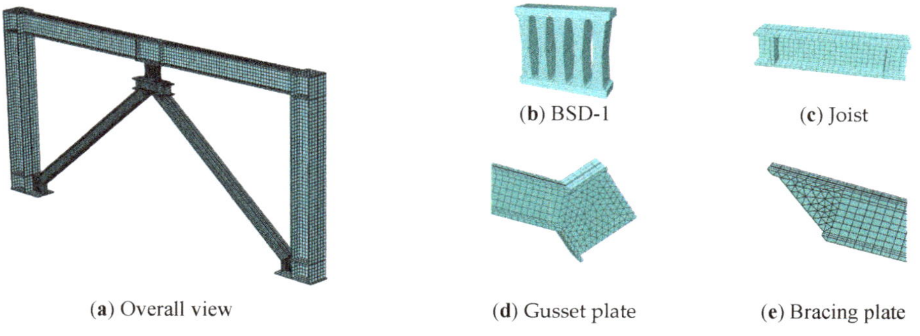

Figure 11. Grid division of BSD-YEBF structure.

3.2. Results Analysis and Comparison

Figure 12 shows the force-displacement curves of samples BASE-Y, BSD-1, and BSD-2. Their hysteresis curves are symmetrical and full, with a shape similar to a shuttle. Specimen BASE-Y experienced premature failure due to buckling of the web in the energy-dissipating beam section, resulting in a continuous decrease in the area enclosed by its hysteresis curve; in contrast, the area enclosed by the hysteresis curve of specimen BSD-1 increased with increasing loading displacement. When loaded to 6 Δ_y, the compressed bracing member experienced bending buckling, which triggered the structure to transition into its secondary defence line. At this point, the steel frame and the buckled bracing jointly bear the load, and the slope of the hysteresis curve outer contour slows but remains full. The hysteresis curve development trend of specimen BSD-2 is similar to that of specimen BSD-1. When loaded to 6 Δ_y, the model failed to converge because significant plastic deformation occurred at the middle horizontal position of the energy-dissipating steel strips. However, it can also be predicted that with the increasing loading displacement, specimen BSD-2 will exhibit bending buckling of the compressed bracing member and a reduction in bearing capacity.

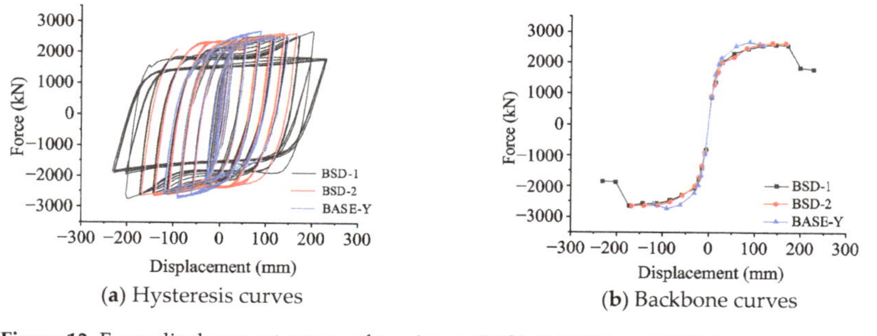

Figure 12. Force-displacement curves of specimens BASE-Y, BSD-1, and BSD-2.

A comparison of the backbone curve in Figure 12b clearly reveals that the BASE-Y sample has the highest slope in the initial stages of loading, indicating a greater bearing capacity at low displacements. After the displacement reaches 3 Δ_y, the slope of the backbone curve begins to decrease until the sample fails. This decline is attributed to the web plate of the energy-dissipating beam section entering the plastic development phase at low deformations due to shear action, leading to local buckling under the restraint of the flanges and stiffeners, which in turn results in a reduction in the bearing capacity. The initial trends of the backbone curves for BSD-1 and BSD-2 are roughly identical. However, upon reaching a loading displacement of 6 Δ_y, specimen BSD-1 experiences buckling of the compressed bracing member, leading to a significant reduction in the slope of the backbone curve.

The detailed mechanical performance indicators are presented in Table 6. The yield displacement (Δ_y), ultimate displacement (Δ_u), yield force (P_y), maximum force (P_{max}), and ductility capacity ($\mu = \Delta_u / \Delta_y$) are as follows: The yield forces of the three types of samples are essentially the same. However, the ductility coefficients of specimens BSD-1 and BSD-2 are increased by 59.57% and 39.47%, respectively, compared with those of sample BASE-Y, whereas their bearing capacities are reduced by 3.26% and 2.78%, respectively. This finding indicates that compared with the traditional YEBFs, the new type of YEBFs equipped with BSD dampers can significantly enhance the ductility of the structure at the expense of a minor reduction in the bearing capacity.

Table 6. Detailed parameters of BASE-Y, BSD-1, and BSD-2.

Specimen	Δ_y/mm	Δ_u/mm	P_y/kN	P_{max}/kN	μ
BASE-Y	28.32	118.45	2142.91	2724.74	4.18
BSD-1	29.77	187.80	2010.10	2635.80	6.67
BSD-2	28.65	167.11	1982.71	2648.83	5.83

4. Analysis of the Hysteresis Performance of the BSDF and TEBF

Previous research has shown that BSD-1 has superior ductility and sustained energy dissipation capabilities compared with BSD-2. Hence, in the modelling endeavours of this section, the specimen designated as BSD-1 was chosen as the representative damper for analysis. Two steel frames, each consisting of three story and one span, were constructed: one equipped with BSD-1 devices (BSDF) and the other a traditional eccentrically bare braced steel frame (TEBF). The dimensions of the beams and columns of the steel frames are consistent with those detailed in Section 3, with a story height of 3000 mm, and the steel material utilized is Q235. Before the experiments, monotonic loading was applied to both BSDF and TEBF to ascertain their yield displacements, which were determined to be 98 mm.

Both frames were subjected to fully fixed boundary conditions at the base of the columns, whereas lateral constraints in the X-direction were applied to both sides of the main beam. Coupling points were established in the Y-direction at the node domain on one side of the 3-story frame, where cyclic horizontal loads were applied. The loading method and grid division were consistent with those described in Section 3. Here, we chose only the grid division, and the boundary condition diagram of the BSDF is provided as an example, as shown in Figure 13.

4.1. *Comparison of Stress Failure MODES*

Figure 14 shows the stress cloud of the BSDF during cyclic loading. When the loading displacement reaches 1 Δ_y, the stress is predominantly concentrated in the steel strips of the BSD, whereas the stress levels in the frame beams, frame columns, and bracing are relatively low. When the load continues to 4 Δ_y, there is a significant increase in stress at the column bases, node domain, and flanges at the ends of the beams in the Y-axis direction, approaching the ultimate strength of the steel material. Notably, the stress increase is most pronounced near the column bases of the Y-axis frame beams on the 2nd story and 3rd story. This is attributed to the fact that during the loading process, the gusset plates are welded to the frame columns and beams, leading to a substantial increase in the flexural stiffness of the frame beams at these locations, causing the plastic hinge that should have appeared at the node position to move towards the midspan of the beam. The stress on the flange of the beam rapidly increases. When loaded to 5 Δ_y, the stress values at the energy-dissipating steel strips are reduced. In contrast, the stress at the outer flanges at the lower ends of the first-story columns significantly increases. After the loading displacement reaches 6 Δ_y, the stress at the column bases further increases and extends continuously along the bracing members. Pronounced plastic hinges appear near the node domains of the Y-axis direction frame beams at each story, and the stress begins to develop from the flanges towards the webs.

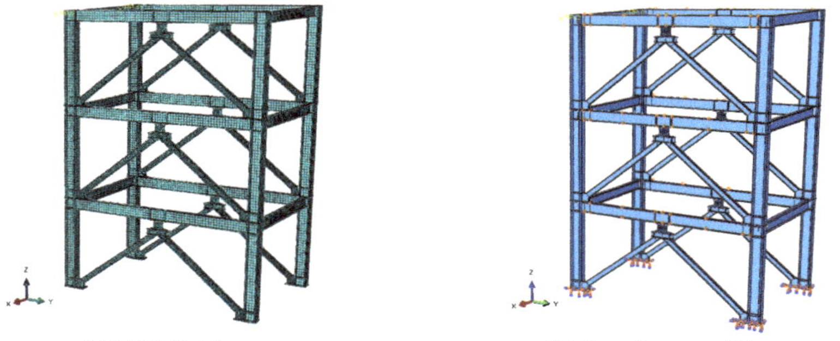

(**a**) Grid division (**b**) Boundary conditions

Figure 13. Grid division and boundary conditions of BSDF.

Figure 15 shows the stress cloud of the TEBF during cyclic loading. When the loading displacement reaches 1 Δ_y, the stress is primarily concentrated in the web region of the eccentrically braced energy-dissipating beam, and the stress near the gusset plates of the Y-axis frame beams also begins to gradually increase. When the load continues to 2 Δ_y, the stress in the web of the energy-dissipating beam approaches the ultimate strength of the steel, and the stress near the gusset plates of the Y-axis frame beams increases to the ultimate strength of the steel and develops towards the centre. When the displacement load reaches 3 Δ_y, the stress at the column bases and bracing feet increases significantly, and the web of the energy-dissipating beam sections on each story undergoes varying degrees of buckling, with the buckling degree being greatest on the 2nd story, followed by the 3rd story, and then the 1st story. When loaded to 4 Δ_y, the web of the energy-dissipating beam on each story experiences significant buckling deformation, and the compressed bracing members also experience out-of-plane instability. The stress at the bracing feet, both ends of the Y-axis frame beams, and the lower flanges at the midspan all reach the ultimate strength of the steel, causing the structure to lose its bearing capacity and fail.

In conclusion, a comparison of the stress clouds of the two steel frames mentioned above revealed that the stresses in the BSDF were primarily concentrated in the node domains, gusset plates, and near the gusset plates of the Y-axis frame beams, with a relatively uniform stress distribution throughout the loading process. In contrast, the TEBF steel framework experienced buckling of the energy-dissipating beam in the mid-to-late stages of loading, leading to structural instability and failure. Moreover, stresses at the energy-dissipating beam, bracing members and node domains significantly exceeded the ultimate strength of the steel material, thus failing to provide sustained energy dissipation capabilities.

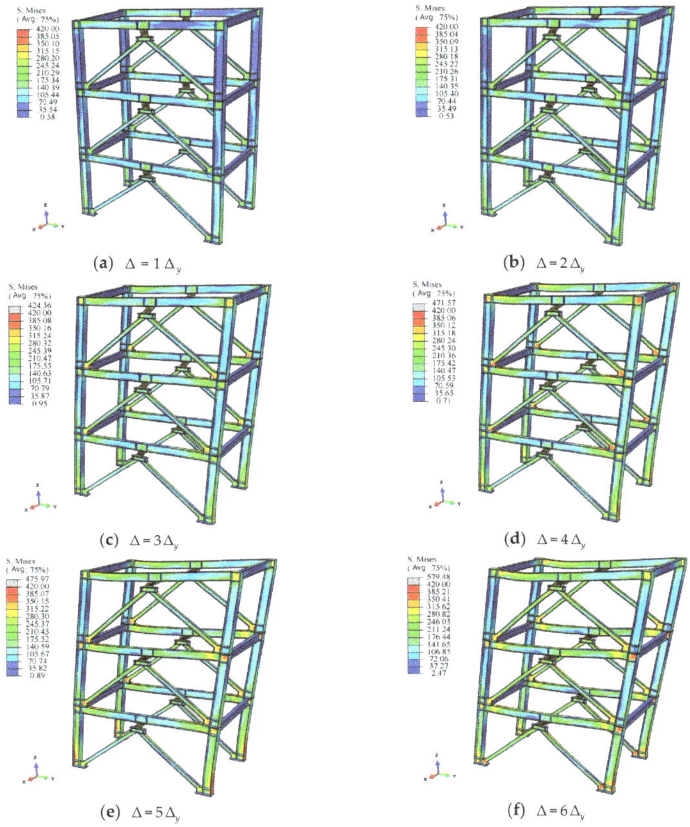

Figure 14. Stress cloud of BSDF during cyclic loading process.

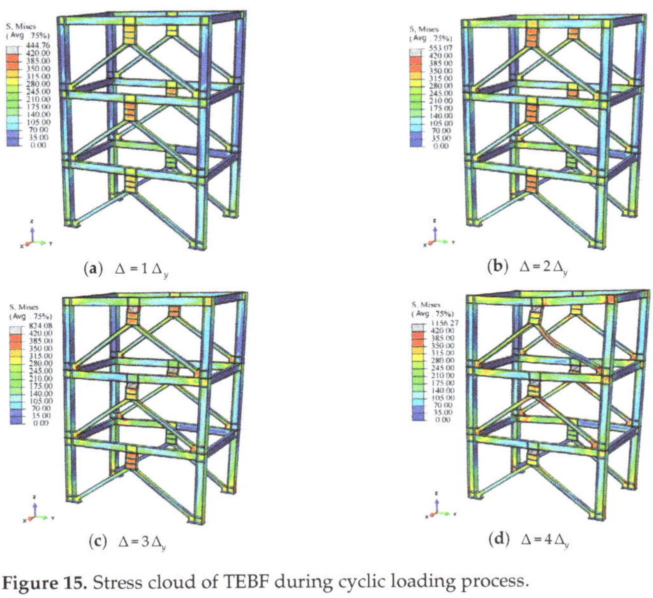

Figure 15. Stress cloud of TEBF during cyclic loading process.

4.2. Comparison of Force-Displacement Curves

Figure 16 shows the force-displacement curves for frames BSDF and TEBF. The hysteresis curve of TEBF approximates a shuttle shape, whereas the hysteresis curve of BSDF resembles a Z-shape. Figure 16a shows that during the initial loading stage, owing to the shear-type energy dissipation beam of the TEBF, it has a high bearing capacity under small deformations, resulting in a larger hysteresis loop envelope area than the frame BSDF. When the loading displacement reaches 150 mm, the hysteresis loop envelope area of the BSDF gradually increases, whereas the TEBF exits operation due to buckling of the energy dissipating beam, so its hysteresis curve envelope area slowly decreases. Figure 16b shows that before attaining the yield displacement, both the energy-dissipating beam and the BSD devices are in the elastic working stage, with their backbone curves essentially being consistent. When the loading displacement reaches 150 mm, the slope of the TEBF backbone curve becomes negative because the web plate of the energy-dissipating beam begins to enter the buckling stage, resulting in a rapid decrease in bearing capacity; conversely, the slope of the BSDF backbone curve consistently increases throughout the loading process, indicative of a stable bearing capability.

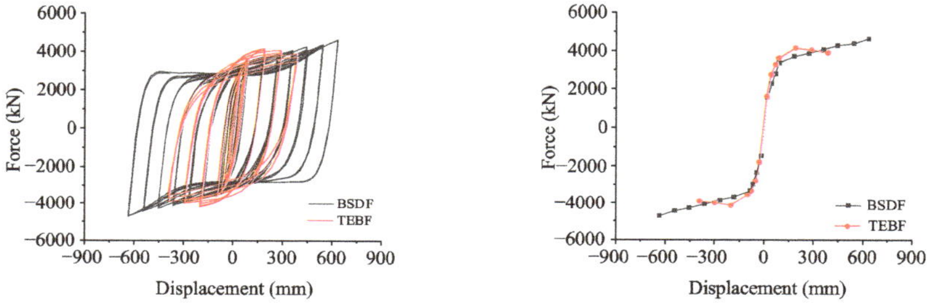

(**a**) Hysteresis curves (**b**) Backbone curves

Figure 16. Force-displacement curves for the BSDF and TEBF frames.

Table 7 shows the relevant indicators of bearing capacity and ductility. Under cyclic loading, the yield-bearing capacity and ultimate bearing capacity of BSDF increased by 3.13% and 11.12%, respectively, compared with those of TEBF, and the ductility coefficient increased by 56.7%. These findings indicate that the YEBFs with BSD devices have a relatively high bearing capacity and ductility.

Table 7. Parameters related to the bearing capacity and ductility.

Specimen	Δ_y/mm	Δ_u/mm	P_y/kN	P_{max}/kN	μ
BSDF	95.21	634.55	3414.07	4619.75	6.66
TEBF	92.00	391.36	3310.32	4157.13	4.25

4.3. Comparison of Stiffness Degradation and Strength Degradation

Figures 17 and 18 illustrate the stiffness degradation and strength degradation curves for BSDF and TEBF, respectively. Both frame types exhibit similar stiffness degradation trends; however, the BSDF frame has an initial stiffness of 7.68% greater than that of the TEBF frame, although the stiffness degradation rate for the BSDF frame is slightly greater than that of TEBF. When the loading displacement is less than 150 mm, the strength of the BSDF frame tends to increase. In contrast, strength degradation begins for the BSDF frame when the loading displacement exceeds 150 mm, whereas the TEBF frame starts to experience strength degradation at a loading displacement of 100 mm.

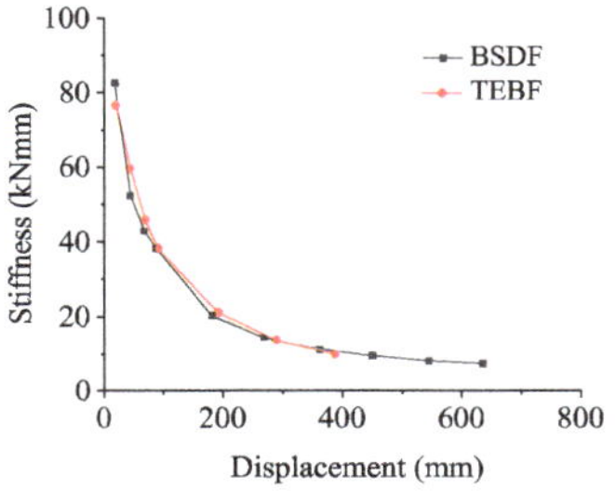

Figure 17. Stiffness degradation curves.

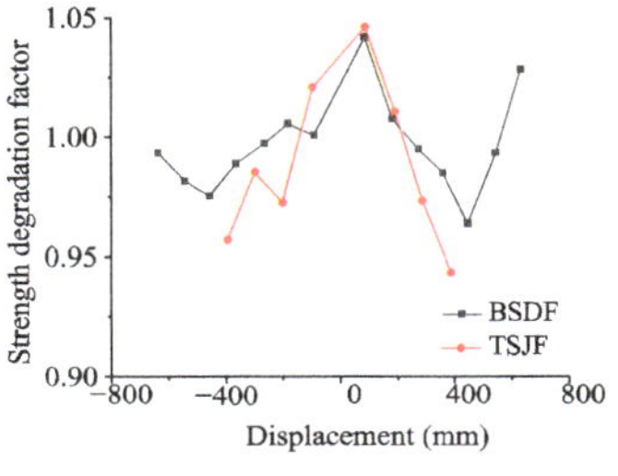

Figure 18. Strength degradation curves.

4.4. Comparison of Energy Dissipation Capacity

Figure 19 provides a comparison of the energy dissipation capacities of the BSDF and TEBF. The comparison is specifically conducted by evaluating the equivalent viscous damping coefficients and energy dissipation at various loading displacements. As illustrated in Figure 19a, under cyclic loading, the equivalent viscous damping coefficients for both frame types initially increase and then decrease, with the inflection point occurring at a displacement of approximately 300 mm. Figure 19b clearly shows that the TEBF results in higher energy dissipation values than the BSDF frame does when the loading displacement is less than 4 Δ_y. However, with further increases in loading displacement, buckling initiates in the web of the energy dissipation beam section of the TEBF, whereas the BSDF demonstrates a sustained capacity for energy dissipation.

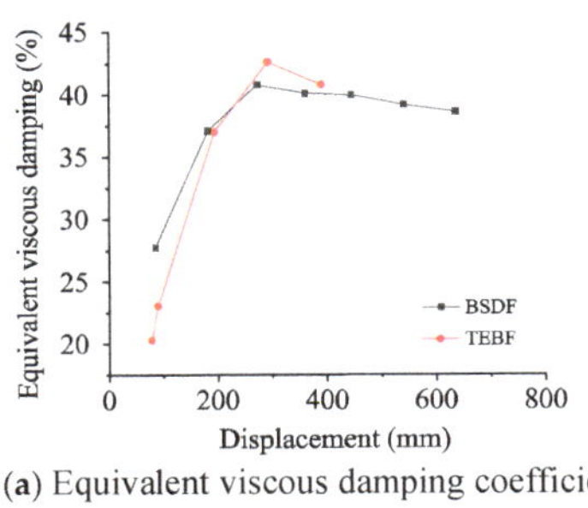

(**a**) Equivalent viscous damping coefficient

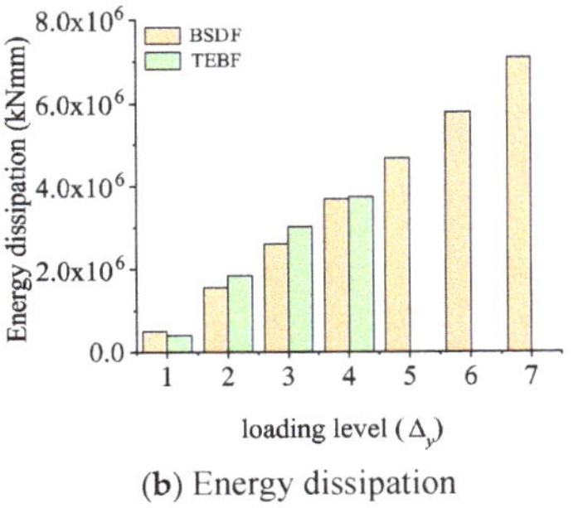

(**b**) Energy dissipation

Figure 19. Comparison of energy dissipation capacity.

As mentioned above, the BSDF has a higher bearing capacity and ductility, as well as a higher energy dissipation capacity than the TEBF.

5. Conclusions

This paper investigated the effect of BSDs on the hysteresis performance of multi-story YEBFs. The existing model was validated through FE analysis. A BSD-YEBFs model was subsequently established, and the seismic performance parameters of multi-story YEBFs under cyclic loading were analysed, with a focus on the impact of BSD devices. The main conclusions obtained are as follows:

1. FE verification was conducted on the samples in the existing single-span, two-story frame test, and the FE simulation and experimental results were roughly the same, with relevant mechanical performance parameter errors within a reasonable range. A correct modelling approach for the YEBFs was provided.
2. Compared with those of BASE-Y, the ductility coefficients of BSD-1 and BSD-2 were increased by 59.57% and 39.47%, respectively, whereas the corresponding bearing capacities were reduced by 3.26% (P_y) and 2.78% (P_{max}), respectively. This finding indicates that the equipment of BSD devices in YEBFs can significantly enhance structural ductility with only a minor sacrifice in bearing capacity.
3. During the cyclic loading process, although the energy dissipation of the TEBF is slightly greater than that of the BSDF in the early and middle stages (displacements less than 4 Δ_y), as loading continues (approximately 4 Δ_y), the web of the energy dissipating beam in the TEBF buckles successively, leading to structural instability and failure. In contrast, the stress distribution in the BSDF remains more uniform throughout the loading process. Furthermore, compared with TEBF, BSDF increased its yield bearing capacity by 3.13%, ultimate bearing capacity by 11.12%, and ductility coefficient by 56.7%. The initial stiffness of the BSDF was also 7.68% higher than that of the TEBF. In summary, YEBFs with BSD devices exhibit superior bearing capacity, ductility, and more sustained energy dissipation capabilities than traditional YEBFs do.

Author Contributions: Conceptualization, C.Z.; methodology, H.Z. (Hong Zheng); software, X.X.; validation, J.Z. and L.J.; formal analysis, C.Z.; investigation, C.Z.; resources, L.J.; data curation, H.W.; writing—original draft preparation, X.X. and H.W.; writing—review and editing, C.Z. and H.W.; visualization, Y.T.; supervision, M.L.; project administration, X.W. and H.Z. (Hai Zhang); funding acquisition, C.Z. All authors have read and agreed to the published version of the manuscript.

Funding: This research received no external funding.

Data Availability Statement: The raw data supporting the conclusions of this article will be made available by the authors on request.

Acknowledgments: Special thanks to H.Z. (Hong Zheng) for theoretical support and X.X. for writing the original draft of this article.

Conflicts of Interest: Authors Jing Zhao and Xiyao Wang were employed by the 3rd Construction Co., Ltd., China Construction 5th Engineering Bureau. Authors Yuxiao Tian, Muhua Liang and Hai Zhang were employed by the China Railway Urban Construction Group Co., Ltd. The remaining authors declare that the research was conducted in the absence of any commercial or financial relationships that could be construed as a potential conflict of interest.

References

1. Chen, X.; Li, S.; Liang, G.; He, M. Numerical Parameter Analysis of High-Strength Steel Frame with Y-Eccentric Brace Using Variable Replaceable Link. *Buildings* **2024**, *14*, 2149. [CrossRef]
2. Wasse, A.D.; Dai, K.; Wang, J.; Sharbati, R. State-of-the-art review: Seismic design and performance assessment of special concentrically braced frames developed for complex industrial building structures. *Int. J. Steel Struct.* **2024**, *24*, 280–295. [CrossRef]
3. Hjelmstad, K.D.; Popov, E.P. Characteristics of eccentrically braced frames. *J. Struct. Eng.* **1984**, *110*, 340–353. [CrossRef]

4. Zhuang, L.; Wang, J.; Nie, X.; Wu, Z. Experimental study on seismic behaviour of eccentrically braced composite frame with vertical LYP steel shear link. *Eng. Struct.* **2022**, *255*, 113957. [CrossRef]
5. Abdelhamid, F.; Yahiaoui, D.; Saadi, M.; Lahbari, N. Lateral reliability assessment of eccentrically braced frames including horizontal and vertical links under seismic loading. *Eng. Technol. Appl. Sci.* **2022**, *12*, 8278–8283. [CrossRef]
6. Xiao, X.; Zhang, P.; Zhang, Q. Seismic Energy Dissipation Analysis of Y And K Type Composite Eccentrically Braced Steel Frames. *IOP Conf. Ser. Earth Environ. Sci.* **2018**, *153*, 032015. [CrossRef]
7. Li, T.; Su, M.; Guo, J.; Zhao, K.; Zhang, Q. Seismic performance evaluation of Y-shaped eccentric brace–high-strength steel frame structure based on remote collaborative hybrid test. *Soil. Dyn. Earthq. Eng.* **2023**, *164*, 107621. [CrossRef]
8. Xu, J.; Wang, S.L.; Mohamed, H.S.; Han, J.; Han, Z.; Tu, J.; Li, Z. Seismic performance of shear energy dissipation beams in D-shaped eccentrically braced steel frames. *J. Constr. Steel Res.* **2021**, *180*, 106584. [CrossRef]
9. Liu, B.; Lu, Y.; Li, W.; Li, J.; Zhao, J.; Wang, S.; Ni, G.; Meng, Q. Study on Seismic Behavior of Different Forms of Eccentrically Braced Steel Frames. *Buildings* **2024**, *14*, 2064. [CrossRef]
10. Zheng, W.; Shen, L. Seismic performance of high strength steel frames with different eccentric braces based on PBSD method. *World Earthq. Eng.* **2020**, *36*, 80–92. (In Chinese)
11. Wang, F.; Su, M.; Hong, M.; Guo, Y.; Li, S. Cyclic behaviour of Y-shaped eccentrically braced frames fabricated with high-strength steel composite. *J. Constr. Steel Res.* **2016**, *120*, 176–187. [CrossRef]
12. Ashrafi, A.; Imanpour, A. Seismic response of steel multi-tiered eccentrically braced frames. *J. Constr. Steel Res.* **2021**, *181*, 106600. [CrossRef]
13. Pourzangbar, A.; Vaezi, M.; Mousavi, S.M.; Saber, A. Effects of brace-viscous damper system on the dynamic response of steel frames. *Int. J. Eng.* **2020**, *33*, 720–731.
14. Piluso, V.; Montuori, R.; Nastri, E.; Paciello, A. Seismic response of MRF-CBF dual systems equipped with low damage friction connections. *J. Constr. Steel Res.* **2019**, *154*, 263–277. [CrossRef]
15. Khazaei, M. Investigation on dynamics nonlinear analysis of steel frames with steel dampers. *Procedia Eng.* **2013**, *54*, 401–412. [CrossRef]
16. Zhao, J.; Tao, M.; Zhuang, L. Development of bent shear panel dampers for eccentrically braced composite frames. *J. Constr. Steel Res.* **2022**, *193*, 107292. [CrossRef]
17. Li, Y.; Wang, W.; Su, S.; Quan, C.; Xu, J.; Jia, Y.; Mi, J. Seismic performance assessment of eccentrically braced steel frame using demountable-metallic-corrugated-shear-panel dampers. *J. Constr. Steel Res.* **2023**, *207*, 107972. [CrossRef]
18. Mirzai, N.M.; Attarnejad, R.; Hu, J.W. Enhancing the seismic performance of EBFs with vertical shear link using a new self-centering damper. *Int. J. Earthq. Eng.* **2018**, *35*, 57–75.
19. Soltani, S.; Javaheri-Tafti, M.R.; Vetr, M.Q. Experimental and analytic study of seismic resistant eccentrically braced frames with dissipative devices. *J. Build. Pathol. Rehabil.* **2025**, *10*, 28. [CrossRef]
20. Ashtari, M.; Lari, S.; Çağlar, H.; Pouraminian, M.; Salmani, R.; Genjaly, A. Development and performance evaluation of an auxetic yield metal damper for enhanced seismic energy dissipation. *Structures* **2025**, *71*, 107892. [CrossRef]
21. Amiri, H.A.; Najafabadi, E.P.; Estekanchi, H.E. Experimental and analytical study of Block Slit Damper. *J. Constr. Steel Res.* **2018**, *141*, 167–178. [CrossRef]
22. Javanmardi, A.; Ibrahim, Z.; Ghaedi, K.; Benisi Ghadim, H.; Hanif, M.U. State-of-the-art review of metallic dampers: Testing, development and implementation. *Arch. Comput. Methods Eng.* **2020**, *27*, 455–478. [CrossRef]
23. Amiri, H.A.; Najafabadi, E.P.; Estekanchi, H.E.; Ozbakkaloglu, T. Performance-based seismic design and assessment of low-rise steel special moment resisting frames with block slit dampers using endurance time method. *Eng. Struct.* **2020**, *224*, 110955. [CrossRef]
24. Mohseni, P.K.; Zahedi-khameneh, A.; Naeemifar, O. Study of the effect of geometric parameters of steel block slit dampers on energy absorption. *Int. J. Steel Struct.* **2020**, *20*, 1069–1079. [CrossRef]
25. Shen, M. Behavior Study of Y-Eccentrically Braced Steel Frames Under Cyclic Load Considering Length Changed of Link Beam. Master's Thesis, Suzhou University of Science and Technology, Suzhou, China, 2008. (In Chinese)
26. *GB 50011-2010*; Code for Seismic Design of Buildings. China Architecture and Building Press: Beijing, China, 2016.
27. *JGJ 99-2015*; Technical Specification for Steel Structure of Tall Building. China Building Industry Press: Beijing, China, 2015.
28. *GB 50017-2017*; Steel Structure Design Standard. China Planning Press: Beijing, China, 2017.
29. Su, M.; Gu, Q.; Shen, L. Finite element analysis of steel members under cyclic loading. *J. Eng. Mech.* **2001**, *18*, 51–59. (In Chinese)

30. *JGJ 101-96*; Specification of Testing Methods for Earthquake Resistant Building. China Academy of Building Research: Beijing, China, 1997.
31. Meng, L. The Analysis and Design Criteria of Y Style Eccentrically Braced Steel Frames under Cyclic Load. Master's Thesis, Xi'an University of Architecture and Technology, Xi'an, China, 2005. (In Chinese)

buildings

MDPI

Article

Experimental Investigation of the Flexural Bearing Capacity of Pull-Up Cantilever Scaffolds with Adjustable Steel Tie Rods

Zhanbo Huang [1,2], Jianjun Yang [1,2], Shizhong Zhou [1,2,*], Chenyang Zhang [1,2] and Jintao Yang [1,2]

[1] School of Civil Engineering, Central South University, Changsha 410075, China; 19118880298@163.com (Z.H.); jjyang@csu.edu.cn (J.Y.)
[2] National Engineering Laboratory for High-Speed Railway Construction, Changsha 410075, China
* Correspondence: 15034738691@163.com

Abstract: This study presents experimental investigations of the flexural bearing capacity of the lower load-bearing structures of pull-up cantilever scaffolds with adjustable steel tie rods. First, load-bearing structures with three different specifications are tested to analyze their primary failure modes and failure mechanisms. The experimental results are then simulated and validated using a finite element analysis (FEA) method. Subsequently, an extended parametric analysis is performed numerically, and a simplified calculation formula is derived through nonlinear curve fitting to evaluate the load-bearing capacity. Finally, the effects of the transverse spacing of the vertical rods, the length of the cantilever beams, and the spacing of the hanging ears on the flexural bearing capacity of the structures are analyzed.

Keywords: pull-up cantilever scaffold; adjustable steel tie rod; flexural bearing capacity

1. Introduction

Scaffolding plays a critical role in building construction [1]. Traditional scaffolding systems, such as floor-standing scaffolding and door-type scaffolding [2,3], provide a certain degree of stability and load-bearing capacity. However, they face issues such as occupying considerable space, long construction periods, high labor costs, and severe material losses. Additionally, due to height limitations, traditional floor-standing scaffolding also involves high risks and complex erection and dismantling processes. In contrast, cantilever scaffolding offers better adaptability and ease of operation, making it more suitable for the complex construction requirements of high-rise buildings [4,5]. Nevertheless, due to technical, organizational, and human factors [6], various accidents frequently occur during the erection and dismantling processes of scaffolding [7]. At present, the normative documents and application standards related to scaffolding remain incomplete [8,9], and standards specifically addressing cantilever scaffolding are particularly scarce. Therefore, further systematic research is urgently needed to lay the foundation for the development of comprehensive technical specifications.

At present, researchers have made advancements in many aspects of scaffolding, such as theoretical modeling, mechanical property determination, engineering monitoring, and construction management methods. In the study of theoretical models of scaffolding, some scholars have provided suggestions for the design of scaffolding through an algorithm for modeling scaffold behavior [10], a method of numerical analysis [11], a approach for topology optimization [12], and an optimization methodology based on genetic algorithms [13]. Zhang et al. [14] and Beale et al. [15] analyzed the reliability of scaffolding systems via a second-order inelastic structural analysis model of scaffolding

https://doi.org/10.3390/buildings15060861

and a nonlinear model of scaffolding. In a study of the mechanical properties of scaffolding, Peng et al. [16] investigated the effect of eccentric loading on the scaffolding system. Zhang et al. [17] investigated the effects of failure modes and different stochastic variables on the variation in the strength of scaffolding structures. The mechanical properties of special types of scaffolds, such as the mortise–tenon steel-tube scaffold [18], the frame-type steel scaffold [19], the pull-up cantilever scaffold [20], the socket-type keyway steel-pipe scaffold [21], and the large-span steel cantilever beam [22], have been studied. In terms of engineering monitoring and construction management methods for scaffolding, Yan et al. [23] explored the construction process and requirements of ultrahigh-tower overhanging scaffolds. Peng et al. [24] investigated and discussed the factors influencing the bearing capacity of single-row steel scaffolds under isolation beams and the strength of steel scaffolds with defects. Zhao et al. [2] systematically investigated the effects of member imperfection and joint stiffness on the ultimate load of door-type modular steel scaffolds. Additionally, scholars have built inspection systems to help inspectors perform scaffold quality assessment through computer vision [25], deep learning [26], and point clouds [27]. In addition, scholars have made limited progress in the study of cantilever scaffolds [28–31], but there is still a large gap in the study of their bearing capacity and other practical applications.

In this paper, a new cantilever scaffold—specifically, a pull-up cantilever scaffold with adjustable steel tie rods—is proposed, building upon traditional cantilever scaffold designs, as shown in Figure 1. The new cantilever scaffold consists of a lower load-bearing structure and an upper scaffold. Its main structural characteristics are as follows:

(1) Lower load-bearing structure: Comprising a cantilevered I-steel beam, oblique tie rods, and adjustable positioning piles, this structure supports and stabilizes the scaffold, transferring the load to the building's load-bearing system.
(2) Cantilever I-steel beam: Connected to the building's load-bearing structure with high-strength bolts and embedded parts, it forms the primary load-bearing element of the scaffold.
(3) Oblique tie rods: These steel rods, positioned between the building's load-bearing structure and the cantilever beam, provide structural stability. The tie rods are adjustable in length to accommodate varying construction requirements.
(4) Adjustable positioning piles and embedded parts: These components anchor the scaffold to the building's load-bearing structure and provide additional stability.
(5) Double-ear hanging rings and hanging ears: Used to connect the oblique tie rods to the building's load-bearing structure and the cantilever beam.

This new pull-up cantilever scaffold with adjustable steel tie rods surpasses traditional cantilever scaffold systems in construction efficiency, cost control, adaptability, and safety, showcasing considerable potential for application. Furthermore, it aligns with the broader principles of sustainable development by promoting more efficient resource use, reducing waste, and improving safety. These advancements not only enhance the operational efficiency of construction projects but also support the goals of sustainability in the construction industry, as detailed in the following advantages [32,33]:

(1) The pull-up cantilever scaffold with adjustable steel tie rods is directly connected to the building's external facade using bolts and double-ear hanging rings, eliminating the need for wall penetration, as shown in Figure 2. This eliminates interference with interior construction, allowing both interior and exterior work to proceed simultaneously. The adjustable rods facilitate flexible modifications according to specific construction needs, allowing workers to rapidly assemble, adjust, and dismantle the scaffold, thereby significantly enhancing overall efficiency.

(2) The steel frame length of the new cantilever scaffold is reduced by more than 50% compared to traditional designs. This not only improves construction efficiency but also shortens project timelines, resulting in substantial cost savings.
(3) The new cantilever scaffold is highly adaptable, enabling rapid adjustments to different construction environments and stages. This flexibility facilitates efficient transitions between stages, enhancing turnover rates and reducing the overall construction cycle.
(4) The robust steel tie rod structure and secure suspension design enhance the stability of the new cantilever scaffold, enabling it to support greater loads while reducing accident risks, thereby improving on-site safety.

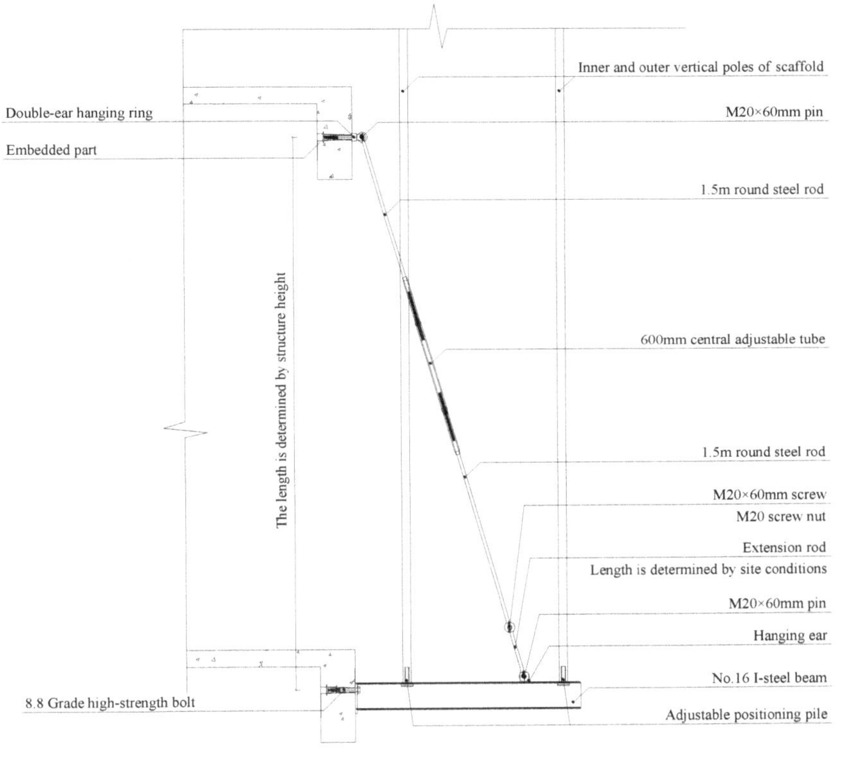

Figure 1. A construction diagram of the pull-up cantilever scaffold with adjustable steel tie rods.

Building upon the current state of research, this study investigates the flexural bearing capacity of the lower load-bearing structure of this new cantilever scaffold through experiments and FEA. First, three groups of load-bearing structures with different specifications are subjected to flexural bearing capacity experiments to analyze their failure modes and bearing mechanisms systematically, with the experimental results simulated and verified using ABAQUS FEA software (version 2023). An extended FEA is subsequently performed to analyze the flexural bearing capacity of various sizes of load-bearing structures commonly used in actual projects, and a simplified formula is derived to calculate their flexural bearing capacity. The effects of the transverse spacing of vertical rods, the lengths of cantilever beams, and the spacing of hanging ears on the flexural bearing capacity of load-bearing structures are also analyzed in this paper. This study lays a theoretical foundation for the application and standardization of pull-up cantilever scaffolds with adjustable steel tie rods and contributes to advancing safety research on temporary structures [9].

(a) (b)

Figure 2. On-site construction diagrams of the cantilevered scaffold: (**a**) a traditional cantilevered scaffold; (**b**) the novel cantilever scaffold, the construction phase in the figure does not yet include the diagonal rod.

2. Flexural Bearing Capacity Experiment

2.1. Experiment Preparation

Three distinct sizes of load-bearing structures were used in the experiments, primarily consisting of components such as cantilever I-beams, oblique tie rods, double-ear hanging rings, high-strength bolts, pins, and gaskets. With the exception of the high-strength bolts, which were made from No. 45 steel [34], all the other components were fabricated from Q235 steel [35]. The cantilever I-beam was fabricated from No. 16 hot-rolled I-steel [36], and its fixed end featured a flat end plate with a thickness of 12 mm. The cantilever lengths of the I-beams for the three test specimens were set at 1.3 m, 2.1 m, and 3.1 m, designated as the L13, L21, and L31 test groups, respectively, with two specimens in each group.

In engineering applications, ten distinct cantilever I-beam lengths, ranging from 1.3 m to 3.1 m, are available for use. Structural differences were observed for cantilever I-beams of varying lengths. Specifically, the test specimens with cantilever lengths of 1.3 m, 1.5 m, and 1.7 m were single tie rod specimens, without extension rods at the ends of the tie rods. The test specimens with cantilever lengths of 1.9 m and 2.1 m were double tie rod specimens, where the inner tie rod lacked an extension rod, but an extension rod was added at the end of the outer tie rod. The test specimens with cantilever lengths of 2.3 m, 2.5 m, 2.7 m, 2.9 m, and 3.1 m were also double tie rod specimens, both of which featured extension rods for the inner and outer tie rods. Due to the constraints of the experimental scale, this study focused on specimens with cantilever lengths of 1.3 m, 2.1 m, and 3.1 m, which were designated as the L13, L21, and L31 test groups, respectively. The selection of these three cantilever lengths was closely aligned with construction practices. The 1.3 m cantilever length is commonly used for single tie rod cantilever scaffolds, typically applied in construction for locations such as balconies, canopies, and bay windows. The 2.1 m and 3.1 m cantilever lengths are commonly used for double tie rod cantilever scaffolds in building projects. Meanwhile, the relative positioning of the inner and outer vertical poles, along with the tie rods, varies for 2.1 m and 3.1 m cantilever scaffolds. Consequently, these two cantilever lengths must be considered separately. Additionally, the 3.1 m cantilever length is the longest used in construction applications. Thus, the three cantilever lengths selected for this study are closely aligned with real-world engineering applications, providing a theoretical foundation for practical use.

Additionally, the relative positions of the hanging ears and vertical poles on the cantilever I-beams with lengths of 1.3 m, 2.1 m, and 3.1 m also differed spatially. The specific structural details are illustrated in Figure 3, where l_0 denotes the length of the cantilever beams, which corresponds to 1.3 m, 2.1 m, and 3.1 m in test groups L13, L21 and L31, respectively; l_1 denotes the transverse spacing of vertical poles, inside and outside, which is set to be 800 mm in all the three groups of specimens. Figure 1 shows that the inner and outer vertical poles are integral components of the upper scaffold structure. During the flexural bearing capacity test of the lower load-bearing structure, the inner and outer vertical poles were replaced with two loading areas at the landing position on the cantilever I-beam to simulate their loading effect, as shown in Figure 3.

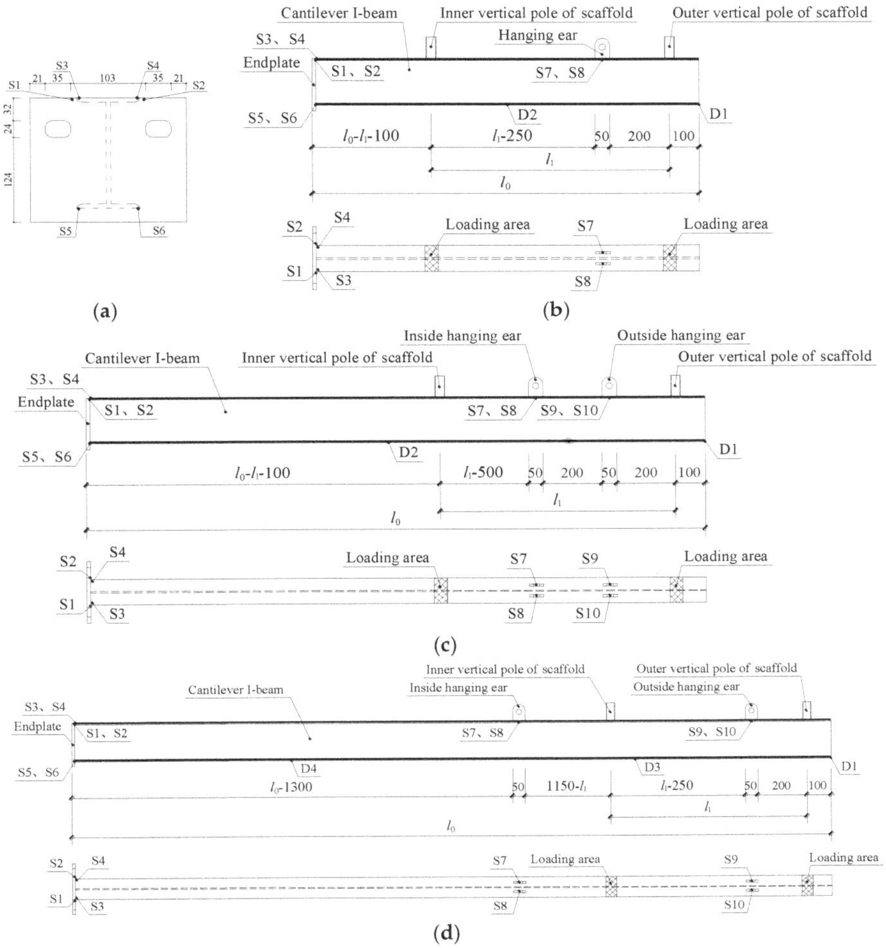

Figure 3. The structure, dimensions, and arrangement of measurement points for the cantilever I-beams of the three test groups. (**a**) The end plates of L13, L21, and L31; (**b**) the cantilever I-beam of L13; (**c**) the cantilever I-beam of L21; (**d**) and the cantilever I-beam of L31.

The oblique tie rod includes three components: a solid round steel rod with a hanging ring at one end and an externally threaded solid round steel rod at the other; a central adjustable steel tube with internal threads at both ends; and an extension rod with solid round steel rods and hanging rings at both ends. The structure of the oblique tie rod is illustrated in Figure 4.

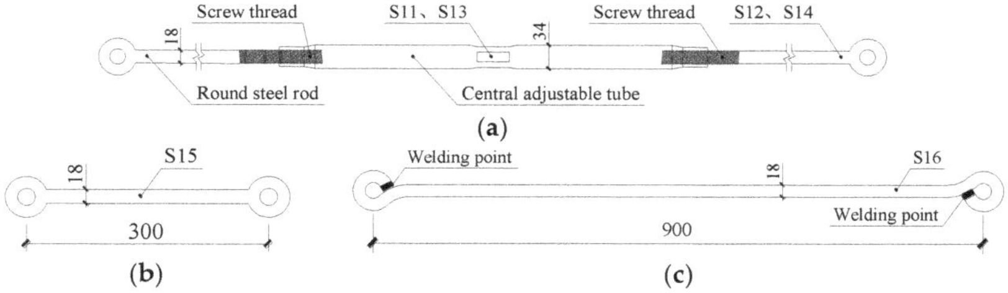

Figure 4. Structure, dimensions, and arrangement of measurement points for the oblique tie rods of the three test groups. (**a**) Adjustable steel tie rod; (**b**) 300 mm extension rod; (**c**) 900 mm extension rod.

The main components of the double-ear hanging ring were two pieces of hanging ear plates and a round steel rod with a diameter of 22 mm, and its construction diagram is shown in Figure 5. The additional accessories used in the experiment included high-strength bolts (22 mm thread diameter), pins (20 mm and 30 mm diameters), and gaskets.

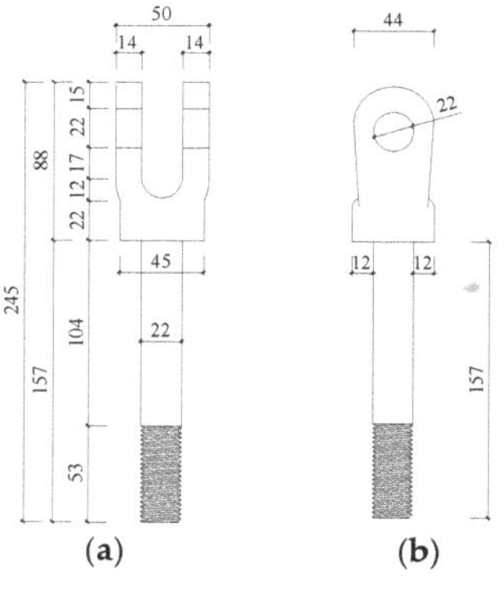

Figure 5. A dimensional diagram of the double-ear hanging ring. (**a**) Front view; (**b**) side view.

From the on-site construction diagram of the new cantilever scaffold shown in Figure 2b, it can be observed that the lower load-bearing structure of the scaffold is arranged continuously in accordance with the layout requirements of the construction site. Due to the constraints of the experimental scale, this study simultaneously tested two load-bearing structures, which closely simulated the actual conditions of the new cantilever scaffold in the construction field. Additionally, there were other considerations for selecting two load-bearing structures for concurrent loading in this experiment. Prior to the experiment, an attempt was made to load a single load-bearing structure. It was found that as the upper load increased, the deflection of the cantilever scaffold also increased, leading to some sliding of the experimental devices fixed at the upper part of the specimen. Moreover, because the tie rods and vertical poles could not be aligned in the same vertical plane and needed to be staggered, the upper fixed end of the tie rods was not directly above the cantilever I-beam, resulting in some out-of-plane deformation of the cantilever I-beam during the experiment. These factors not only led to larger experimental errors but also introduced significant safety risks. Therefore, prior to the formal experiment, improvements were made to the experimental setup. The experiment used three distribution beams to load two specimens of the same specification simultaneously, thereby minimizing the impact of these factors on the experiment and ensuring the proper loading of the specimens.

The experimental device layout for each test group is shown in Figure 6, with two specimens from each test group being loaded simultaneously. Prior to testing, the cantilever I-beam was aligned horizontally in both the width and length directions during installation. Next, the first- and second-level distribution beams were positioned. The two first-level distribution beams were placed on the inner and outer loading areas of the cantilever I-beams, oriented perpendicular to them. The second-level distribution beam was then placed atop the two first-level distribution beams, centered between the cantilever I-beams. Finally, the hinge support, jack, and force sensor were installed at the center of the second-level distribution beam.

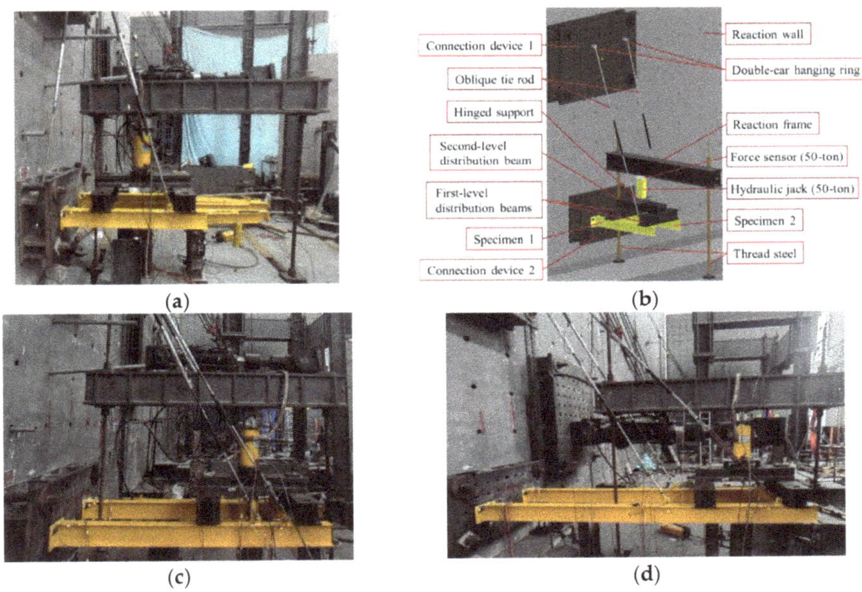

Figure 6. Experimental device layout. (**a**) L13 test group; (**b**) 3D model visualization of the L13 test group; (**c**) L21 test group; (**d**) L31 test group.

The specimen was secured to the laboratory reaction wall by connecting devices 1 and 2. The double-ear hanging rings were affixed to connecting device 1. The end plates of the cantilever I-beams in each test group were secured to connecting device 2 using high-strength bolts. The hanging rings at one end of the tie rods in each test set were connected to the double-ear hanging rings with pins, while the rings at the other end were attached to the hanging ears on the cantilever I-beams with pins.

As the inner and outer vertical poles were erected on the load-bearing frame, the attachment points at the upper ends of the tie rods were typically staggered relative to the vertical poles. As shown in Figure 7, installation position 1 corresponds to the double-ear hanging rings of the L13 test group and the double-ear hanging rings at the upper ends of the outer tie rods in the L21 and L31 test groups, while installation position 2 corresponds to the double-ear hanging rings at the upper ends of the inner tie rods in the L21 and L31 test groups. The distance from the double-ear hanging ring installation position to the end plate installation position was 3 m, approximately corresponding to the height of a single building floor.

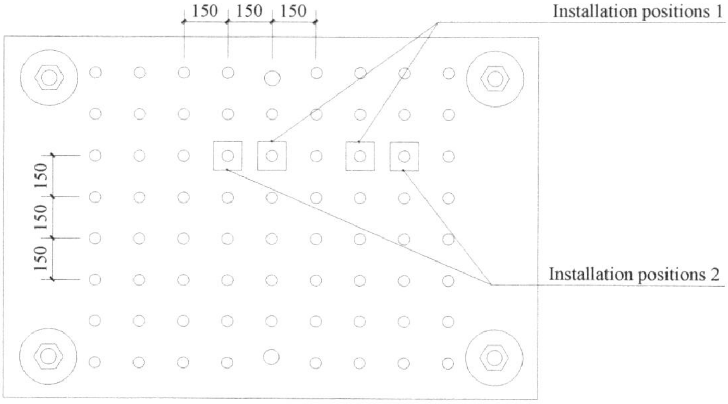

Figure 7. Installation positions of the double-ear hanging rings for each test group on connection device 1.

Each specimen in test group L13 was equipped with a single tie rod directly connected to the hanging ear on the I-beam. Each specimen in test group L21 was equipped with two tie rods: the inner tie rod was connected directly to the inside hanging ear, while the outer tie rod was connected to the outside hanging ear via a 300 mm long extension rod. Each specimen in test group L31 was equipped with two tie rods: the inner tie rods connected to the inside hanging ear via a 300 mm long extension rod, while the outer tie rods connected to the outside hanging ear via a 900 mm long extension rod. The tie rods and extension rods were joined using pins.

Before testing, the material properties of the flanges, end plates, and tie rods were individually evaluated using three specimens for each component type. The tensile test results for the cantilever I-beam flange and end plate revealed an average yield strength of 323.3 MPa and an average ultimate strength of 452.7 MPa. However, the stress–strain curves of the tie rods did not show a distinct yielding plateau, with an average ultimate strength of 548.1 MPa.

2.2. Experimental Loading System

FEAs were performed prior to the experiments to estimate the specimens' ultimate flexural bearing capacity. Table 1 lists the loading stages for the three specimen groups, with values representing the loads applied by the hydraulic jack.

Table 1. Load at graded loading of the specimens for each test group (unit: kN).

Loading Stage	L13	L21	L31
1 [1]	3.6	3.6	3.6
2	30	30	20
3	30	30	20
4	30	30	20
5	5	30	20
6 [2]	5	5	5
N [3]	263.6	288.6	178.6

[1] Weights of the distribution beams and the hydraulic jack. [2] The load applied at each stage following the sixth stage until the specimen fails. [3] The load stage at the failure of the specimens.

Prior to each set of tests, preloading was conducted to assess the loading conditions of the two specimens. The tie rods were adjusted to ensure both specimens experienced identical loading conditions. Thus, during the experiment, the loading conditions for both specimens were nearly identical, and it was uncommon for one specimen to fail while the other did not. If one specimen failed, no further load could be applied, and the test was deemed a failure. The specimens were then re-prepared for another round of testing. The experimental load was applied using a manual hydraulic pump. The specimens were preloaded to minimize the effect of mounting voids on the experimental results. The formal loading process followed a force-controlled staged procedure. When yielding at strain measurement points or bending of the double-ear hanging rings occurred during the experiment, the applied load was reduced to 5 kN at each subsequent stage until specimen failure. The load application rate at the jack was controlled to 5 ± 1 kN per minute during both the preloading and formal loading phases of the experiments.

2.3. Arrangement of Measurement Points

Figures 3 and 4 illustrate the arrangement of the displacement meters and strain gauges for each test group. Displacement meters, labeled D1 through D4, were installed below the lower flange of the cantilever I-beam in each test group. Specifically, displacement meter D1 was positioned at the cantilever end of the I-beam, D2 at the mid-span, D3 between the inner and outer hanging ears, and D4 between the inner hanging ear and the end plate.

Strain gauges, labeled S1 to S8, were affixed to the cantilever I-beams of the specimens in each test group. S1 and S2 were mounted on the end plate, near the left and right sides of the upper flange. S3 and S4 were positioned on the left and right sides of the upper surface of the upper flange, close to the end plate. S5 and S6 were affixed to the left and right sides of the lower surface of the lower flange, near the end plate. S7 to S10 were positioned on the left and right sides of the lower surface of the upper flange, close to the locations of the hanging ears.

Strain gauges S11 to S14 were affixed to the oblique tie rods in each test group. Specifically, S11 and S13 were mounted at the midpoint of the central adjustable rod, while S12 and S14 were placed on the round steel rod near the end of its hanging ring. Additionally, S15 was affixed at the midpoint of the 300 mm extension rod, and S16 was mounted on the 900 mm extension rod near the end of its hanging ring.

2.4. Experimental Observations

Each test group comprised two specimens, resulting in a total of four loading areas. The load applied by the hydraulic jack was referred to as the total load, while the load applied to the loading areas was referred to as the single-point load.

During testing for test group L13, at the fourth loading stage (with a total load of 93.6 kN and a single-point load of 23.4 kN), the double-ear hanging rings began to bend downward, which corresponds to the normal serviceability limit state of the specimens, as shown in Figure 8a. In the subsequent loading stages, the total load was reduced to 5 kN. When the total load reached 143.6 kN, the center of the end plate of the cantilever I-beam exhibited outward bending in the transverse direction, as shown in Figure 8b. As the loading continued, no additional significant phenomena occurred other than increased bending deformation of the double-ear hanging rings and the end plates. The loading process continued until the total load reached 263.6 kN (with a single-point load of 65.9 kN), at which point the double-ear hanging rings completely failed, marking the load-bearing capacity limit state of the specimen. The specimens could no longer sustain the load, marking the end of the experiment.

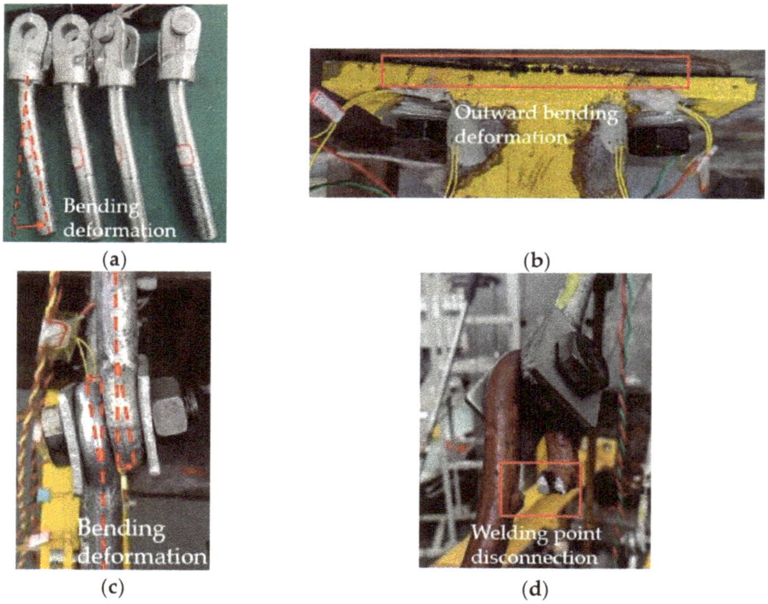

Figure 8. Failure diagrams for each test group. (**a**) The force bending of the double-ear hanging ring in test group L13; (**b**) the force bending of the end plate of test group L13; (**c**) the force bending of the hanging ring connecting the outer tie rod to the extension rod in test group L21; (**d**) disconnection at the welding point on the hanging ring of the extension rod of test group L31.

During testing for test group L21, at the fifth loading stage (with a total load of 123.6 kN and a single-point load of 30.9 kN), the double-ear hanging rings at the upper end of the outer tie rods began to bend downward, which corresponds to the normal serviceability limit state of the specimens, as shown in Figure 8a. When the total load reached 183.6 kN, the double-ear hanging rings at the upper end of the inner tie rods also began to bend downward. Simultaneously, the hanging rings connecting the outer tie rods to the extension rods deformed, as shown in Figure 8c. The loading process continued until the total load reached 288.6 kN (with a single-point load of 72.15 kN), at which point the double-ear hanging rings completely failed, marking the load-bearing capacity limit state of the specimen. The specimens could no longer sustain the load, marking the end of the experiment.

During testing for test group L31, at the fifth loading stage (with a total load of 83.6 kN and a single-point load of 20.9 kN), the double-ear hanging rings began to bend downward, which corresponds to the normal serviceability limit state of the specimens, as shown in Figure 8a. When the total load reached 123.6 kN, the hanging rings where the tie rods connected to the extension rods experienced bending deformation. At a total load of 178.6 kN (with a single-point load of 44.65 kN), a welding point at the hanging rings of the 900 mm long extension rods disconnected, as shown in Figure 8d. This marked the load-bearing capacity limit state of the specimen. The specimens could no longer sustain the load, marking the end of the experiment.

The failure modes of the specimens from the three test groups reveal that the double-ear hanging rings and extension rods are the weakest components in the lower load-bearing structures of pull-up cantilever scaffolds with adjustable steel tie rods. Specifically, the bending of the double-ear hanging rings, as shown in Figure 8a, results in increased vertical displacement of the cantilever I-beams, thereby elevating the stress at both the fixed end of the I-beam and the double-ear hanging ring attached to the wall. This creates stress concentrations, which, in turn, reduce the ultimate load-bearing capacity of the lower load-bearing structures under the serviceability limit state. During testing, the welded joint of the 900 mm extension rod failed, as shown in Figure 8d, resulting in a brittle failure that poses a considerable risk to the safe use of this novel cantilever scaffold. Furthermore, during testing, the 300 mm extension rod exhibited bending at the end ring, as shown in Figure 8c, which caused stress concentration at the connection between the tie rod and the extension rod, thereby diminishing the load-bearing capacity of the tie rods.

2.5. Stress Analysis

Based on data from the material property tests, the average yield strength of the specimens was 323.3 MPa, with similar stress trends observed across all test groups. Within the same test group, the highest stresses were recorded at the end plate measurement points, followed by those on the oblique tie rods, with the lowest stresses observed near the hanging ear locations. Since the specimens in these three test groups had different structural arrangements, primarily differing in the positions of the hanging ears (i.e., the connection points of the tie rods to the cantilever I-beams) and the vertical poles, this led to differences in the distribution of load on the end plates and tie rods during the load-bearing process. For test groups L21 and L31, the stresses at the outer tie rods consistently exceeded those at the inner tie rods. The highest stresses among measurement points S1 to S6 on the end plates occurred at S1 and S2. Among measurement points S11 to S16 for each test group, the highest stresses were observed at points S13 and S14 on the 18 mm round steel rods of the outer tie rods. The maximum stress data from the end plate and tie rod measurement points of the two specimens in each test group were used to construct the load–stress diagrams presented in Figure 9.

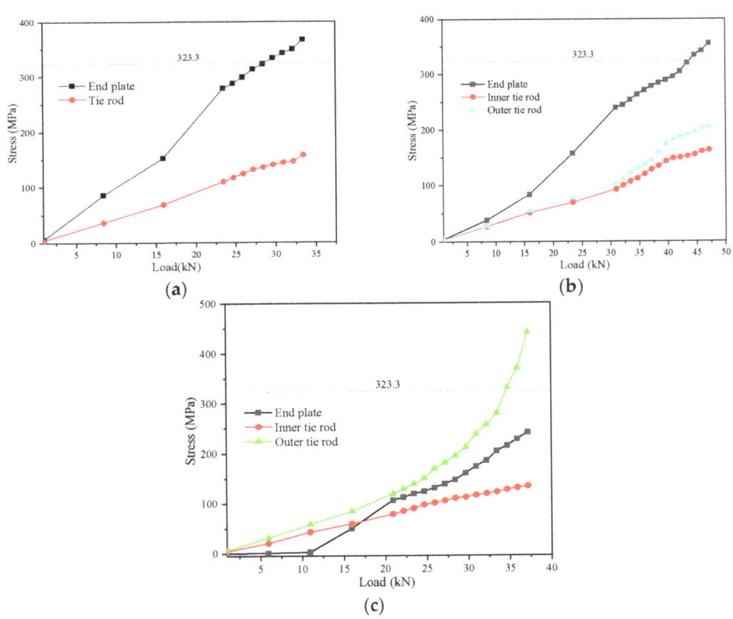

Figure 9. Load–stress diagrams representing the maximum stress measurement points on the end plates and the tie rods for each test group. (**a**) Test group L13; (**b**) test group L21; (**c**) test group L31.

At the fixed ends of each test group, the cantilever I-beams were welded to the end plates, which provided effective restraints. As shown in Figure 9, in test groups L13 and L21 the end plates of the cantilever I-beams yielded first, with yield loads of 29.65 kN and 44.65 kN, respectively. In test group L31, the specimens yielded first at the outer tie rods with a yield load of 35.9 kN. The load–stress curves at the maximum stress measurement points of the end plates for all three test groups are shown in Figure 10a. The maximum stress at the end plates decreased as the length of the cantilever I-beams increased under equal loading conditions.

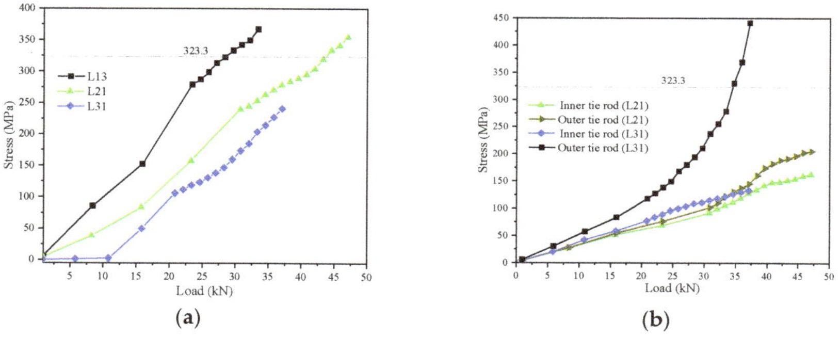

Figure 10. Comparison of load–stress curves. (**a**) Load–stress curves for the maximum stress measurement points at the end plates of the cantilever I-beams for each test group; (**b**) load–stress curves for the maximum stress measurement points at the tie rods for test groups L21 and L31.

A comparison of the load–stress curves at the maximum stress measurement points for the inner and outer tie rods in test groups L21 and L31 is shown in Figure 10b. The maximum stresses of the tie rods gradually increased with increasing length of the cantilever I-beams under the same applied load. Moreover, the stresses of the outer tie rods always exceeded those of the inner tie rods. The spacing between the lower suspension points of the inner and outer tie rods, which was also the spacing between the inside and outside hanging ears, was greater in test group L31 than in test group L21, resulting in a more noticeable difference in the stresses of the inner and outer tie rods between the two test groups.

Analysis of the stress data at the cantilever I-beam ends and tie rods revealed that with longer cantilever I-beams, the load borne by the cantilever I-beam ends decreased, while the load needed to yield the ends increased. Concurrently, the load shared by the tie rods increased, and the stress at the tie rods increased when the specimens yielded. Throughout the loading process, the stresses at the outer tie rods consistently exceeded those at the inner tie rods. Additionally, as the distance between the inside and outside hanging ears increased, the stress difference between the inner and outer tie rods also increased.

2.6. Displacement Analysis

The maximum displacement was recorded at the D1 displacement meter positioned at the cantilever end of the cantilever I-beam among the four displacement meters (D1–D4). As the applied load increased during the test, the bending deformation of the double-ear hanging ring caused a significant increase in the vertical displacement of the cantilever I-beam. Therefore, the vertical displacement analysis for each test group utilized data from the D1 displacement meter, focusing on measurements recorded prior to the bending deformation of the double-ear hanging rings.

The curves showing the relationship between the maximum displacement and applied load for the two specimens in each test group are presented in Figure 11a. As shown in

Figure 11a, the vertical displacement at the cantilever end progressively increased with increasing cantilever I-beam length across the test groups, with maximum displacements of 4.31 mm, 9.03 mm, and 14.65 mm for test groups L13, L21, and L31, respectively.

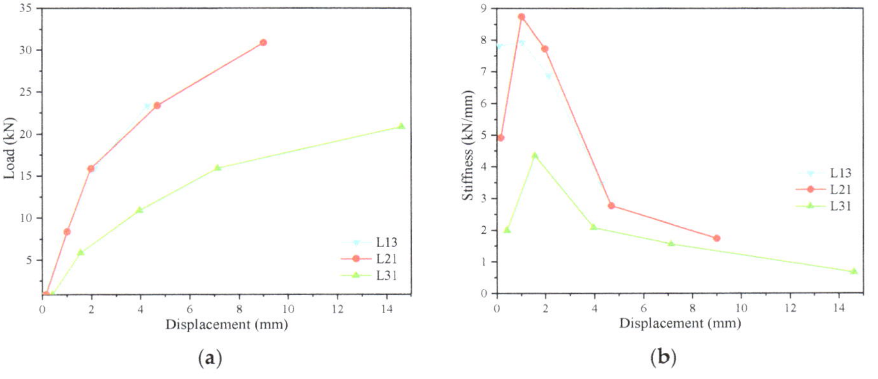

Figure 11. The relationship diagram of the displacement, load, and stiffness of the cantilever end of the cantilevered I-beam in each test group. (**a**) Displacement–load curves; (**b**) displacement–stiffness curves.

The stiffness degradation curve was measured using secant stiffness. Figure 11b illustrates the relationship between the maximum displacement and secant stiffness. The specimen in the L13 test group initially maintained a relatively stable stiffness of approximately 7.8 kN/mm, which then decreased to 5.42 kN/mm before the double-ear hanging ring experienced bending deformation. In the L21 and L31 test groups, the stiffness of the specimens initially increased, reaching 8.077 kN/mm and 3.69 kN/mm, respectively. Subsequently, the stiffness gradually decreased, and before the double-ear hanging ring underwent bending deformation, the stiffness reduced to 3.42 kN/mm and 1.43 kN/mm, respectively.

3. Finite Element Analysis

3.1. Finite Element Modeling

The load-bearing structure of the pull-up cantilever scaffold with adjustable steel tie rods is analyzed using ABAQUS FEA software. In the finite element model, the analysis step was selected as "Static, General," and the geometric nonlinearity option (NLGEOM = ON) was enabled to account for large deformation effects in the structure. Figure 12 shows the finite element model (FEM) for the L13 test group.

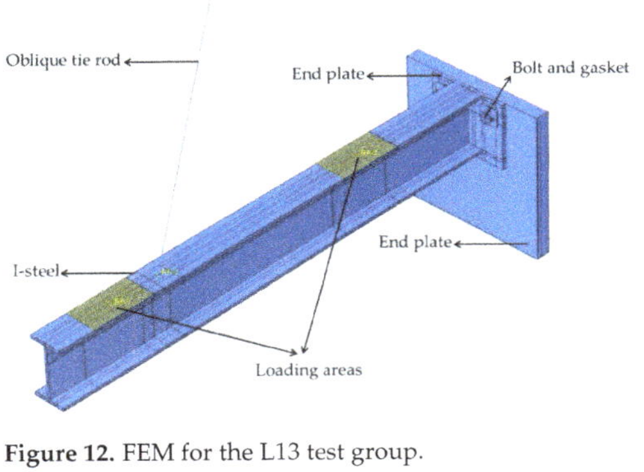

Figure 12. FEM for the L13 test group.

3.1.1. Material Properties

Based on the results of the material property test, the specimen's yield strength is 323.3 MPa, its Poisson's ratio is 0.3, the elastic modulus is 206 GPa, and the mass density of the steel is 7850 kN/m^3. The material model selected was an isotropic hardening elasto-plastic model. The stress–strain relationship of the steel follows a bilinear model, as shown in Figure 13.

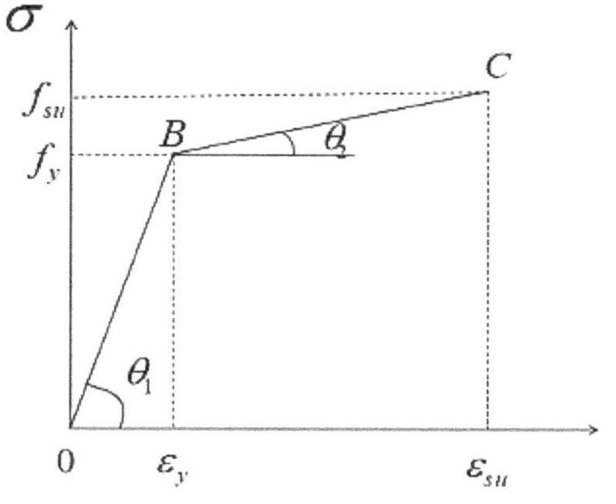

Figure 13. Stress–strain curve of Q235 steel.

3.1.2. Element Selection

The main components of the FEM include I-beams, end plates, tie rods, double-ear hanging rings, bolts, and gaskets. The I-beams, end plates, bolts, and gaskets were modeled using the solid element C3D8R. The tie rods, subjected only to axial forces, were modeled using the truss element T3D2. The double-ear hanging rings, subjected to a combination of downward shear and outward tensile forces, were modeled using the beam element B31. The reaction wall, which is primarily used to restrain the end plate without participating in stress or displacement analysis, was modeled as a rigid body.

3.1.3. Interaction, Contact, and Boundary Condition

The components of the FEMs were assembled to replicate the experimental setup. Tie constraints were applied between the I-beams and the end plates, as well as between the bolts and the reaction wall. The interactions between the end plates, bolts, gaskets, and reaction wall were defined by face-to-face contact constraints, with "hard contact" in the normal direction and a "penalty" function friction model in the tangential direction, with a friction coefficient of 0.4. The constraints between the hanging rings on the tie rods and the hanging ears of the I-beams, as well as between the hanging rings on the tie rods and the double-ear hanging rings on the reaction wall, were modeled as hinged connections (MPC-pin). The reaction wall and the double-ear hanging rings on it are fully fixed as shown in Figure 14.

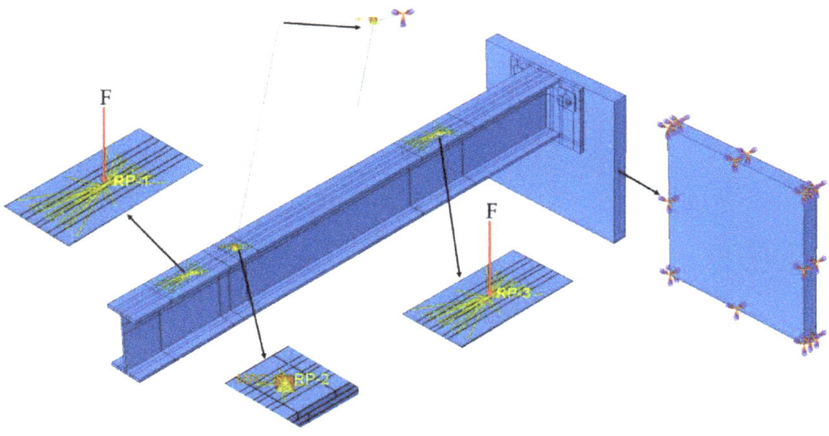

Figure 14. Load application and boundary condition diagram.

3.1.4. Mesh Division

The FEM utilized a structured adaptive meshing technique. Based on the proportions of the overall model size, the global mesh size was set to 25 mm, while a local mesh size of 2 mm was applied to critical areas.

3.1.5. Load Application

The load application method in the FEM followed the same methodology as in the experiments, employing a force-controlled staged loading approach. The magnitudes of the loads at each stage were identical to those applied in the experiments, as specified in Table 1. Two loading areas were coupled at their midpoints, referred to as coupling points. Downward concentrated forces were subsequently applied to these coupling points, as shown in Figure 14. The simulated loading was terminated when the FEMs for each test group reached their ultimate flexural bearing capacity under normal use conditions, corresponding to the bending deformation of the double-ear hanging rings observed in the experiments.

3.2. Finite Element Analysis Results

A comparative analysis between the FEM data and experimental results is presented in this section. To ensure consistency, the FEM maximum stress and displacement values were evaluated at locations corresponding to the maximum strain and displacement measurement points in the experimental model. The experimental results indicated that the maximum stress on the cantilever I-beams for each test group occurred on the end plates near the left and right sides of the upper flange, the maximum stress on the inner and outer tie rods occurred on the 18 mm round tie rods, and the maximum vertical displacement for each test group was observed at the cantilever end of the cantilever I-beams.

The load–stress and load–displacement comparison curves for the experiment and FEA were derived from data collected before the bending deformation of the double-ear hanging rings in each test group, as shown in Figure 15. When the double-ear hanging rings began to bend in each test group, the single-point loads on the loading areas were 23.40 kN, 30.90 kN, and 20.90 kN for test groups L13, L21, and L31, respectively.

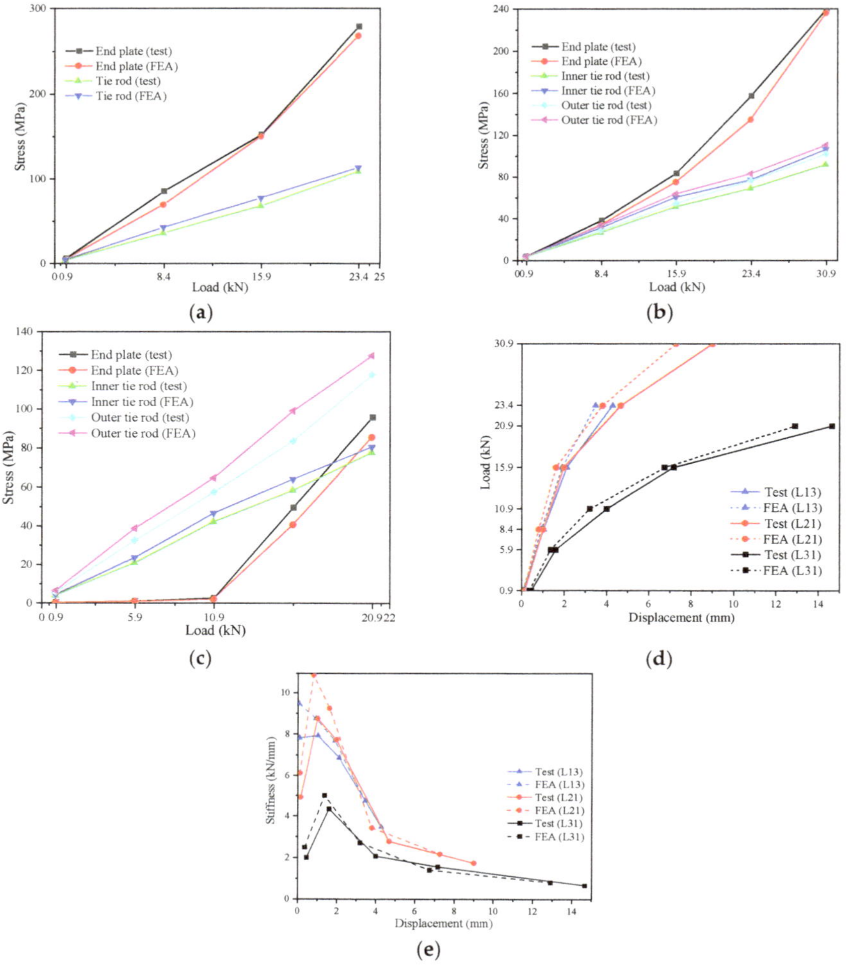

Figure 15. Comparison of test results and FEA results for each test group. (**a**) L13 stress comparison; (**b**) L21 stress comparison; (**c**) L31 stress comparison; (**d**) L13, L21, and L31 displacement comparison; (**e**) L13, L21, and L31 secant stiffness comparison.

At this stage, the maximum stress values on the end plates and tie rods, the maximum displacement values at the cantilever ends, and the secant stiffness were compared, as summarized in Table 2. The stress values are expressed in MPa, and the displacement values are expressed in mm; the secant stiffness values are expressed in kN/mm.

Table 2. Comparison of the stress and displacement results between the experimental and FEA results for each test group.

Categories	Test Groups	Test Values	FEA Values	Errors (%)
End plate stresses	L13	279.15	268.03	3.98
	L21	239.70	236.16	1.48
	L31	95.84	85.53	10.76
	Mean			5.41
	Standard deviation			3.92

Table 2. *Cont.*

Categories	Test Groups	Test Values	FEA Values	Errors (%)
Tie rod stresses	L13	108.96	113.76	4.40
	L21	102.02	110.591	8.41
	L31	118.02	127.60	8.12
	Mean			6.98
	Standard deviation			1.83
Cantilever end displacements	L13	4.31	3.51	18.63
	L21	9.03	7.30	19.17
	L31	14.65	12.9	11.95
	Mean			16.58
	Standard deviation			3.28
Secant stiffness	L13	3.47	4.75	36.89
	L21	1.74	2.17	24.71
	L31	0.67	0.81	20.90
	Mean			27.50
	Standard deviation			6.78

3.3. Error Analysis

Figure 15a–c and Table 2 indicate that the stresses at the end plates of the test models slightly exceeded those predicted by the FEMs, while the stresses at the tie rods of the test models were slightly lower. The maximum error in the stress comparison was 10.76%. Similarly, Figure 15d and Table 2 show that the vertical displacements of the test models were marginally larger than those of the FEMs, with a maximum displacement error of 19.17%. Figure 15e and Table 2 indicate that the stiffness variation in the test models in each group closely corresponded to that of the FEMs throughout the entire test process (before the bending deformation of the double-ear hanging rings). However, during the early loading stage, a noticeable discrepancy was observed between the two.

The primary reason for the differences between the FEA results and the experimental results is the idealization of components and interactions in the FEMs, which neglects certain influencing factors present in the actual test models. For example, bending deformations of the double-ear hanging rings were observed during the tests, and relative displacements might have occurred between the central adjustable steel tie rods and the round steel rods. Additionally, localized stresses on the hanging ears and hanging rings, as well as gaps in component contact, might have been present during the tests.

These factors contributed to an increase in the displacement at the cantilever ends of the cantilever I-beams. The increase in vertical displacement at the cantilever ends resulted in higher stresses at the end plates of the cantilever I-beams. The increased stress at the end plates indicated that the load carried by the fixed ends of the cantilever I-beams increased, thereby reducing the load carried by the tie rods and lowering the stress at the tie rods. This explains why the experimental results revealed higher stresses at the end plates, lower stresses at the tie rods, and greater vertical displacements at the cantilever ends than did the FEA results.

Despite some differences between the experimental data and the FEA results, the data trends remained consistent, with relatively small errors observed. Therefore, the load application, boundary conditions, and constraints on the interactions between different components in the FEM were considered reasonable. This demonstrates that the FEA closely aligns with the experimental process, ensuring high reliability and accuracy.

4. Flexural Bearing Capacity Analysis

The specimens with three different cantilever lengths in the experimental section cannot be directly compared due to the varying arrangements of tie rods and hanging ears. Therefore, in this section, finite element simulations were conducted using Abaqus software (version 2023) for cantilever I-beams with ten distinct lengths, ranging from 1.3 m to 3.1 m, commonly used in practical engineering. The simulations aimed to determine the maximum flexural bearing capacity of the entire structure at the point of yield. Additionally, a comparative analysis of relevant parameters, such as cantilever length, vertical pole spacing, and hanging ear spacing, was performed to assess their impact on the flexural bearing capacity of the load-bearing structure. This also provides valuable insights for the research and enhancement of similar cantilever scaffolds.

Five types of inner and outer vertical pole spacings are commonly used: 700 mm, 750 mm, 800 mm, 850 mm, and 900 mm. For clarity, the FEM with a cantilever beam length of 1300 mm and an inner and outer vertical pole spacing of 750 mm is labeled F2-1300-750, with other models following the same naming convention.

4.1. Load-Bearing Structures with Cantilever Lengths of 1.3, 1.5, and 1.7 m

The load-bearing structures with cantilever lengths of 1.3 m, 1.5 m, and 1.7 m were three distinct types of structures, each equipped with a single tie rod. A simplified diagram illustrating these models is presented in Figure 3a as an example. In the figure, l_0 denotes the cantilever length of the cantilevered I-beams for different models: 1.3 m, 1.5 m, and 1.7 m, respectively. For each model, the distance from the hanging ear to the cantilever end of the I-beam was 325 mm, and the distance from the outer vertical pole to the cantilever end was 100 mm. The inner and outer vertical pole spacings, l_1, were assigned five values ranging from 700 mm to 900 mm, as previously described.

In the FEMs discussed in this section, the end plates at the fixed ends of the cantilevered I-beams yielded first. At the yielding point, the maximum flexural bearing capacity of the load-bearing structure (F_0) was equal to the maximum flexural bearing capacity of the I-beam (F_1). The results, summarized in Table 3, indicate that the load-bearing structure F2-1300-900 exhibited the highest flexural bearing capacity, sustaining a single-point load of 28.06 kN.

Table 3. Maximum flexural bearing capacity of load-bearing structures with cantilever beam lengths of 1.3 m, 1.5 m, and 1.7 m.

Names	$F_0 = F_1$ (kN)	Names	$F_0 = F_1$ (kN)	Names	$F_0 = F_1$ (kN)
F2-1300-700	24.51	F2-1500-700	25.77	F2-1700-700	27.20
F2-1300-750	24.96	F2-1500-750	25.92	F2-1700-750	27.28
F2-1300-800	26.00	F2-1500-800	26.15	F2-1700-800	27.32
F2-1300-850	26.86	F2-1500-850	26.52	F2-1700-850	27.45
F2-1300-900	28.06	F2-1500-900	27.03	F2-1700-900	27.68

4.2. Load-Bearing Structures with Cantilever Lengths of 1.9 and 2.1 m

The load-bearing structures with cantilever lengths of 1.9 m and 2.1 m were two distinct types of structures, each equipped with double tie rods. A simplified diagram illustrating these models is presented in Figure 3b as an example. In the figure, l_0 denotes the cantilever length of the cantilevered I-beams for different models: 1.9 m and 2.1 m, respectively. For each model, the distance from the outer hanging ear to the cantilever end of the I-beam was 325 mm, and the distance between the inner and outer hanging ears was 250 mm. The inner and outer vertical pole spacings, l_1, were assigned five values ranging from 700 mm to 900 mm, as previously described.

In the FEMs discussed in this section, the outer tie rods yielded first. At the yielding point, the maximum flexural bearing capacity of the load-bearing structure (F_0) was equal to the maximum flexural bearing capacity of the outer tie rod (F_2). The results, summarized in Table 4, indicate that the load-bearing structure F2-1900-900 exhibited the highest flexural bearing capacity, sustaining a single-point load of 55.73 kN.

Table 4. Maximum flexural bearing capacity of load-bearing structures with cantilever beam lengths of 1.9 m and 2.1 m.

Names	$F_0 = F_2$ (kN)	Names	$F_0 = F_2$ (kN)
F2-1900-700	51.12	F2-2100-700	49.47
F2-1900-750	52.35	F2-2100-750	50.56
F2-1900-800	53.51	F2-2100-800	51.64
F2-1900-850	54.60	F2-2100-850	52.70
F2-1900-900	55.73	F2-2100-900	53.76

4.3. Load-Bearing Structures with Cantilever Lengths of 2.3, 2.5, 2.7, 2.9, and 3.1 m

The load-bearing structures with cantilever lengths of 2.3 m, 2.5 m, 2.7 m, 2.9 m, and 3.1 m were five distinct types of structures, each equipped with double tie rods. A simplified diagram illustrating these models is presented in Figure 3c as an example. In the figure, l_0 denotes the cantilever length of the cantilevered I-beams for different models: 2.3 m, 2.5 m, 2.7 m, 2.9 m, and 3.1 m, respectively. For each model, the distance from the outer hanging ear to the cantilever end of the I-beam was 325 mm, and the distance between the inner and outer hanging ears was 950 mm. The inner and outer vertical pole spacings, l_1, were assigned five values ranging from 700 mm to 900 mm, as previously described.

In the FEMs discussed in this section, the outer tie rods were the first to yield. At the yielding point, the maximum flexural bearing capacity of the load-bearing structure (F_0) was equal to the maximum flexural bearing capacity of the outer tie rod (F_2). The results, summarized in Table 5, indicate that the load-bearing structure F2-2300-900 exhibited the highest flexural bearing capacity, sustaining a single-point load of 37.28 kN.

Table 5. Maximum flexural bearing capacity of load-bearing structures with cantilever beam lengths of 2.3 m, 2.5 m, 2.7 m, 2.9 m, and 3.1 m.

Names	$F_0 = F_2$ (kN)	Names	$F_0 = F_2$ (kN)	Names	$F_0 = F_2$ (kN)
F2-2300-700	33.95	F2-2500-700	33.54	F2-2700-700	33.14
F2-2300-750	34.76	F2-2500-750	34.35	F2-2700-750	33.92
F2-2300-800	35.59	F2-2500-800	35.17	F2-2700-800	34.70
F2-2300-850	36.51	F2-2500-850	35.99	F2-2700-850	35.50
F2-2300-900	37.28	F2-2500-900	37.26	F2-2700-900	36.33
F2-2900-700	33.07	F2-3100-700	32.74	/	/
F2-2900-750	33.43	F2-3100-750	33.11	/	/
F2-2900-800	34.51	F2-3100-800	34.29	/	/
F2-2900-850	35.29	F2-3100-850	35.08	/	/
F2-2900-900	36.07	F2-3100-900	35.88	/	/

4.4. Flexural Bearing Capacity Calculation Formula

In this subsection, data from extended FEAs were nonlinearly curve-fitted using Origin (version 2021) data analysis software to derive a simplified formula for calculating the flexural bearing capacity of load-bearing structures. The fitting was conducted using the Levenberg–Marquardt [37] iterative algorithm, which is a widely used nonlinear least squares method. By combining the experimental and simulation results, three key variables were identified: the length of the cantilever beams, the spacing of the hanging ears, and the transverse spacing of the vertical poles, as these significantly influenced the flexural bearing capacity.

The variables were defined as follows: x_1 (mm) was the length of the cantilever beams; x_2 (mm) was the spacing of the hanging ears, where $x_2 = 0$ for structures with a single tie rod; x_3 (mm) was the transverse spacing of the vertical poles; F_G (N) was the flexural bearing capacity of the load-bearing structures derived from the fitting; and F_0 (N) was the flexural bearing capacity of the load-bearing structures obtained from the FEA. Additionally, in each model, the distance from the outer vertical pole to the cantilever end was 100 mm, and the distance from the outside hanging ear to the cantilever end was 325 mm. The fitting process started by analyzing the relationships between any two variables and the flexural bearing capacity before incorporating all three variables together.

$$F_G = 17.9023x_1 + 144.0528x_2 + 14.4073x_3 - 0.0052x_1^2 - 0.1740x_2^2 + 0.0091x_1x_2 + 0.0032x_2x_3 \quad (1)$$

Finally, the simplified formula for the flexural bearing capacity of load-bearing structures derived from the fitting is presented in Equation (1). As shown in Table 6, the ratio of the fitted values to the FEA values for the flexural bearing capacity of the structures ranged from 0.94 to 1.04, with an average of 1.00 and a standard deviation of 0.02. The small error between the fitted and FEA values demonstrates that the derived formula predicts the flexural bearing capacity of load-bearing structures with high accuracy.

Table 6. Comparison between the FEA results and fitting results for the flexural bearing capacity of load-bearing structures.

x_1-x_2-x_3	F_0	F_G	F_G/F_0	x_1-x_2-x_3	F_0	F_G	F_G/F_0
1300-0-700	24,510	24,591.17	1.00	2300-900-700	33,950	33,470.52	0.99
1300-0-750	24,960	25,311.53	1.01	2300-900-750	34,760	34,333.49	0.99
1300-0-800	26,000	26,031.90	1.00	2300-900-800	35,590	35,196.46	0.99
1300-0-850	26,860	26,752.26	1.00	2300-900-850	36,510	36,059.44	0.99
1300-0-900	28,060	27,472.63	0.98	2300-900-900	37,280	36,922.41	0.99
1500-0-700	25,770	25,266.60	0.98	2500-900-700	33,540	33,716.49	1.01
1500-0-750	25,920	25,986.97	1.00	2500-900-750	34,350	34,579.46	1.01
1500-0-800	26,150	26,707.33	1.02	2500-900-800	35,170	35,442.44	1.01
1500-0-850	26,520	27,427.70	1.03	2500-900-850	35,990	36,305.41	1.01
1500-0-900	27,030	28,148.06	1.04	2500-900-900	37,260	37,168.38	1.00
1700-0-700	27,200	25,527.03	0.94	2700-900-700	33,140	33,547.46	1.01
1700-0-750	27,280	26,247.40	0.96	2700-900-750	33,920	34,410.43	1.01
1700-0-800	27,320	26,967.76	0.99	2700-900-800	34,700	35,273.40	1.02
1700-0-850	27,450	27,688.13	1.01	2700-900-850	35,500	36,136.38	1.02
1700-0-900	27,680	28,408.49	1.03	2700-900-900	36,330	36,999.35	1.02
1900-200-700	51,120	51,141.84	1.00	2900-900-700	33,070	32,963.42	1.00
1900-200-750	52,350	51,893.89	0.99	2900-900-750	33,430	33,826.40	1.01
1900-200-800	53,510	52,645.95	0.98	2900-900-800	34,510	34,689.37	1.01
1900-200-850	54,600	53,398.01	0.98	2900-900-850	35,290	35,552.34	1.01
1900-200-900	55,730	54,150.06	0.97	2900-900-900	36,070	36,415.31	1.01
2100-200-700	49,470	50,937.94	1.03	3100-900-700	32,740	31,964.38	0.98
2100-200-750	50,560	51,689.99	1.02	3100-900-750	33,110	32,827.35	0.99
2100-200-800	51,640	52,442.05	1.02	3100-900-800	34,290	33,690.33	0.98
2100-200-850	52,700	53,194.11	1.01	3100-900-850	35,080	34,553.30	0.98
2100-200-900	53,760	53,946.16	1.00	3100-900-900	35,880	35,416.27	0.99

5. Parametric Finite Element Analysis

Based on experimental studies and finite element analyses, this section presents a parametric FEA of key parameters influencing the mechanical performance of pull-up cantilever scaffolds with adjustable steel tie rods. The analysis employs the one-variable method and focuses on load-bearing structures commonly used in engineering applications. Specifically, the effects of the transverse spacing of vertical poles, the length of cantilever beams, and the spacing of hanging ears on the mechanical properties of load-bearing structures were investigated. During the parametric FEA, the steel yield strength was uniformly set to 215 MPa. Through simulation loading, load–stress and load–displacement curves for each model were obtained and analyzed, providing a reference for the design, calculation, and further research of cantilever scaffolds.

5.1. Influence of the Transverse Spacing of Vertical Rods on the Mechanical Properties of Load-Bearing Structures

Load-bearing structures with cantilever lengths of 1.3 m, 1.5 m, and 1.7 m from Section 4.1 were selected to study the influence of transverse spacing on their mechanical properties. In these models, the transverse spacing of the vertical poles was varied, while all other parameters remained constant. The load–stress curves are shown in Figure 16, and the load–displacement curves are shown in Figure 17. The simulation results are summarized in Table 7, where F_0, σ_2, and D_1 denote the maximum flexural bearing capacity of the load-bearing structure, the maximum stress on the tie rod, and the maximum displacement at the cantilever end, respectively, at the yielding point of the end plate.

Table 7. FEA results for load-bearing structures with different transverse spacings of vertical poles.

Names	F_0 (kN)	σ_2 (MPa)	D_1 (mm)
F2-1300-700	24.51	172.52	3.86
F2-1300-750	24.96	167.13	3.76
F2-1300-800	26.00	161.23	3.66
F2-1300-850	26.86	156.21	3.53
F2-1300-900	28.06	151.30	3.45
F2-1500-700	25.77	186.91	4.14
F2-1500-750	25.92	181.97	4.04
F2-1500-800	26.15	177.14	3.94
F2-1500-850	26.52	172.40	3.86
F2-1500-900	27.03	167.79	3.78
F2-1700-700	27.20	198.52	4.37
F2-1700-750	27.28	193.95	4.27
F2-1700-800	27.32	189.46	4.17
F2-1700-850	27.45	185.05	4.08
F2-1700-900	27.68	180.73	4.00

The results in Figures 16 and 17 and Table 7 show that, for load-bearing structures with the same cantilever length, an increase in the transverse spacing of vertical poles improved the flexural bearing capacity. Among these, the structure F2-1300-900 exhibited the highest flexural bearing capacity, sustaining a single-point load of 28.06 kN.

Additionally, for load-bearing structures with identical cantilever lengths and applied loads, an increase in transverse spacing reduced the tie rod stress and decreased the cantilever end displacement. These results indicated that, within a certain range of transverse spacing, smaller spacings (i.e., closer inner vertical poles to the cantilever beam's fixed ends) decreased the load shared by tie rods and increased the load borne by end plates.

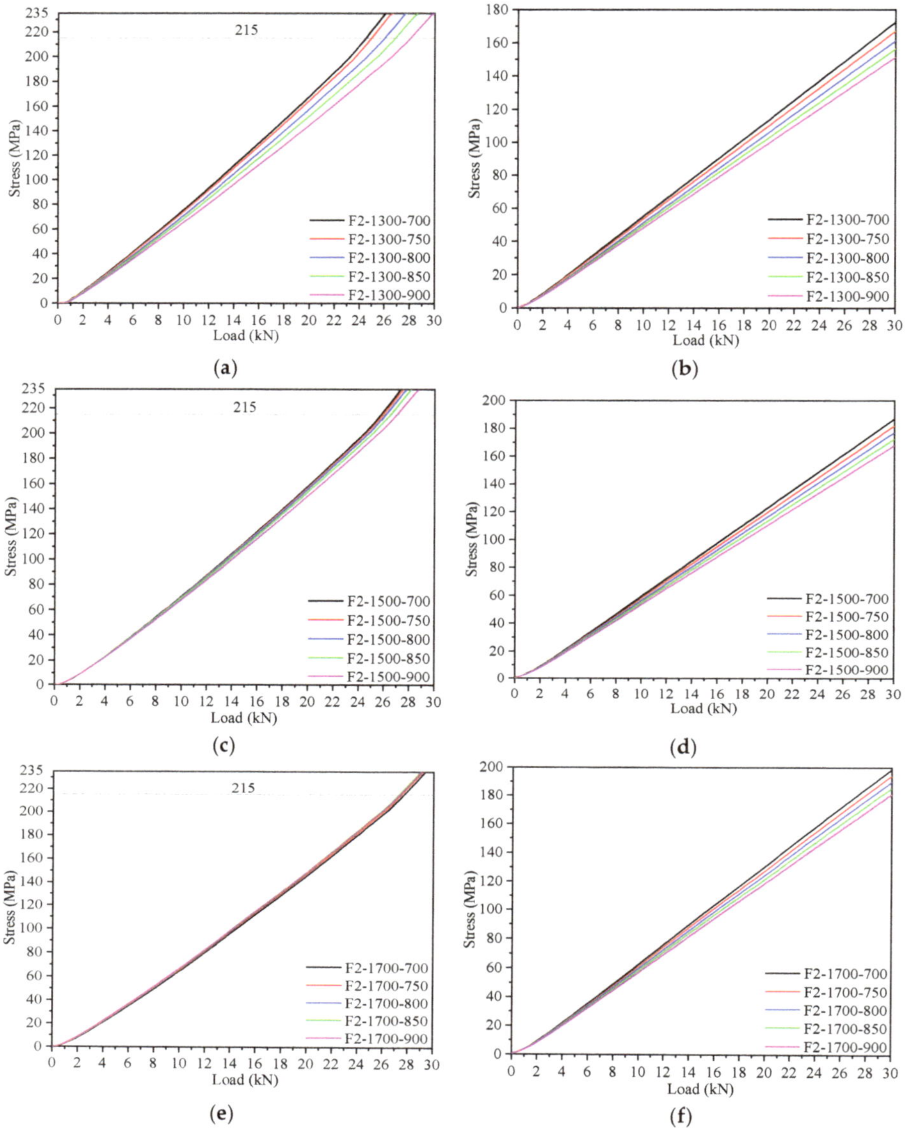

Figure 16. Load–stress curves for load-bearing structures with cantilever lengths of 1.3 m, 1.5 m, and 1.7 m. (**a**) End plate for the 1.3 m model; (**b**) tie rod for the 1.3 m model; (**c**) end plate for the 1.5 m model; (**d**) tie rod for the 1.5 m model; (**e**) end plate for the 1.7 m model; (**f**) tie rod for the 1.7 m model.

In addition, Table 7 and Figure 18 show that with an increase in the transverse spacing of the vertical poles under the same applied load, the flexural bearing capacity of the load-bearing structure with a 1.3 m cantilever length increased by 14.5%. In comparison, the flexural bearing capacity of the 1.7 m model increased by only 1.8%, indicating that variations in the transverse spacing had a greater effect on the flexural bearing capacity of the 1.3 m cantilever structure than on that of the 1.7 m cantilever structure.

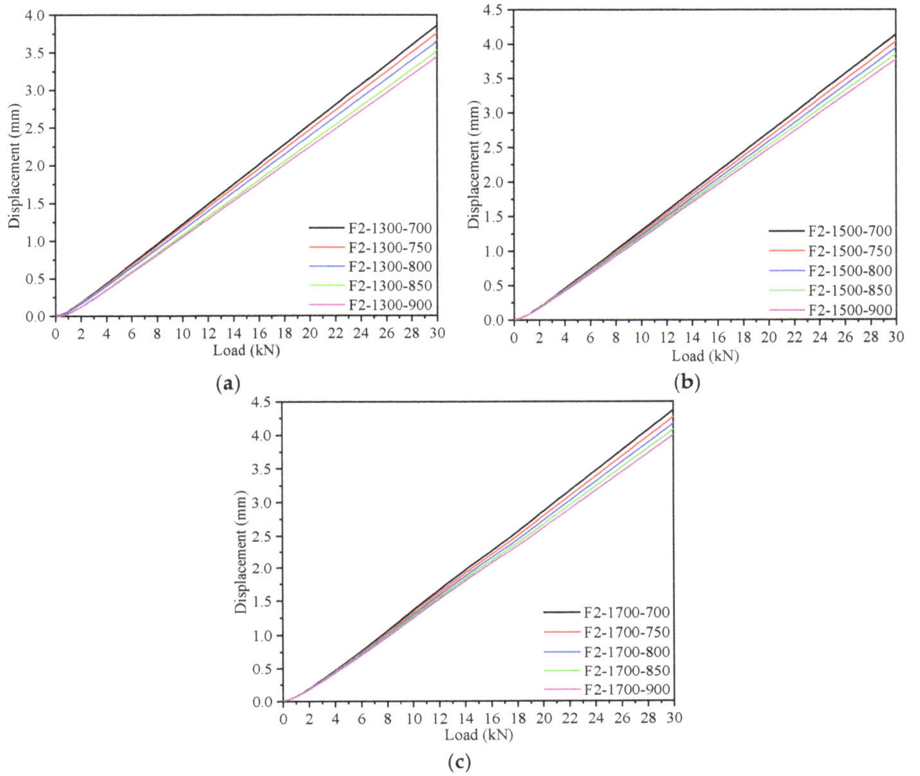

Figure 17. Load–displacement curves at the cantilever ends of load-bearing structures with cantilever lengths of 1.3 m, 1.5 m, and 1.7 m. (**a**) The 1.3 m model; (**b**) the 1.5 m model; (**c**) the 1.7 m model.

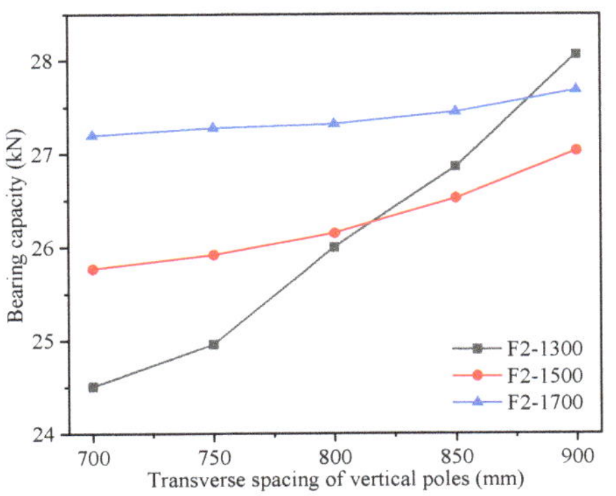

Figure 18. Comparison of the flexural bearing capacity of load-bearing structures with cantilever lengths of 1.3 m, 1.5 m, and 1.7 m.

5.2. The Influence of the Length of Cantilever Beams on the Mechanical Properties of the Load-Bearing Structures

The load-bearing structures with cantilever lengths of 2.3 m, 2.5 m, 2.7 m, 2.9 m, and 3.1 m from Section 4.3 were selected to investigate the effects of cantilever beam length on the mechanical properties of the load-bearing structures. In these models, all the parameters except the cantilever beam length were kept constant, with the transverse spacing of the vertical poles set to 800 mm. The maximum stresses on the I-beams in models F2-2300-800 and F2-2500-800 occurred at the outside hanging ears, while in the other models, they occurred at the inside hanging ears.

Figure 19a shows that the maximum stresses on the I-beams for cantilever lengths of 2.3 m and 2.5 m were nearly identical under the same load, whereas those for cantilever lengths of 2.7 m, 2.9 m, and 3.1 m increased with increasing cantilever length. Figure 19b shows that the outer tie rods yielded first during the loading process, bearing greater loads than the inner tie rods did. Additionally, as cantilever beam length increased under the same applied load, the loads borne by both the inner and outer tie rods increased, with the load increase on the outer tie rods being smaller than that on the inner tie rods. Figure 19c shows that under the same applied load, as the cantilever beam length increased, the displacements at the cantilever ends increased, and the slopes of the load–displacement curves showed an increasing trend at the yielding points of the outer tie rods.

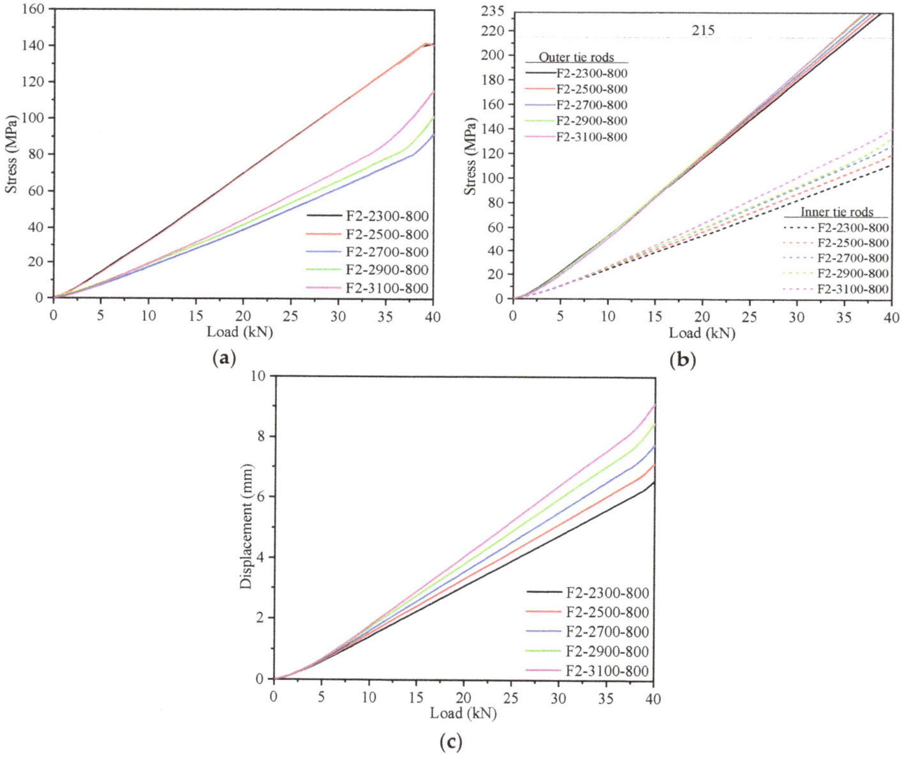

Figure 19. Load–stress and load–displacement curves for load–bearing structures with cantilever beams of different lengths. (**a**) Load–stress curves for the I-beams; (**b**) load–stress curves for the inner and outer tie rods; (**c**) load–displacement curves for the cantilever ends.

A further analysis of the results is shown in Table 8, where F_0, σ_3, σ_1, and D_1 represent the maximum flexural bearing capacity of the load-bearing structure, the maximum stress on the inner tie rod, the maximum stress on the end plate, and the maximum displacement at the cantilever end, respectively, at the yielding point of the outer tie rod. Table 8 reveals that increasing the cantilever beam length from 2.3 m to 3.1 m had a minimal effect on the flexural bearing capacity of the load-bearing structure, which decreased by

3.6%. The change in length had a smaller effect on the stresses on the outer tie rods but a greater effect on the stresses on the inner tie rods. Additionally, the increase in cantilever length significantly influenced the positions of the maximum stresses on the I-beams and noticeably increased the displacements at the cantilever ends, which rose by 37.8%.

Table 8. FEA results for load-bearing structures with cantilever beams of different lengths.

Names	F_0 (kN)	σ_3 (MPa)	σ_1 (MPa)	D_1 (mm)
F2-2300-800	35.59	110.96	141.68	6.44
F2-2500-800	35.17	118.05	140.55	6.98
F2-2700-800	34.70	125.28	89.07	7.60
F2-2900-800	34.51	130.69	98.16	8.28
F2-3100-800	34.29	138.07	112.58	8.88

5.3. The Influence of the Spacing of Hanging Ears on the Mechanical Properties of Load-Bearing Structures

A load-bearing structure with a cantilever length of 1.9 m was selected to study the effects of the spacing of the hanging ears on the mechanical properties of load-bearing structures. The structural layout diagram is shown in Figure 20, where l_2 represented the distance between the inside and outside hanging ears, which was set to 100 mm, 200 mm, 300 mm, and 400 mm in the FEMs. Meanwhile, the distances from the outside hanging ears to the cantilever ends were fixed at 325 mm, with all other parameters unchanged. The four models were named 1900-800-J100, 1900-800-J200, 1900-800-J300, and 1900-800-J400. The FEA results are presented in Figure 21 and Table 9.

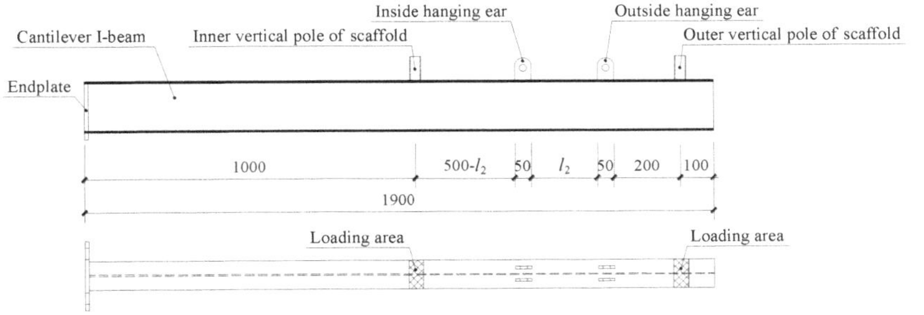

Figure 20. Structural diagram of cantilever I-beams with different spacing of hanging ears.

Table 9. FEA results of the load-bearing structures with different spacings of hanging ears.

Names	F_0 (kN)	σ_1 (MPa)	D_1 (mm)
1900-800-J100	56.40	218.73	5.14
1900-800-J200	53.51	178.84	5.28
1900-800-J300	49.97	150.02	5.41
1900-800-J400	46.83	129.42	5.46

Figure 21 shows that during the loading process, the outer tie rods consistently carried more load than did the inner tie rods. Under the same applied load, as the spacing between the inside and outside hanging ears increased, the load shared by the inner tie rods decreased, the load carried by the outer tie rods increased, the difference in the loads shared by the two tie rods increased, and the displacement at the cantilever ends gradually increased. When the outer tie rods yielded, the load–stress curves of the inner tie rods experienced abrupt changes with significant stress increases, indicating stress redistribution throughout the entire structural system.

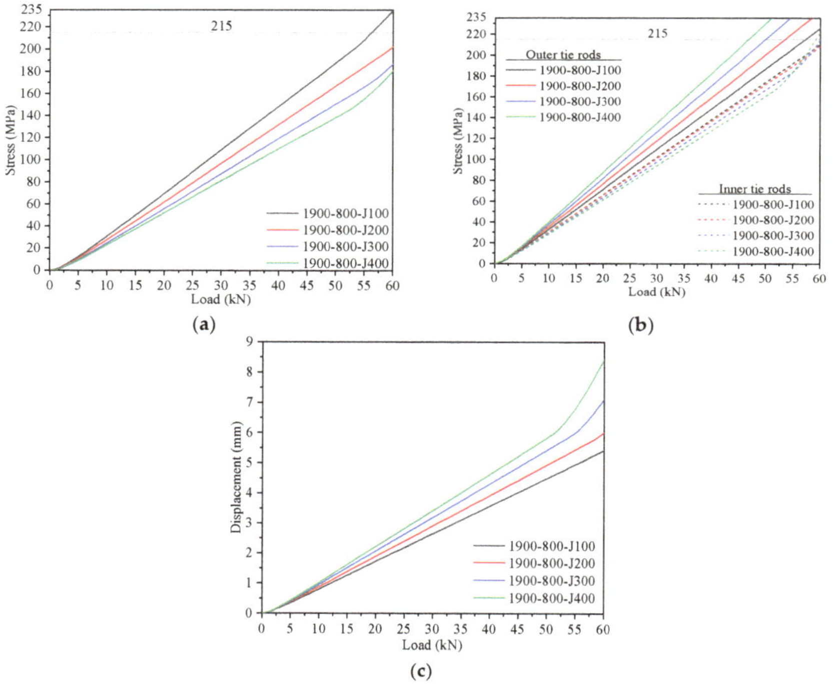

Figure 21. Load–stress curves and load–displacement curves for load-bearing structures with different spacings of hanging ears. (**a**) The load–stress curves of the I-beams; (**b**) the load–stress curves of the inner and outer tie rods; (**c**) the load–displacement curves of the cantilever ends.

The FEA results for load-bearing structures with different spacings of hanging ears are presented in Table 9, where F_0, σ_1, and D_1 represent the maximum flexural bearing capacity of the load-bearing structure, the maximum stress on the end plate, and the maximum displacement at the cantilever end, respectively, at the yielding point of either the outer tie rods or the end plates. For the load-bearing structure with a hanging ear spacing of 100 mm (1900-800-J100), the end plate yielded first, whereas for the other three load-bearing structures, the outer tie rods were the first to yield. The flexural bearing capacity of the load-bearing structure with a spacing of 100 mm between the inner and outer hanging ears was 20.4% greater than that of the structure with a spacing of 400 mm.

In summary, within a certain range of spacing between the inside and outside hanging ears, a smaller spacing led to a more balanced load distribution between the inner and outer tie rods, ensuring better utilization of the tensile capacities of both tie rods. This also increased the overall flexural bearing capacity of the load-bearing structures and reduced the displacement at the cantilever ends.

6. Conclusions

In this paper, the mechanical properties of three sets of pull-up cantilever scaffolds with adjustable steel tie rods of varying specifications were investigated through both experimental and numerical methods. Based on experimental validation, the study investigated the transverse spacing of vertical rods, length of cantilever beams, and spacing of hanging ears on the load-bearing capacity of structural systems. The primary findings are outlined as follows:

(1) The main damage modes of three sets of pull-up cantilever scaffolds include bending damage to the double-ear hanging ring, outward bending at the center of the end plate, bending deformation of the hanging ring connecting the tie rod and the extension rod, and disconnection at the welding point on the hanging ring of the extension rod. As the cantilever beam length increases, the loads borne by the tie rods progressively increase, whereas those carried by the end plates decrease correspondingly. In load-bearing structures featuring two tie rods, the outer tie rod consistently bears a greater load than the inner tie rod does. Additionally, increasing the spacing of the hanging ears amplifies the stress differences between the two tie rods.

(2) The results of the finite element model closely match the experimental results, demonstrating the high reliability of the FEM. An extended FEA is subsequently conducted using ABAQUS to evaluate the bearing capacities of 10 different specifications of load-bearing structures commonly used in practical engineering. The nonlinear curve fitting of the extended FEA data simplifies the formula for estimating the bearing capacities of load-bearing structures. The minimal discrepancy between the fitted and FEA values demonstrates the high accuracy of the derived formula in predicting the flexural bearing capacity of load-bearing structures.

(3) Parametric FEA is employed to further investigate the effects of the transverse spacing of vertical rods, the length of cantilever beams, and the spacing of hanging ears on the flexural bearing capacity of load-bearing structures. For the load-bearing structures with cantilever lengths of 1.3 m, 1.5 m, and 1.7 m featuring single tie rods, the flexural bearing capacity increases as the transverse spacing of the vertical poles widens within a specific range. For the load-bearing structures with cantilever lengths of 2.3 m to 3.1 m featuring double tie rods, increasing the cantilever beam length within a specific range slightly reduces the flexural bearing capacity while significantly increasing the maximum displacement at the cantilever ends. For the load-bearing structures with a cantilever length of 1.9 m featuring double tie rods, reducing the spacing of the hanging ears results in a more balanced load distribution between the inner and outer tie rods, ensuring the better utilization of their tensile capacities, increasing their overall flexural bearing capacity, and reducing deflections at the cantilever ends.

Author Contributions: Z.H., conceptualization, methodology, software, formal analysis, investigation, writing—original draft preparation, and writing—review and editing. J.Y. (Jianjun Yang)., conceptualization, methodology, validation, resources, funding acquisition, and Writing—review and editing. S.Z., conceptualization, supervision, and writing—review and editing. C.Z. and J.Y. (Jintao Yang), investigation, validation, and supervision. All authors have read and agreed to the published version of the manuscript.

Funding: This research was financially supported by the Hunan Province Engineering Construction Local Standard Making (Revision) Plan Project (Grant number: BZ20220005). The sponsor is Hunan Pankou Leasing Co., Ltd.

Data Availability Statement: Data are contained within the article.

Acknowledgments: The authors wish to acknowledge the National Engineering Laboratory's technical support for High-Speed Railway Construction of the School of Civil Engineering, Central South University.

Conflicts of Interest: The authors declare no conflicts of interest. The authors declare that this study received funding from Hunan Pankou Leasing Co., Ltd. The funder designed and provided the original test specimens but was not involved in the study design, collection, analysis, interpretation of data, the writing of this article or the decision to submit it for publication.

References

1. Sakhakarmi, S.; Park, J.; Asce, A.M.; Cho, C.; Student, P.D. Enhanced Machine Learning Classification Accuracy for Scaffolding Safety Using Increased Features. *J. Constr. Eng. Manag.* **2019**, *145*, 04018133. [CrossRef]
2. Zhao, Z.; Chen, Z. Analysis of Door-Type Modular Steel Scaffolds Based on a Novel Numerical Method. *Adv. Steel. Constr.* **2016**, *12*, 316–327. [CrossRef]
3. Bin, L.; Bo, Z. Parameterized Computing System on Structure of Lift Scaffolding Attached to the Wall. *Appl. Mech. Mater.* **2012**, *130–134*, 3972–3977.
4. Wu, K.; de Soto, B.G.; Adey, B.T.; Zhang, F. BIM-Based Estimation of Vertical Transportation Demands during the Construction of High-Rise Buildings. *Autom. Constr.* **2020**, *110*, 102985. [CrossRef]
5. Laleicke, P.F.; Cimino-Hurt, A.; Gardner, D.; Sinha, A. Comparative Carbon Footprint Analysis of Bamboo and Steel Scaffolding. *J. Green Build.* **2015**, *10*, 114–126. [CrossRef]
6. Hoła, B.; Nowobilski, T.; Woźniak, Z.; Białko, M. Qualitative and Quantitative Analysis of the Causes of Occupational Accidents Related to the Use of Construction Scaffoldings. *Appl. Sci.* **2022**, *12*, 5514. [CrossRef]
7. Cho, C.; Kim, K.; Park, J.; Cho, Y.K. Data-Driven Monitoring System for Preventing the Collapse of Scaffolding Structures. *J. Constr. Eng. Manag.* **2018**, *144*, 04018077. [CrossRef]
8. Chuang, K.-Y.; Tserng, H.-P. A Study of Automated Intelligent Monitoring of Scaffolding Safety for Construction Projects. *J. Chin. Inst. Eng.* **2024**, *47*, 626–642. [CrossRef]
9. Jin, Z.; Gambatese, J. Exploring the Potential of Technological Innovations for Temporary Structures: A Survey Study. *J. Constr. Eng. Manag.* **2020**, *146*, 04020049. [CrossRef]
10. Prabhakaran, U.; Beale, R.G.; Godley, M.H.R. Analysis of Scaffolds with Connections Containing Looseness. *Comput. Struct.* **2011**, *89*, 1944–1955. [CrossRef]
11. Pieńko, M.; Błazik-Borowa, E. Numerical Analysis of Load-Bearing Capacity of Modular Scaffolding Nodes. *Eng. Struct.* **2013**, *48*, 1–9. [CrossRef]
12. Cai, Q.; Ma, J.; Xie, Y.M.; San, B.; Zhou, Y. Topology Optimization and Diverse Truss Designs Considering Nodal Stability and Bar Buckling. *J. Constr. Steel Res.* **2025**, *224*, 109128. [CrossRef]
13. Song, S.; Zhao, Y.; Liang, F.; Guo, H.; Zhang, T.; Li, P.; Xiong, G. Optimization of Lower Suspension Point Position in Attached Cantilever Scaffold. *Buildings* **2024**, *14*, 2592. [CrossRef]
14. Zhang, H.; Chandrangsu, T.; Rasmussen, K.J.R. Probabilistic Study of the Strength of Steel Scaffold Systems. *Struct. Saf.* **2010**, *32*, 393–401. [CrossRef]
15. Beale, R.G. Scaffold Research—A Review. *J. Constr. Steel Res.* **2014**, *98*, 188–200. [CrossRef]
16. Peng, J.-L.; Chen, K.-H.; Chan, S.-L.; Chen, W.T. Experimental and Analytical Studies on Steel Scaffolds under Eccentric Loads. *J. Constr. Steel Res.* **2009**, *65*, 422–435. [CrossRef]
17. Zhang, H.; Rasmussen, K.J.R.; Ellingwood, B.R. Reliability Assessment of Steel Scaffold Shoring Structures for Concrete Formwork. *Eng. Struct.* **2012**, *36*, 81–89. [CrossRef]
18. Hongbo, L.; Yuan, Z.; Zhiua, C.; Qun, L. Structural Performance and Design Method of New Mortise–Tenon Full Steel-Tube Scaffold. *Adv. Steel Constr.* **2018**, *14*, 291–307.
19. Peng, J.-L.; Wang, P.-L.; Chan, S.-L.; Wu, P.-K. Experimental Study on Load-Bearing Capacities of Frame-Type Scaffolds Used in Precast Construction. *Int. J. Steel Struct.* **2020**, *20*, 400–414. [CrossRef]
20. Liu, H.; Wang, X.; Li, Z.; Zhong, Z. Comparative Study on Ultimate Stress Test and COMSOL Analysis of Steel Beam of Pull-Up Cantilever Scaffold. *Adv. Transdisc. Eng.* **2022**, *31*, 525–531.
21. Zhang, C.; Yang, J.; Jiang, L.; He, Y. Experimental Studies and Finite Element Analysis of Socket-Type Keyway Steel Pipe Scaffolding. *Buildings* **2024**, *14*, 245. [CrossRef]
22. Liu, F.; Yang, J.; Wang, J.; Hu, C. Study on Application of Long Cantilever Support Frame. *Constr. Technol.* **2013**, *42*, 25–27. (In Chinese)
23. Yan, X.M. Upper Cantilevered Scaffold Construction Technology Analysis of Ultra-High Tower. *Adv. Mater. Res.* **2012**, *446–449*, 3698–3701. [CrossRef]
24. Peng, J.-L.; Ho, C.-M.; Lin, C.-C.; Chen, W.-F. Load-Carrying Capacity of Single-Row Steel Scaffolds with Various Setups. *Adv. Steel Constr.* **2015**, *11*, 185–210.
25. Kim, A.; Lee, K.; Lee, S.; Song, J.; Kwon, S.; Chung, S. Synthetic Data and Computer-Vision-Based Automated Quality Inspection System for Reused Scaffolding. *Appl. Sci.* **2022**, *12*, 10097. [CrossRef]
26. Dzeng, R.-J.; Cheng, C.-W.; Cheng, C.-Y. A Scaffolding Assembly Deficiency Detection System with Deep Learning and Augmented Reality. *Buildings* **2024**, *14*, 385. [CrossRef]
27. Zhao, J.; Chen, J.; Liang, Y.; Xu, Z. Feature Selection-Based Method for Scaffolding Assembly Quality Inspection Using Point Cloud Data. *Buildings* **2024**, *14*, 2518. [CrossRef]

28. Chen, L.; Zhao, Q. The Application Comparison of Bottom Anchored Cable-Stayed Cantilever Scaffold in Building Assembly. In Proceedings of the 2018 7th International Conference on Energy and Environmental Protection (ICEEP 2018), Shenzhen, China, 4–15 July 2018; Atlantis Press: Shenzhen, China, 2018.
29. Huang, Z.; Chen, Q.; Ran, X.; Lan, S.; Wang, Z. Construction Technology of Cantilever Scaffold of Bolt Connection. *Constr. Technol.* **2018**, *47*, 419–425. (In Chinese)
30. Chen, N.; Yin, F.; Shi, W. Construction Technology of the Support Frame with Combination of I Steel Inclined Support and Steel Cantilever Beam. *Constr. Technol.* **2015**, *44*, 88–90. (In Chinese)
31. Wang, S.; Wu, Z.; Liu, Y. Study on the Design and Calculation of Cantilever Scaffold-Type Steel Cantilever Beam. *Build. Struct.* **2023**, *53*, 1150–1154. (In Chinese)
32. Qi, J.; Shi, J.; Huang, H.; Wu, H. Application of Suspended Section Steel Scaffold Buckled with Lever TypeBasket to High-Rise Building Construction. *Build. Constr.* **2012**, *34*, 435–437. (In Chinese)
33. Mao, J.; Guo, X.; Wu, F.; Wang, D.; Li, H. Research and Application of Basket Pull Rod Cantilever Scaffold Construction. *Build. Constr.* **2021**, *43*, 2101–2103. (In Chinese)
34. *GB/T 699-2015*; Quality Carbon Structure Steels. Chinese Standards Institution: Beijing, China, 2015.
35. *GB/T 700-2006*; Carbon Structural Steels. Chinese Standards Institution: Beijing, China, 2006.
36. *GB/T 706-2016*; Hot Rolled Section Steel. Chinese Standards Institution: Beijing, China, 2016.
37. Ranganathan, A. The Levenberg-Marquardt Algorithm. *Tutorial LM Alg.* **2004**, *11*, 101–110.

buildings

MDPI

Article

The Synergy and Accumulation Model for Analysis (SAMA): A Novel Approach to Transforming Risk Analysis in Construction with a Focus on the Deepwater Horizon Disaster Case

Elias Medaa [1], Ali Akbar Shirzadi Javid [1,*], Hassan Malekitabar [1] and Saeed Banihashemi [2]

[1] School of Civil Engineering, Iran University of Science and Technology (IUST), Tehran 13114-16846, Iran; medaa_e@civileng.iust.ac.ir (E.M.); malekitabar@iust.ac.ir (H.M.)
[2] School of Built Environment, University of Technology Sydney, Sydney, NSW 2007, Australia; saeed.banihashemi@uts.edu.au
* Correspondence: shirzad@iust.ac.ir

Abstract: Risk analysis is critical for preventing catastrophic failures in complex systems, as exemplified by the Deepwater Horizon disaster, a stark reminder of systemic vulnerabilities in offshore drilling operations, where inadequate appraisal of overlapping failures led to severe environmental and human losses. This study addresses the absence of a predictive framework capable of capturing cumulative risk interactions across both time stages and defensive layers. To fill this gap, and by drawing on prior frameworks such as the Swiss Cheese Model (SCM) and the Risk Matrix (RM), as well as critiques of their limitations, we introduce the Synergy and Accumulation Model for Analysis (SAMA). This model defines project life-cycle stages and risk recipients, characterizes each risk by four parameters (the focus of impact, suddenness, frequency, and effectiveness), and calculates horizontal (RF_h) and vertical (RF_v) risk factors. We applied SAMA to fifteen identified failure modes of the Macondo well, categorizing them across two time stages (operational and construction) and four defensive layers. Horizontal analysis revealed that the regulatory-laws layer accumulated the highest risk factors, RFh_{1laws} = 129.25 during the operational stage and RFh_{2laws} = 95.98 during the construction stage. Vertical analysis showed that the safety objective experienced the greatest systemic vulnerability, with $RF_{vsafety}$ = 135.8 across ten overlapping risks, followed by the quality objective at $RF_{vquality}$ = 128.39. These findings demonstrate SAMA's enhanced capability to identify critical collapse paths often overlooked by conventional models. For researchers, SAMA offers a transparent, parameter-driven methodology applicable across engineering and construction domains. For industry stakeholders, regulators, project managers, and safety engineers, this model provides actionable insights to prioritize resource allocation and strengthen specific defensive layers, thereby enhancing both preventive planning and resilience against future disasters.

Keywords: risk analysis; defensive layers; collapse paths; recipients of risks; Synergy and Accumulation Model for Analysis (SAMA); Deepwater Horizon disaster

1. Introduction

Accidents occur when control over risks is lost and can stem from deficiencies in the defense system, noncompliance with regulations, unsafe behaviors, or unforeseen circumstances [1,2]. The scholarly consensus affirms that accidents occurring within complicated systems result from a series of interrelated factors, their amalgamation giving rise to risks and subsequent accidents within them [3].

Consequently, accident and risk factors within systems can be classified into two primary categories: (1) Human factors, which often serve as the root cause of major accidents or are linked to their causation [4], and (2) Factors encompassing system tracking and policies, wherein the influence of challenges of the physical, political, and economic work environment becomes apparent, alongside inherent vulnerabilities [5].

The literature emphasizes the significance of classifying and organizing risks. Such endeavors help direct efforts and resources toward addressing the following with their prioritization and consequences [5].

The advancement of techniques across all industries has led to the introduction of multiple variables and standards in the domain. Consequently, scholars concur that the complexity of projects positively correlates with the level of uncertainty involved [6]. As a result, difficulties have intensified in estimating the requisite variables for risk analysis and management and comprehending the interconnectedness and interdependence among transactions and variables. These concerns have become a source of apprehension for risk management professionals and project managers alike [7].

This line of contemplation assumes a pivotal role within the present article, as it underscores the necessity of dissecting risks based on their impact on essential resources or the entities referred to as recipients of risks. This analytical framework predominantly aligns with the discourse on resource study and management in engineering projects, specifically construction projects and other spheres of public administration within diverse high-risk industries. Additionally, this approach emphasizes the component influenced by the risk, namely the recipients of risks, rather than solely focusing on the risk itself.

Several models exhibit deficiencies in integrating human and system interdependencies among the existing array of approaches. Furthermore, these models confront substantial obstacles in effectively addressing the dynamic systems, encompassing the dynamic interplay of internal system dynamics, software architectures, various process variables, and the key human dimension [8]. For example, post-occupancy evaluation (POE) offers a framework for integrating user feedback into building performance and management, enhancing the overall system effectiveness [9,10]. A primary challenge encountered in safety, risk prediction, and risk analysis is the absence of a comprehensive model encompassing the desired attributes within these disciplines. Accordingly, the principal objective of this article is to introduce a novel semi-quantitative model for risk analysis that aspires to tackle the amalgamation and accumulation of risks. Initially, this model will be tailored to serve the field of construction risk analysis. Subsequently, it will be linked to other management domains under construction management's umbrella.

According to Eccleston and Doub [11], risk assessment can be defined as the systematic procedure of identifying, evaluating, accepting, avoiding, and managing risks. Hence, a crucial aspect of risk analysis lies in selecting a model suitable for risk assessment [5].

In recent years, the construction industry has adopted more advanced risk assessment methods to deal with the complexities of modern projects. Although traditional methods are effective, they fail to capture the dynamic interactions and accumulation of risks over time. New approaches, such as graph convolutional networks, have been used to model construction safety risks better and to identify patterns in accident data. [12]. Monte Carlo simulations have been used to quantify risks and to account for uncertainties in probability and impact assessments [7]. Probabilistic digital twins have been developed to incorporate different sources of uncertainty in geotechnical design and construction processes to enhance decision-making [13]. Building on these advancements, the Synergy and Accumulation Model for Analysis (SAMA) introduces a dual-axis framework that captures both the temporal and structural dimensions of risk. By quantifying how risks accumulate and interact across different system layers and over time, SAMA is a comprehensive risk assessment approach and a new tool in the construction industry risk analysis toolbox.

The principal aim underlying the development of a systematic approach to risk management is to augment the efficacy of risk management practices by identifying and analyzing risk accumulation. The present study endeavors to accomplish this objective by subjecting two fundamental risk management models, the Swiss Cheese Model (SCM) and Risk Matrix (RM), to critical evaluation. Specifically, the research strives to address the primary criticisms and limitations associated with both models, with the ultimate aim of proffering a superior alternative in the form of the Synergy and Accumulation Model for Analysis (SAMA).

2. Literature Review

This article draws inspiration from two prominent frameworks: the Swiss Cheese Model (SCM) and the Risk Matrix (RM). A comprehensive examination highlights the limitations inherent in these models, motivating the development of an integrated approach.

2.1. Why RM and SCM?

A comparative study references substantial scientific articles and publications emphasizing various risk analysis and management models. Table 1 presents a detailed comparison of widely recognized risk analysis models as classified by referenced studies [6,14–16].

A review of Table 1 reveals that each model typically adheres to a specific analytical approach, lacking the flexibility to integrate multiple methodologies. Furthermore, an examination of reviewed scientific articles indicates that most existing models overlook time as a critical factor in risk identification. These models often treat risk analysis as a perpetual, iterative process, focusing on the repetition of newly identified risks while neglecting the inherent risks within the system.

Given the simplicity and widespread adoption of the Swiss Cheese Model (SCM) and the Risk Matrix (RM), along with the significant criticisms they have faced, synthesizing these models' strengths became essential. From the comparative analysis presented, it is clear that these models complement each other's shortcomings. This realization inspired the creation of a model that builds on their capabilities while addressing their known limitations.

Consequently, the proposed model aims to provide users with greater flexibility, more comprehensive results, and broader analytical methods within a unified framework.

2.2. Literature on the Swiss Cheese Model (SCM)

The failures often arise from interactions among various factors at each system level [3,17].

In his seminal work, Reason [14] introduced a model that employs slices of Swiss cheese to illustrate human errors and failures within defense barriers. The perforations in the cheese symbolize the weaknesses and gaps in the defense barriers [1]. Since its inception in the 1990s, this model has engendered a paradigm shift in safety sciences and accident studies [15].

Table 1. Comparison table of the most popular risk analysis models.

The Method	The Standard Provided for Comparison												
	Qualitative Analysis	Quantitative Analysis	Semi-Quantitative Analysis	Graphical Representation	Theoretical Representation	Focuses on One Risk	Focuses on a Group of Risks Together	Focuses on the System's Risk	Focuses on a Person's Risk	Explains the Intertwining of Risks	Difficult to Apply	Easy to Apply	Used in Construction Projects
Swiss Cheese Model	*			*			*	*		*		*	*
The Risk in Early Design (RED)	*				*	*		*			*		*
Event Tree			*	*			*	*	*			*	*
Fishbone Diagram	*			*			*	*	*				*
Fault Tree Analysis	*	*		*		*			*		*		*
Layer of Protection Analysis			*		*	*		*				*	*
Risk Matrices			*		*	*			*			*	*
Failure Mode and Effects Analysis	*				*	*		*			*		*

* Indicates that this standard is available in this model.

Reason identified multiple causes for forming these perforations and their contribution to accidents or failures. These causes encompass inherent system conditions, including inadequate design, suboptimal procedures, insufficient training, and decisions made by human factors influencing the system [16,18,19].

Another pattern leading to the formation of these perforations is the presence of an active defect or failure that breaches one of the defensive layers within the system [20]. Notably, Reason et al. [3] emphasized that the consecutive alignment of perforations can give rise to accidents that jeopardize the system and its operations.

Nevertheless, Reason did not explicate the selection of the locations for these perforations or the mechanisms governing their opening and closure. He merely proposed preventive measures without delineating a mechanism or law for manipulating the occurrence and closure of the perforations [19,21]. The flexibility inherent in representing accidents through this model serves as a primary incentive for its utilization by numerous safety researchers across various academic disciplines [22–24].

Through an exhaustive examination of pertinent scholarly literature, a comprehensive compilation of key criticisms of the SCM was formulated, culminating in a structured inventory of critical aspects as delineated in Table 2 below:

Table 2. Criticism of the Swiss Cheese Model.

Category	Source	Quotes
Criticisms related to the holes	[25,26]	Failure to locate cheese holes accurately—failure to determine the nature of holes and their interrelationships
	[3]	The holes' location, cause, and change—the mechanism of aligning the holes—focus on barriers rather than risks.
Criticisms related to relationships and etiology	[27]	Failure to take into account the interrelationship between the causative factors
	[28]	Ignore the reasons behind the events
	[1]	Lack of understanding of causal links, regulatory, and local—over-general
Criticisms related to defenses	[29]	Inconsistency in defenses and their independence

2.3. Literature on Risk Matrix Model (RM)

Risk matrices (RMs) are a valuable graphical tool employed for risk representation within the realm of safety science. The underlying framework of RMs utilizes a bitmap approach, relying on two fundamental components: probability and consequences [8,30–32]. Probability indicates the likelihood of potential consequences, with estimations derived from available information about previous or similar risks. This likelihood value is represented within a range from 0 to 1, effectively reflecting the existing state of uncertainty [33,34]. On the other hand, consequences pertain to the impact of risks on diverse objectives, wherein each objective possesses its unique level of urgency. For instance, in the context of preserving human lives, the consequences would manifest as the number of lives lost or injured [35,36].

This section provides an overview of the Risk Matrix (RM) limitations as highlighted in prior scholarly investigations. Cox [37] brought attention to the issue of RM ranges and central bias, wherein individuals tend to favor average values to avoid extreme responses—another concern, as noted by Thomas et al. [30], is the potential problems stemming from the color schemes employed in the Risk Matrix. Furthermore, the literature has extensively discussed the problem of inverting the order in the Risk Matrix, with researchers attributing this issue to erroneous calculations of risk consequences and probabilities [38]. Additionally, several studies have indicated that utilizing a Risk Matrix can result in misleading decisions, inappropriate resource allocation, and potential harm to organizations [31,38]. The following Table 3 outlines a selection of criticism points raised in previous research:

Table 3. Criticisms of the Risk Matrix.

Category	Criticisms Related to Results				Criticisms Related to Estimates	
Criticisms Source	Risk-Ranking Errors	Risk-Ranking Reversals	Accessing the Wrong Actions	Not Accounting for the Accumulation of Risks	Uncertainty in Estimating Probability and Consequences	Color Classification Problem
[37]	*	*	*		*	*
[30]					*	*
[39]					*	
[40]					*	
[41]					*	
[38]		*		*		
[42]		*				
[43]		*	*	*	*	
[44]			*			*
[45]	*		*		*	
[31]			*		*	

* Indicates that this Criticism is mentioned in this reference.

2.4. Conceptual Framework

Drawing on the strengths and addressing the limitations of the Swiss Cheese Model and the Risk Matrix, we propose an integrated framework that (1) embeds cumulative risk interactions across distinct life cycle stages, (2) captures the vertical layering of organizational defenses, and (3) each risk is defined by four defining parameters (focus of impact, suddenness, frequency, and effectiveness). From these parameters, two complementary metrics emerge: horizontal risk factors (RF_h), which quantify how individual hazards accumulate over time, and vertical risk factors (RF_v), which measure the overlap of vulnerabilities across regulatory, technical, managerial, and human-behavioral layers. By unifying temporal progression with defensive layering, this framework synthesizes existing theories into a cohesive structure that guides the development of our hypotheses on stage-specific risk escalation, layer-specific collapse paths, and their joint influence on overall system failure potential.

3. Methodology

Through an extensive review of relevant scholarly literature concerning the Swiss Cheese Model and Risk Matrix, along with a thorough examination of the critical points presented in Tables 2 and 3, the Synergy and Accumulation Model for Analysis (SAMA) aims:

To offer a methodical framework to elucidate the concept of overlapping risks;

To shift the focus of risk management from a mere interpretation and analysis of risk to the consideration of the risk recipient, SAMA aims to enhance the efficacy and applicability of risk management practices;

To mitigate uncertainties associated with the estimation of risk parameters, this goal is achieved by introducing novel parameters that draw upon the historical background of the risk and its corresponding consequences.

Ultimately, the primary objective of our novel model is to establish a robust connection between the outputs generated by the model and the outcomes of its analyses across various domains, encompassing comprehensive management in the construction industry.

These articulated goals serve as a foundation for developing the proposed model, which aims to address the limitations of previous approaches while providing a comprehensive framework for risk analysis and management.

3.1. Conceptual Theoretical Principles of SAMA

Our proposed model focuses on identifying essential conditions associated with overlapping risks to ascertain the most critical scenario. Specifically, we compile risks that impact a singular recipient within a specific time stage, subsequently evaluating the level of riskiness resulting from their overlapping and grouping. Moreover, the perspective of the individual responsible for risk analysis is considered a significant contributing factor. Figure 1 presents the mechanics underlying the SAMA workflow.

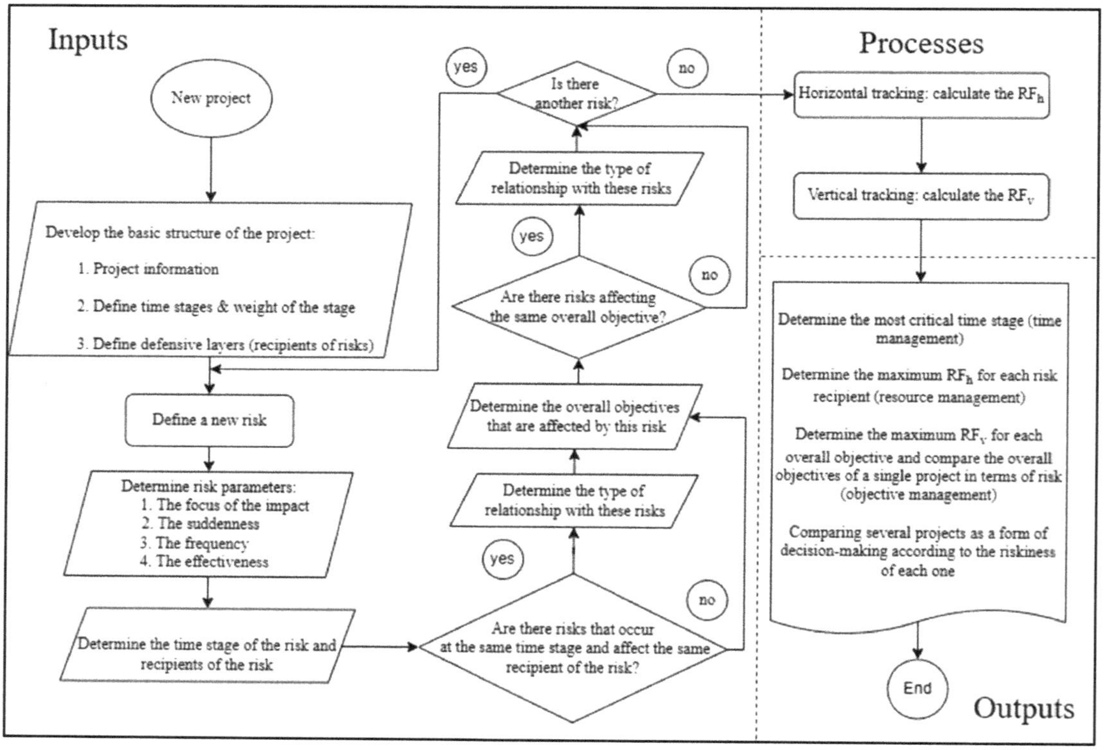

Figure 1. Flowchart of the workflow of SAMA.

Detailed elucidation of the model is expounded upon in subsequent sections, which include three phases: inputs, processes, and outputs.

3.2. Input Phase

3.2.1. Developing the Basic Structure of the Studied Case

Within every risk analysis model, various organizational factors play a pivotal role in enhancing the model's performance [46]. The Synergy and Accumulation Model for Analysis (SAMA) shares similarities with other established risk analysis models in this regard. To operationalize the model, the user first specifies the project's temporal stages. Subsequently, within each stage, the pertinent individuals or entities responsible for assuming the associated risks are referred to as "defensive layers". It is conceivable to allocate weight to time stages or defensive layers within the project, should such allocation be deemed essential for risk monitoring and the commensurate elevation of attention levels.

3.2.2. The New Definition of Constituent Risk Parameters

In light of the fact that each novel model introduces new parameters designed to align with the model's application and contribute to minimizing uncertainty, a crucial question arises: Why do we confine the definition of risk to predetermined parameters? Is it not true that these parameters should be flexible options that users of the model can freely utilize, tailored to the specific requirements of their respective projects? In the ensuing section, we put forth a set of parameter recommendations that have been discerned through an extensive review of pertinent literature on the subject matter. We contend that these suggested parameters align aptly with delineating risks inherent to the building and construction management domain.

The second step determines the value associated with the consequences of risk. To circumvent this issue, we combine the influence of two factors that exhibit a direct proportionality to the consequences of risk. These factors are denoted as "the focus of the impact (F)" and "the suddenness (S)" see Figure 2, with respective ranges of [1, 10] and [0.1, 1]. A comprehensive breakdown of these factors is presented in Table 4.

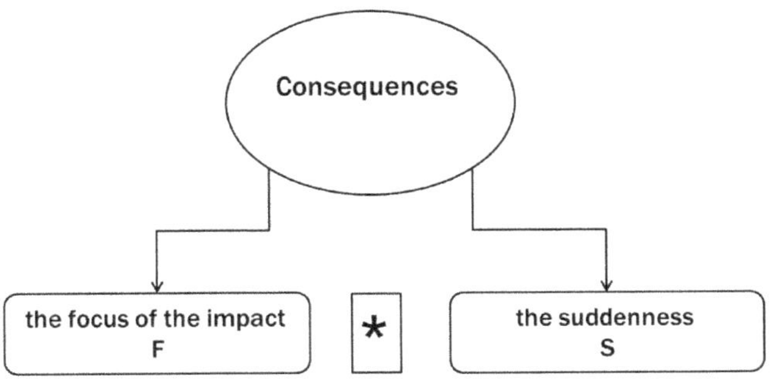

Figure 2. Consequences' definition parameters.

A higher F corresponds to a greater magnitude of consequences. Similarly, a reduced level of S indicates more noticeable warning signs, subsequently reducing the risk's consequences. The interplay between these factors can be mathematically expressed as Equation (1), yielding a range of consequences between [0.1, 10].

$$Consequences = F * S = [0.1, 10] \quad (1)$$

The third step in the risk assessment process involves estimating risk probability. To enhance the accuracy of this estimation, a proposed solution is to incorporate two essential factors: frequency and effectiveness. Frequency, denoted as Fr, represents the number of repetitions for this risk within a specific timeframe (which may include the project duration or the investment period, at the discretion of the risk manager) and is taken from historical records of such risk. The range for Fr is defined as [0.1, 1]. On the other hand, the second factor, effectiveness, measures the extent to which risks influence the project's original objectives, referred to as effectiveness, and is denoted as E. The range for E is specified as [1, 10], as shown in Figure 3. Further elaboration on these factors is provided in Table 4.

Consequently, the probability of risk can be determined by multiplying the values of Fr and E, resulting in Equation (2):

$$probability = Fr * E = [0.1, 10] \quad (2)$$

Table 4. Define new parameters, their description, and details of their value.

Parameter's Name	Value from-to	Value Details	Description	References and Semantic Sentences
The focus of the impact	1–2.5	The effect includes a quarter of the statistical community or less	It indicates the extent of the impact of the risk on the studied statistical community, and it is left within a single or wide range	"The focus of the impact was on the procurement of . . ." [47] "It is found that the focus of the impact of housing prices and . . ." [48] "Which the main inductive focus of the impact belongs . . ." [49] "Strategic refers to the focus of the impact . . ." [50]
	2.6–5	The effect includes one-quarter to one-half of the statistical community		
	5.1–7.5	The effect includes one-half to three-quarters of the statistical community		
	7.6–10	The effect includes three-quarters of the whole of the statistical community		
The suddenness	0.1–0.25	We are clearly alert to the occurrence of danger through warning signs	Indicates the extent of warning of the occurrence of risk through warning signs that warn of imminent risk	"Any suddenness in the later stages of the development system can . . ." [51] "The suddenness of impact in many disasters and the urgency . . ." [52]
	0.26–0.5	Warning signs can alert to the occurrence of danger		
	0.51–0.75	There are warning signs, but they are not very clear		
	0.76–1	There are no warning signs at all		
The frequency	1–2.5	The frequency of the occurrence of the risk is less than the frequency mentioned in historical records	It refers to the frequency with which a particular risk occurs during a specified period. This is related to previous studies on the history of the risk and its impact on the statistical community studied.	Frequency spillovers between global Green Bonds (GBs), WTI oil, and G7 stock markets using the time–frequency . . . [53] performance with regard to the frequency of supply . . . [54] We provide a detailed application to multivariate high-frequency . . . [55] the expected frequency or predict the frequency . . . [56]
	2.6–5	The frequency of the occurrence of the risk is close to or equal to the frequency mentioned in the historical records		
	5.1–7.5	The frequency of the occurrence of the risk is more than the frequency mentioned in historical records but less than twice that frequency		
	7.6–10	The frequency of the occurrence of the risk is more than twice the frequency mentioned in historical records		
The effectiveness	0.1–0.25	Risk exhibits a constant effect intensity during its repetition within one time period	Indicates the fluctuation of the intensity for this risk over one time period	Purpose To evaluate the effectiveness of a . . . [57] The effectiveness of current guidance and practice . . . [58] The study design was based on previous evidence for the effectiveness of . . . [59] Examining the effectiveness of risk elicitations . . . [60]
	0.26–0.5	Risk exhibits a little change in effect intensity during its recurrence within one time period		
	0.51–0.75	Risk exhibits a variable effect intensity during its recurrence within one time period		
	0.76–1	Risk exhibits a significantly variable effect intensity during its recurrence within one time period		

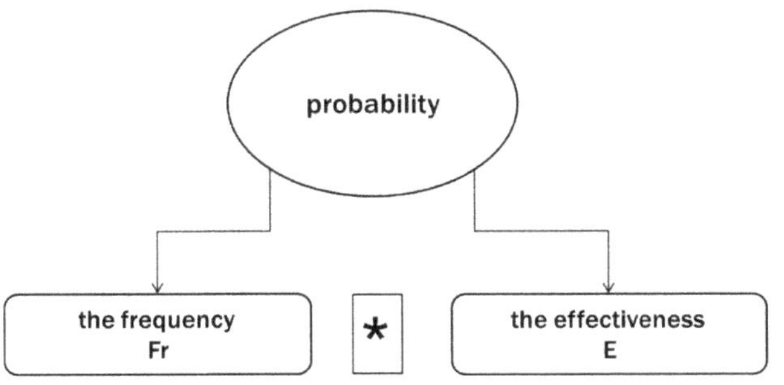

Figure 3. Probability definition parameters.

When a new risk emerges within a particular defensive layer and time frame, assessing its relationship with previously identified risks within the same context becomes crucial. Specifically, the team must ascertain whether these risks exhibit an overlapping relationship, if one's impact encompasses that of another, or if their influences remain entirely distinct and unrelated. Answering this question is pivotal for ensuring a precise and comprehensive risk assessment and aids in identifying pathways of risk accumulation.

To enable a comprehensive tracking and analysis of risks, the team must establish the main objectives affected by each studied risk. Furthermore, they must determine the relationship type with other risks impacting the same objective. This information serves as a critical component in the vertical analysis of risks and informs the decision-making processes.

To support the validity and comparability of the risk factor calculations, the ranges of all parameters (focus of impact, suddenness, frequency, and effectiveness) are defined in a manner that normalizes their combined outputs within a final risk factor scale of [0.1, 100] [37]. Such bounds ensure a consistent mathematical structure and facilitate risk ranking across diverse projects and contexts. The lower bounds (e.g., 0.1 for suddenness and frequency) allow the model to account for minimal or rare risks. In contrast, the upper bounds (e.g., 10 for focus on impact and effectiveness) represent the most severe or influential risks. Without universal, exact thresholds, they are adjusted based on theoretical modeling needs and logical differentiation between low and high-risk scenarios. This range provides a practical resolution for analysis and ensures that risk values remain interpretable, scalable, and analytically tractable throughout the framework of SAMA.

3.3. Processes Phase

3.3.1. Horizontal Tracking and Analysis

By applying the previous stage, the model is configured to work. Subsequently, risks are introduced individually, carefully considering their interrelation with previously defined risks within the project. These risks are identified within distinct "layers", representing defensive layers or recipients of risks across the project's life cycle. To effectively manage these risks, it is crucial to thoroughly examine the collapse paths of each defensive layer and ascertain the most critical line of collapse by evaluating the accumulation of risks and their mutual impact. Subsequent sections will elaborate more comprehensively on the mechanism, overlay, and associated conditions and types. This methodology draws inspiration from the rupture patterns observed in bolted steel plates within steel connections, where the breaking line is determined based on the applied tensile forces, signifying the breakdown concerning the number of holes present in the slice.

Overlapping risks share several features, including temporal congruence and similar recipients of risks. They might also have conflicting effects. The expert's perspectives and insights are valuable in establishing whether a given risk overlaps with or remains distinct from other risks. Subsequently, we present a comprehensive set of recommendations and restrictions to elucidate the underlying mechanism governing the phenomenon of overlapping risks:

- The horizontal analysis and evaluation of risks are predicated on identifying and assessing common risks during a specific phase in the project life cycle. This approach involves studying the impact of these risks within the designated stage, primarily focusing on their effects on project resources as recipients of risk. Consequently, the method facilitates the identification and cumulative impact of shared risks in terms of both temporal occurrence and resource utilization.
- The proposed model exclusively considers risks with negative implications while disregarding those associated with positive outcomes.
- For horizontal analysis, segmented objectives or any other suitable delineation that aligns with the project's requirements may be employed rather than relying exclusively on resource-related considerations.
- Risks that manifest a similar impact can be grouped in the same risk path. Therefore, risks can be aggregated within a collapse paths mechanism when they affect a specific type of resource or recipient.
- The collapse path mechanisms shall be identified within each resource segment of a given project life cycle phase. Subsequently, the most optimistic and pessimistic scenarios are selected based on the following criteria:
    - The highest horizontal Risk Factor (Max RF_h) determines the most optimistic scenario;
    - The lowest horizontal risk factor (Min RF_h) determines the most pessimistic scenario.

Thus, the equation by which the horizontal risk factor "RF_h" for any horizontal risk path is calculated is as follows in Figure 4 and Equation (3):

$$RF_h = \mathrm{f}(RF_i) = RF_1 + RF_2 + \ldots + RF_n \quad (3)$$

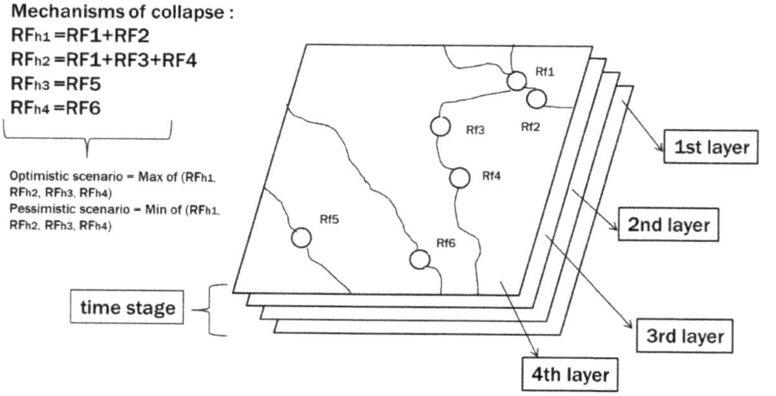

Figure 4. Horizontal analysis and tracking mechanism in SAMA.

3.3.2. Vertical Tracking and Analysis:

The analysis commences by tracing the vertical trajectory within the model, adhering to the subsequent steps:

- The progression of the project, embraced by defensive layers or stages in the Swiss cheese model, necessitates an exact definition based on the project life cycle. If warranted by the project's significance and magnitude, a solitary stage may be subdivided into several segments corresponding to the specific project.
- Determining the significance of the defensive layers is paramount. The impact diagram and the cost associated with responding to risks, as expounded in the relevant literature [61], indicate that addressing risks may prove more arduous in the project's later stages than in the early stages. The model advocates assigning greater weight to the defensive layers, which exhibit a higher risk factor whenever a stage progresses within the project life cycle. Estimating the value of this weight, ascribed to the defensive layer, necessitates considering the project's importance, scope, precision required in risk analysis, and overall project duration, as shown in Figure 5.
- Certain risks solely impact a singular stage and do not reverberate across other stages. Consequently, these risks should be excluded from the analysis of the vertical risk trajectory unless they act as causative agents or catalysts for generating additional risks.
- There are risks that directly and primarily affect the project's ultimate objectives, manifesting across all stages of the project life cycle. These risks align with the Swiss cheese model and warrant separate examination, resulting in the derivation of a final vertical risk factor (RF_v) as Equation (4).

$$RF_v = W_i * RF_i \tag{4}$$

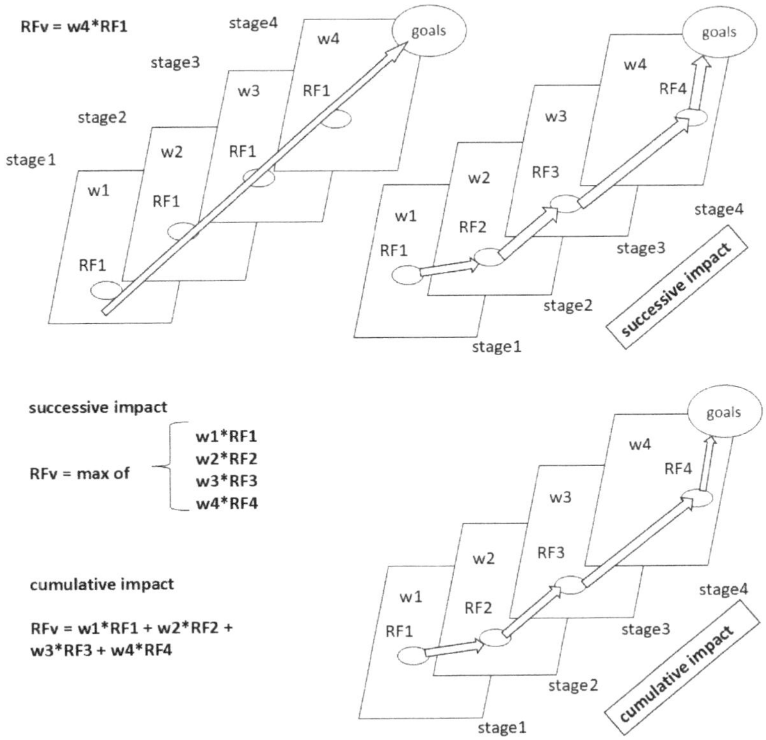

Figure 5. Vertical analysis and tracking mechanism in SAMA.

- Additionally, certain risks recur across multiple layers or demonstrate associations with risks in other layers. These risks may engender vertical overlap, cumulative effects, or cascading impacts, wherein the initial risk begets subsequent risks, and so forth.
- In the case of cumulative risks, it is logical to posit that the vertical risk factor is the aggregate of the risk factors, as in Equation (5), duly considering the weight assigned to each stage during the risk factor computation.

$$RF_v = W_1 * RF_1 + W_2 * RF_2 + \ldots + W_n * RF_n \quad (5)$$

- When confronted with a situation involving successive impacts of risks, wherein the effect of one risk is implicitly embedded in the effect of another, careful consideration is required. For instance, let us consider the delay in the team's arrival at the work site and the subsequent delay in delivering essential materials needed. Both cases result in the same effect. Traditionally, when multiple risks are present, the prevailing practice involves attributing the vertical risk factor to be equivalent to the final risk within the sequence, typically occurring during the later stages of the project's life cycle. However, our proposed model challenges this convention by advocating for the assignment of the vertical risk factor based on the risk factor exhibiting the most significant influence within the analyzed series, as shown in Equation (6).

$$RF_v = max(W_1 * RF_1, W_2 * RF_2, W_3 * RF_3, \ldots, W_n * RF_n) \quad (6)$$

3.3.3. Assessing Risk Interactions

After calculating each risk's RF_h and RF_v values, those pairs or groups of risks that share the same time stage and defensive layer are to be reviewed. For each grouping, the risks are determined whether they:

- Synergize: Two or more risks amplify one another's effects. In order to check their cumulative impact, their RF_h or RF_v scores must be combined.
- Cancel out: certain risks may counteract each other. The combined score is adjusted downward to reflect the reduction, although ignoring the cancellation effect is safer.
- Overlap without synergy: When risks arise from a similar root cause but do not materially interact with one another, only the higher of the individual scores is counted to avoid double-counting.

This expert judgment step ensures that the model captures actual interactive behavior, whether risks magnify, neutralize, or coexist, without relying on rigid formulas.

3.4. Output Phase

3.4.1. Outputs and Implications Derived from the Computation of the Horizontal Risk Factor RF_h

The horizontal risk factor (RF_h) serves as an evaluative metric, illustrating the cumulative and integrated impact of risks inherent in a specific time stage of the project life cycle for a particular risk recipient. Subsequently, we delineate the conclusions and implications inherent to the RF_h calculation, which are particularly pertinent to the risk management specialist scrutinizing the project and its associated risks:

1. RF_h serves as a quantitative indicator of the risk level associated with a given time stage in the project life cycle. Consequently, it apprises management of the imperative need to intensify focus on risk management activities, particularly in stages characterized by heightened risk, thereby advancing the overall state of risk management.

2. The computation of RF_h facilitates the determination of resource management direction during a specific stage of the project life cycle. It provides insights into the requisite maximum and minimum resource allocations essential for effectively managing anticipated project risks at that stage. Consequently, the outcomes of the new model SAMA are linked to resource management, specifically if the defensive layers employed in the model encompass project resources.
3. When employing staged objectives as defensive layers at each time stage, RF_h aids in identifying the most hazardous risk combinations and synergy scenarios that threaten the staged objectives based on each time stage. By comparing these scenarios, predicated on the risk of attaining staged objectives as delineated by SAMA, it plays a pivotal role in diverse domains such as time management, quality management, and the strategic management of phased plans within the scrutinized project.

3.4.2. Outputs and Implications Derived from the Computation of the Vertical Risk Factor RF_v

The vertical risk factor (RF_v) elucidates the cumulative and synergistic effects of risks associated with the project's ultimate goals, which are predetermined at the initiation of SAMA for risk analysis. Subsequently, it also explains the conclusions and implications of the RF_v calculation, which are particularly pertinent to the risk management specialist scrutinizing the project and its associated risks:

1. Assessment of Impact Magnitude and Identification of At-Risk Goals: The utilization of RF_v provides a means to gauge the magnitude of impact, thereby facilitating the evaluation of the extent to which identified risks or the synergy and accumulation of risks influence the project's ultimate goals. These goals encompass crucial elements such as time, cost, and quality. By discerning the potential impact of risks on these vital goals, RF_v emerges as a valuable instrument for identifying specific areas necessitating concentrated efforts to mitigate risks or allocate additional resources for their resolution. This, in itself, constitutes a form of goal management.
2. Facilitating project comparisons: RF_v contributes to the project selection by enabling a comparative analysis of different projects regarding their respective risk levels. Projects with lower risk levels may be perceived as more desirable investments than those with higher ones.

In summary, RF_v emerges as a valuable tool for comprehensively assessing and managing risks within projects, enabling project managers to make informed decisions and effectively prioritize risk mitigation measures.

4. Case Study

4.1. Explanation of the Case Study

The Deepwater Horizon disaster (Macondo well) on 20 April 2010, which resulted in the death of 11 individuals and the release of 53,000 barrels of oil per day, stands as a stark reminder of the catastrophic consequences of inadequate risk management in offshore drilling operations. From the outset of the Macondo well project, risks, uncertainties, and hazards were neither properly assessed nor managed [62].

The primary flaws leading to this disaster can be traced to critical lapses in both technical and organizational risk management practices. Notably, the failure of the cementing process, essential for sealing the well, was a significant factor. Halliburton's cement job exhibited known stability issues, which were neither adequately tested nor addressed [63]. Compounding this, the Blowout Preventer (BOP), designed as a fail-safe mechanism, failed due to multiple deficiencies, including a miswired solenoid and a dead battery in the control pod. These issues were overlooked during the maintenance and testing phases [64].

Moreover, the rig crew misinterpreted the negative pressure test, a critical step for verifying well integrity, leading to the erroneous conclusion that the well was secure despite clear signs of pressure anomalies [65]. Additionally, the crew either missed or misinterpreted several warning signs of a kick, and appropriate actions were not taken promptly [66]. These technical failures were exacerbated by significant lapses in human and organizational factors, including poor communication and decision-making processes among BP, Transocean, and Halliburton. The pressure to complete the well on schedule led to disregarding safety protocols and adopting undue shortcuts [65]. These interconnected failures underscore the profound deficiencies in risk assessment and management practices that prevailed at the time. This disaster emphasizes the need for continuous improvement in risk management practices to prevent similar catastrophic events in the future [67].

In this article, we will use the example of the Deepwater Horizon disaster to demonstrate the shortcomings of the Risk Matrix and the Swiss Cheese Model in accurately explaining risk analysis for such complex projects. These models fail to account for all the scenarios that may accumulate to culminate in a disaster. In contrast, the Synergy and Accumulation Model for Analysis (SAMA) will be applied to illustrate how effectively it aligns with the actual events of the incident.

4.2. The Risks of the Macondo Well

The Deepwater Horizon disaster resulted from multiple critical failures, including the Blowout Preventer (BOP) malfunction and an improperly executed cement seal, which allowed hydrocarbons to leak. Misinterpretation of a negative pressure test, coupled with inadequate well design, further contributed to the incident. These technical failures were exacerbated by poor risk assessment and management, delayed emergency response, and a series of poor decisions by the involved organizations. Human and organizational errors, regulatory failures, inadequate equipment maintenance, communication breakdowns, and insufficient crew training played significant roles in the catastrophe.

The following are the most prominent risks identified in the literature addressing the Deepwater Horizon incident:

- Cementing Process Failures: The cement job on the Macondo well, intended to seal the well from hydrocarbon zones, suffered significant failures. Post-incident investigations revealed that the cement slurry design and testing were flawed [63]. Halliburton, the contractor responsible for the cementing, was aware of the cement's instability but did not adequately communicate these concerns, nor were they incorporated into the risk assessments [65].
- Blowout Preventer (BOP) Reliability: The BOP, a critical safety device designed to seal the well in case of uncontrolled pressure, was fraught with design and maintenance issues. Investigations revealed that the BOP had a dead battery and a miswired solenoid, which prevented it from functioning correctly during the blowout [64]. The compounded technical failures highlighted significant gaps in risk management practices.
- Kick Detection and Response: Kick detection, the process of identifying uncontrolled hydrocarbon flow, is critical in well operations. On the day of the blowout, there were clear signs of a kick that the crew either misinterpreted or ignored, leading to a delayed response [65]. The failure to emphasize the urgency and protocols for kick detection and response was a critical oversight.
- Negative Pressure Test Interpretation: The negative pressure test, used to ensure well integrity, was misinterpreted by the rig crew. Conflicting pressure readings during the test were disregarded, leading to the erroneous conclusion that the well was

secure [68]. This false sense of security directly resulted from not recognizing the criticality of correctly interpreting negative pressure tests.

- Emergency Disconnect System (EDS) Activation: The EDS, designed to allow the rig to disconnect from the well in case of a blowout, failed during the incident. Attempts to activate the EDS were unsuccessful, exacerbating the disaster [66]. The underestimation of the EDS's critical role and potential failure modes highlighted significant gaps in risk management.

4.3. Risk Matrix Analysis Shortcomings for the Deepwater Horizon Disaster

The Deepwater Horizon disaster, which occurred on 20 April 2010, exemplifies how the traditional Risk Matrix approach inadequately assessed and mitigated critical risks, ultimately contributing to one of the worst environmental catastrophes in history. To analyze the risks using the Risk Matrix, it is necessary to identify the detailed risks associated with the main groups mentioned above. This process involves determining the impact and likelihood of each risk identified in the literature examining the Deepwater Horizon disaster. These studies sought to understand the causes of this catastrophic incident and to explain them through various risk analysis models. Table 5 below presents the risks that contributed to the Deepwater Horizon disaster, as assessed using the Risk Matrix [69]. Figure 6 also shows the Risk Matrix for the risks studied in the Deepwater Horizon disaster.

Table 5. Risk Matrix Table for Deepwater Horizon.

#	Code	Risk	Impact	Likelihood	Mitigation Failure
1	R_1	Blowout Preventer (BOP) Failure	Catastrophic	Likely	BOP was not maintained properly and failed to seal the well in an emergency
2	R_2	Cement Seal Failure	Catastrophic	Likely	The cement job at the well bottom failed to isolate hydrocarbons
3	R_3	Misinterpretation of Pressure Test	Severe	Likely	Negative pressure test results were misinterpreted, indicating no issues when there were multiple issues
4	R_4	Inadequate Well Design	Severe	Possible	Design did not account for the potential high pressure and gas influx
5	R_5	Failure in Risk Assessment and Management	Severe	Likely	Risks were underestimated, and proper assessments were not conducted
6	R_6	Emergency Response Failure	Severe	Likely	Emergency response plans were not effective, leading to delayed containment efforts
7	R_7	Poor Decision-Making	Severe	Likely	Multiple poor decisions and tradeoffs led to the disaster
8	R_8	Human and Organizational Malfunctions	Severe	Likely	Organizational and human errors significantly contributed to the failures
9	R_9	Regulatory Failures	Severe	Possible	Regulatory bodies did not enforce proper safety measures and regulations
10	R_{10}	Equipment Maintenance	Major	Likely	Equipment was not maintained or tested according to industry standards
11	R_{11}	Communication Failures	Major	Possible	Poor communication between the parties involved in the operation and management
12	R_{12}	Training Deficiencies	Major	Possible	The crew was not adequately trained for emergencies

Risk Matrix for Deepwater Horizon

Impact \ Likelihood	Rare	Unlikely	Possible	Likely
Moderate				
Major			R_{11} R_{12}	R_{10}
Severe			R_4 R_9	R_3 R_5 R_6 R_7 R_8
Catastrophic				R_1 R_2

Figure 6. Risk Matrix Analyses for Deepwater Horizon.

The results of the Risk Matrix in the Deepwater Horizon case highlight a prioritization of risks for treatment but fail to address critical systemic issues, such as inadequate management flow, delayed response to risks, and ineffective handling of complaints. This analysis reveals several limitations of the Risk Matrix approach. For instance, the concentration of risk parameter estimates in terms of impact and likelihood tends to fall within a specific range, making it challenging to distinguish between negligible risks and those with high likelihood but low impact. This can lead to overly polarized assessments. Additionally, the complexity of ranking risks by their impact and likelihood coefficients is evident, as many risks occupy similar positions within the classification system, thereby complicating prioritization.

4.4. Swiss Cheese Model Analysis of the Deepwater Horizon Incident

The Swiss Cheese Model categorizes errors into Latent Failures and Active Failures. Additionally, unsafe actions contribute to disasters by creating significant vulnerabilities within the whole defense system, leading to potentially catastrophic accidents [70]. Furthermore, we consider failures in protection systems originally designed to serve as safety and prevention mechanisms. In some instances, these systems can become primary contributors to disasters, exacerbating the severity of the incident by acting counter to their intended operational function. Accordingly, we will categorize these risks from the perspective of the Swiss cheese model as follows:

- Latent Failures (Organizational and Regulatory Weaknesses):
    1. Regulatory Oversight: Inadequate regulatory oversight and enforcement by the Minerals Management Service (MMS). The MMS had conflicts of interest, as it was both promoting offshore drilling and responsible for its regulation, so the MMS's oversight was not strict in enforcing regulatory requirements on ongoing offshore operations.

2. Safety Culture: BP's organizational culture prioritized cost-cutting and speed over safety. BP's main focus was meeting financial targets, leading to compromised safety standards and inadequate risk management practices.

- Active Failures (Errors and Violations):
  3. Decision-Making Errors: Critical decisions were made without adequate risk assessment. Decisions such as using a single long string casing instead of a safer liner/tieback system were made to save time and money despite being riskier.
  4. Operational Deviations: Failure to adhere to established procedures and best practices. The negative pressure test, a crucial step in determining well integrity, was misinterpreted, leading to the incorrect assumption that the well was secure.
- Preconditions for Unsafe Acts (Human and Environmental Factors):
  5. Crew Training and Competence: The rig crew lacked adequate training and preparedness. Crew members were not sufficiently trained to handle the negative pressure test and other critical procedures, resulting in misjudgments and errors.
  6. Communication Failures: Poor communication between BP, Transocean, and Halliburton. Significant lapses in communication regarding changes in drilling plans and risk assessments resulted in misunderstandings and a lack of coordinated response.
- Defenses in Depth (Safety Systems and Equipment):
  7. Blowout Preventer (BOP) Failures: The BOP, a critical safety device, failed to function correctly. The BOP had several design and maintenance issues, including a dead battery and a miswired solenoid valve, which prevented it from sealing the well effectively.
  8. Well Design Flaws: Design choices increased the risk of a blowout. The well design included fewer barriers to hydrocarbon flow, and the decision to use a long string casing instead of a liner/tieback system increased the likelihood of a blowout.

4.5. SAMA Analysis of the Deepwater Horizon Incident

The Deepwater Horizon incident analysis begins by applying the Synergy and Accumulation Model for Analysis (SAMA), defining the time stages. The stages of risk analysis for the Macondo well before the incident can be divided into two phases: Operational and Construction. By reviewing the company's previous incidents and conducting interviews in earlier studies with officials and workers involved with the Macondo well, defensive layers (recipients of the risks) were identified for each phase. These layers include protection procedures and tools, regulatory laws, structural integrity, and the project's primary objectives to ensure quality implementation and occupational safety.

This approach provides the foundational structure for collecting the necessary data for the risk analysis of the Macondo well and for interpreting the Deepwater Horizon incident. It is important to note that the numerical data for risk parameters presented in this analysis are approximate and were derived from reviewing scientific articles on related topics. The primary aim of these estimates is to demonstrate the SAMA mechanism and clarify the form of the results produced through its application.

Table 6 outlines SAMA inputs used in the Deepwater Horizon analysis, including the identification of influential risks, their parameters, calculation of risk factors, identification of the affected targets, time stages, and the recipients of the risks (defensive layers).

Table 6. SAMA inputs used in the Deepwater Horizon analysis.

The Studied Risks	Coding	Parameters							Objectives		Time Stages		Defensive Layers (the Recipients of the Risks)		
		The Focus of the Impact	Suddenness	Frequency	Effectiveness	Consequences	Probability	Risk Factor	Safety	Quality	1-Operational	2-Construction	Protection Procedures and Tools	Regulatory Laws	Structural Structure
Blowout Preventer (BOP) Failure	RF_1	10	0.8	1	1	8	1	8	*		*	*	1 + 2		
Cement Seal Failure	RF_2	8	0.3	4	0.6	2.4	2.4	5.76	*		*	*	1 + 2		
Misinterpretation of Pressure Test	RF_3	6	0.2	8	0.8	1.2	6.4	7.68	*	*	*		1		
Inadequate Well Design	RF_4	9	0.3	1	1	2.7	1	2.7		*	*	*			1 + 2
Failure in Risk Assessment and Management	RF_5	8	0.9	7	0.9	7.2	6.3	45.36	*	*	*	*		1 + 2	
Emergency Response Failure	RF_6	9	0.2	7	0.6	1.8	4.2	7.56	*		*			1	
Poor Decision-Making	RF_7	8	0.8	8	0.4	6.4	3.2	20.48		*	*	*		1 + 2	
Human and Organizational Malfunctions	RF_8	7	0.5	5	0.5	3.5	2.5	8.75		*	*			1	
Regulatory Failures	RF_9	5	0.1	9	0.6	0.5	5.4	2.7		*	*	*		1 + 2	
Neglecting Equipment Maintenance	RF_{10}	8	0.3	9	0.9	2.4	8.1	19.44	*	*	*		1		
Communication Failures	RF_{11}	4	0.3	7	0.8	1.2	5.6	6.72		*	*			1	
Training Deficiencies	RF_{12}	3	0.2	2	0.8	0.6	1.6	0.96	*	*	*		1		
Weak Safety Culture	RF_{13}	8	0.7	7	0.7	5.6	4.9	27.44	*		*	*		1 + 2	
Unsatisfactory Concrete Tests	RF_{14}	7	0.4	2	0.6	2.8	1.2	3.36	*	*		*	2		
Operational Deviations	RF_{15}	8	0.2	8	0.8	1.6	6.4	10.24	*	*	*			1	

* Indicates that this risk affects this Objective and affects the specified time stage.

Furthermore, many risk paths were identified by meticulously examining the interrelationships between risks associated with the same time stage and the risk recipient. These risk paths signify the amalgamation of diverse risks and are visually depicted in Figure 7.

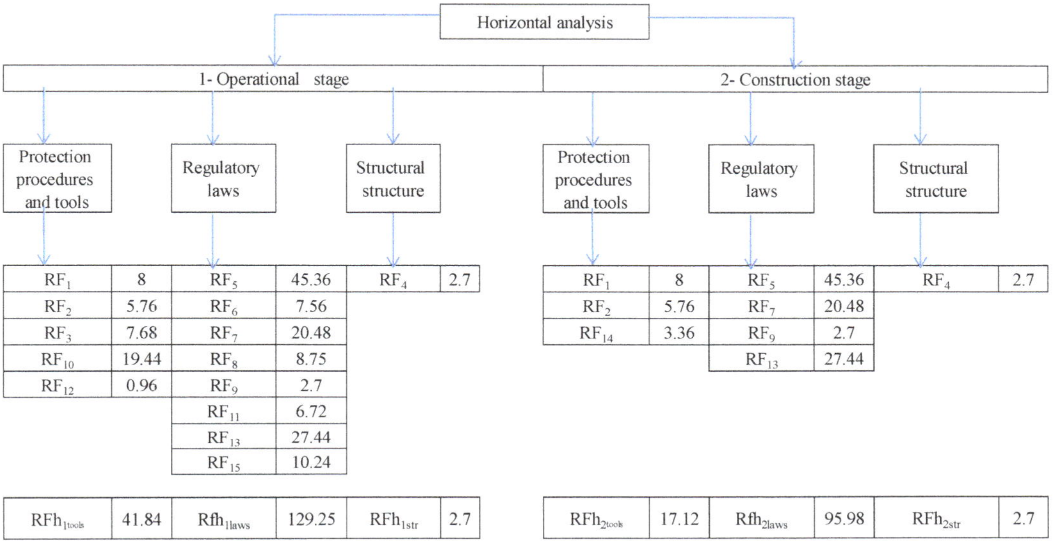

Figure 7. Risk accumulation paths according to horizontal analysis.

The horizontal analysis process of SAMA commences with a comprehensive assessment of each layer, wherein potential risks and their interconnections are meticulously examined. This analytical endeavor enables the identification of horizontal pathways of risk or collapse paths for each layer. This process is subsequently replicated for all existing layers, culminating in the computation of the horizontal risk factor for each layer, as shown in Table 7.

Table 7. Total risk factor for risk paths in horizontal analysis.

Time Stage	Defensive Layers	Coding	Horizontal Paths	The Value
Operational stage	Protection procedures and tools	$RF_{h1tools}$	$RF_1 + RF_2 + RF_3 + RF_{10} + RF_{12}$	41.84
	Regulatory laws	Rf_{h1laws}	$RF_5 + RF_6 + RF_7 + RF_8 + RF_9 + RF_{11} + RF_{13} + RF_{15}$	129.25
	Structural structure	RF_{h1str}	RF_4	2.7
Construction stage	Protection procedures and tools	$RF_{h2tools}$	$RF_1 + RF_2 + RF_{14}$	17.12
	Regulatory laws	Rf_{h2laws}	$RF_5 + RF_7 + RF_9 + RF_{13}$	95.98
	Structural structure	RF_{h2str}	RF_4	2.7

Additionally, a vertical analysis is conducted by scrutinizing each risk's characteristics and its overall objective. Throughout this process, the objectives impacted by the studied risks are considered. When common factors emerge among the risks, they are assigned to the same vertical collapse paths. The vertical risk factor for each path can be accurately calculated by identifying all vertical paths, as demonstrated in Figure 8 and Table 8.

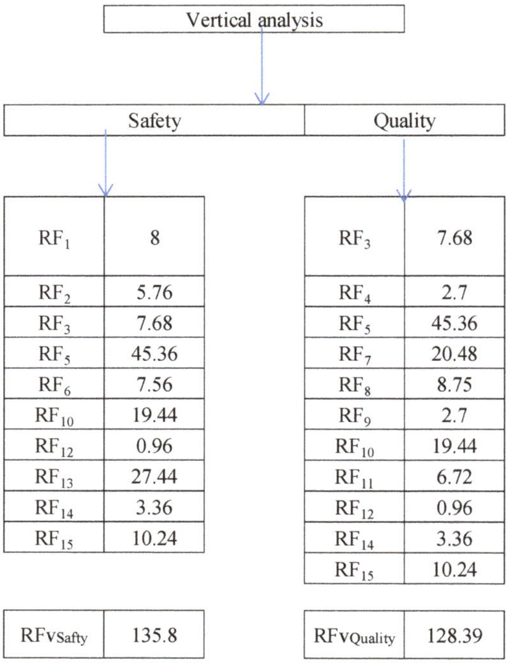

Figure 8. Risk accumulation paths according to vertical analysis.

Table 8. Total risk factor for risk paths in vertical analysis.

	The Objective	Coding	Vertical Paths	The Value
The Objective	Safety	RF_{vSafty}	$RF_1 + RF_2 + RF_3 + RF_5 + RF_6 + RF_{10} + RF_{12} + RF_{13} + RF_{14} + RF_{15}$	135.8
	Quality	$RF_{vQuality}$	$RF_3 + RF_4 + RF_5 + RF_7 + RF_8 + RF_9 + RF_{10} + RF_{11} + RF_{12} + RF_{14} + RF_{15}$	128.39

5. Findings

Applying SAMA to the Deepwater Horizon incident provides a comprehensive understanding of how interacting and overlapping risks contributed to systemic failure. This dual-perspective model has horizontal analysis, which looks at the accumulation of risks over the project timeline, and vertical analysis, which looks at the synergy and accumulation of risks on specific defensive layers associated with key project objectives.

Horizontal Analysis: The horizontal part of the SAMA framework shows that the defensive layer of regulatory laws had the highest risk accumulation across both major time stages, operational and construction. During the operational phase, this layer was exposed to eight interrelated risk events, resulting in a horizontal risk factor of $Rfh_{1laws} = 129.25$. Four notable risks were identified in the construction phase, resulting in a horizontal risk factor of $Rfh_{2laws} = 95.98$.

These values show that regulatory oversight was highly vulnerable throughout the project's life. The accumulation of risks in this layer means a persistent failure to translate legal frameworks into operational controls. The data also show a broader inability of the regulatory system to adapt to dynamic conditions in offshore drilling environments. The horizontal findings show a synergistic buildup of risks that were not just isolated to a stage but compounded across project phases, indicating a form of temporal risk entrenchment.

Vertical Analysis: The vertical analysis reveals the synergy and accumulation of risks on key project objectives, highlighting the most vulnerable collapse paths within the system. The safety objective stands out as the most affected, with 10 overlapping risks resulting in a vertical risk factor of $RF_{vSafety} = 135.8$. This is not just the number of risk events but the convergence of those risks on a single critical outcome: human life and operational integrity. The quality objective follows closely with 11 overlapping risks and a vertical risk factor of $RF_{vQuality} = 128.39$. This synergy and accumulation of risks in the quality layer means failures in construction integrity, process verification, and engineering compliance, which indirectly compromise safety performance.

The vertical results show the systemic interdependence between project layers and objectives. Rather than isolated weaknesses, these figures show how multiple risks converged through interrelated pathways to undermine core functions. Through vertical risk mapping, SAMA identifies critical defensive breakdowns where synergistic effects amplify the potential for collapse. This vertical perspective is particularly powerful in showing how risks with moderate individual impact can collectively escalate into severe systemic vulnerabilities when they accumulate across defensive layers.

These findings demonstrate SAMA's strength: it can quantify not only the presence of risk but also the structured accumulation and synergistic interaction of risk, as well as how failure modes propagate over time.

The insights from both horizontal and vertical analysis inform practical recommendations for addressing risk at systemic and organizational levels:

1. Regulatory Laws (Horizontal Track): The case study shows a persistent weakness in regulatory enforcement and communication. The accumulation of risks in this layer means there is no feedback mechanism, administrative delays, and a lack of operational adaptation to regulatory expectations. To address this, one must implement continuous monitoring systems, dynamic regulatory frameworks that evolve with project conditions, and integration between field operations and legal compliance teams. A real-time regulatory risk tracking system could have flagged the inconsistencies and prompted preventive action before the disaster struck. This issue is consistent with the U.S. report of the Chemical Safety and Hazard Investigation Board [64], which addressed this disastrous incident and its causes.
2. Safety (Vertical Track): The synergy and accumulation of risks on the safety objective shows a major systemic failure—defective cement seals, postponed maintenance, and inadequate emergency protocols aligned to create a high-risk environment. Post-incident reports show that despite knowing of these failures, the decision-makers did not act [64]. Strengthening this layer means proactive maintenance protocols, cross-verifying critical safety systems, and a culture of accountability where safety is prioritized even under cost or schedule pressure.

6. Discussion

While the Risk Matrix (RM) and the Swiss Cheese Model (SCM) are industry standards for risk assessment, our SAMA framework addresses their most significant limitations by introducing a parametric, multi-dimensional approach. We summarize the key criticisms of RM and SCM alongside SAMA's solutions and results below:

1. Overlapping Probability–Impact Ranges and Color-Based Classification:

Criticism (RM): Traditional risk matrices have overlapping probability–impact cells and use color coding (e.g., green/yellow/red), which introduces ambiguity and ranking bias.

SAMA Solution: By defining four independent risk parameters: focus of impact, suddenness, frequency, and effectiveness, SAMA replaces arbitrary color bands with numerical horizontal and vertical risk factors. These metrics stratify risk pathways into equal, non-overlapping intervals so each scenario gets an unambiguous, mathematically aggregated score. This numeric approach eliminates subjective interpretation and enables precise prioritization of hazards, reducing inconsistency in color-based designs.

2. Ignoring Synergy and Accumulation of Risks:

Criticism (RM): Risk matrices treat each hazard in isolation, ignoring how multiple moderate risks can combine to cause a significant failure.

SAMA Solution: Through its horizontal and vertical analysis, SAMA maps how risks converge into collapse paths. Horizontal factors (RF_h) quantify risk buildup across lifecycle stages; vertical factors (RF_v) capture the synergy and accumulation of failures across layers. This two-dimensional mapping reveals systemic vulnerabilities that single-point models miss, highlighting critical collapse paths for proactive mitigation.

3. "Holes" Without Time Factor or Quantification:

Criticism (SCM): SCM's "holes in layers" metaphor effectively illustrates defensive breakdowns, but it lacks quantification and overlooks the project timeline, making it difficult to manage risk proactively.

SAMA Solution: SAMA reframes layers as dynamic risk recipients and computes risk factors for each defensive layer at each project stage. Tracking RF_h within phases and RF_v across layers turns static "holes" into quantified vulnerability profiles. The model enables the prediction of risk trajectories at each time stage and for every risk recipient.

4. Event-Centric Focus vs. Recipient-Centric Framework:

Criticism (SCM): SCM is event-centric, focusing on post-incident interpretation rather than aligning defenses to protect project objectives.

SAMA Solution: SAMA's recipient-centric approach prioritizes safeguarding defined objectives—safety, quality, cost, and schedule, rather than just explaining past failures. This means risk management efforts focus on preserving the integrity of critical system functions. By aligning risk assessment with resource, objective, time, and financial management, SAMA provides a proactive roadmap for strengthening defenses and allocating mitigation efforts.

SAMA's modular structure allows seamless embedding into widely used management platforms:

- Enterprise Risk Management (ERM) Suites: Export RF_h/RF_v tables from SAMA into ERM dashboards (e.g., as custom metrics in Power BI or Tableau) for continuous monitoring and analysis.
- Project Management Tools: Link stage-specific RF_h values to scheduling software (e.g., Microsoft Project or Primavera) so that timelines automatically flag high-risk tasks for additional review.
- Maintenance and Compliance Systems: Feed vertical RF_v scores into Computerized Maintenance Management Systems (CMMS) to prioritize work orders where synergy of failures heightens potential impact.

Organizations can integrate our model without replacing existing workflows by mapping SAMA's outputs onto familiar data fields such as Risk Score, Control Priority, and Review Date. This integration transforms SAMA from a standalone analytical exercise into a live decision-support module that augments, rather than disrupts, current management systems.

7. Conclusions

This study introduces and validates the Synergy and Accumulation Model for Analysis (SAMA), a new parameter-driven framework inspired by the Risk Matrix (RM) and the Swiss Cheese Model (SCM). SAMA was developed to understand risk overlaps and to predict shared risks by defining discrete life cycle stages and defensive layers. Each risk is characterized by four parameters—focus of impact, suddenness, frequency, and effectiveness—and aggregates into horizontal (RF_h) and vertical (RF_v) risk factors. This systematic approach addresses the limitations of traditional models by quantifying how risks accumulate over time and interact across system defenses. The key contributions of this research are:

1. Robust Framework: SAMA provides a structured schema where layers are weighted recipients of risk, allowing for precise classification based on temporal and systemic attributes.
2. Dual Analytical Approach: Horizontal analysis identifies the most hazardous paths within each time stage. In contrast, vertical analysis uncovers the synergy and accumulation of risks that, although temporally distinct, jointly threaten core objectives.
3. Risk Manager's Role: The model incorporates expert judgment to evaluate the interrelationships between risks and to identify the highest-risk collapse paths, thereby integrating rational oversight into an otherwise quantitative process.

When applied to the Deepwater Horizon case study, SAMA found the Regulatory Laws layer had RF_h values of 129.25 (operational) and 95.98 (construction), and the safety objective had an RF_v of 135.8 due to 10 overlapping risks. These numbers indicate that SAMA can capture and address cumulative risk interactions more effectively than traditional models.

SAMA provides actionable guidance for various management domains. In resource management, decision-makers can allocate resources to the stages and layers with the most synergy of risks, preventing shortages or overstaffing. In time management, project schedulers receive an early warning of temporal risk bottlenecks and can adjust timelines and inspection intervals. In cost management, financial controllers can model the cumulative impact of layered risks on budget projections and prioritize investments in high-leverage controls. In quality and compliance management, auditors can focus their review on the defense layers with the highest accumulation of risk and ensure regulatory compliance before minor lapses escalate. Finally, in safety and maintenance management, safety officers and engineers can target overlapping failure modes, such as cement integrity and equipment maintenance, for simultaneous remediation and turn SAMA's insights into action plans. Risk assessments can move from past incident mapping to predictive synergistic analysis of future threats. SAMA enables users to see collapse paths, cross-functional controls, and system resilience across all aspects of project management. In future studies, we encourage applying the SAMA framework to additional sectors or case studies. Such applications would validate the model's adaptability and contribute to its refinement through cross-contextual comparison.

Author Contributions: Conceptualization, E.M. and A.A.S.J.; methodology, E.M. and H.M.; data collection, E.M. and S.B.; validation, E.M. and H.M.; investigation, A.A.S.J. and S.B.; resources, H.M.; writing—original draft preparation, E.M. and H.M.; writing—review and editing, A.A.S.J. and S.B.; supervision, A.A.S.J. and H.M. All authors have read and agreed to the published version of the manuscript.

Funding: This research received no external funding.

Data Availability Statement: The data generated or analyzed during this study are available from the corresponding author upon reasonable request.

Conflicts of Interest: The authors declare no conflicts of interest.

Abbreviations

The following abbreviations are used in this manuscript:

SAMA	The Synergy and Accumulation Model for Analysis
RM	The Risk Matrix
SCM	The Swiss Cheese Model
BP	British Petroleum
BOP	Blowout Preventer
EDS	Emergency Disconnect System

References

1. Larouzee, J.; Le Coze, J.-C. Good and bad reasons: The Swiss cheese model and its critics. *Saf. Sci.* **2020**, *126*, 104660. [CrossRef]
2. Stein, J.E.; Heiss, K. The Swiss cheese model of adverse event occurrence—Closing the holes. In *Seminars in Pediatric Surgery*; Elsevier: Amsterdam, The Netherlands, 2015.
3. Reason, J.; Hollnagel, E.; Paries, J. Revisiting the Swiss cheese model of accidents. *J. Clin. Eng.* **2006**, *27*, 110–115.
4. Haghighattalab, S.; Chen, A.; Fan, Y.; Mohammadi, R. Engineering ethics within accident analysis models. *Accid. Anal. Prev.* **2019**, *129*, 119–125. [CrossRef] [PubMed]
5. Altabbakh, H.M. Risk Analysis: Comparative Study of Various Techniques. Ph.D. Thesis, Missouri University of Science and Technology, Rolla, MO, USA, 2013.
6. Jafarnejad, A.; Ebrahimi, M.; Abbaszadeh, M.A.; Abtahi, S.M. Risk management in supply chain using consistent fuzzy preference relations. *Int. J. Acad. Res. Bus. Soc. Sci.* **2014**, *4*, 77.
7. Yazdani, M.; Abdi, M.R.; Kumar, N.; Keshavarz-Ghorabaee, M.; Chan, F.T. Improved decision model for evaluating risks in construction projects. *J. Constr. Eng. Manag.* **2019**, *145*. [CrossRef]
8. Qazi, A.; Dikmen, I. From risk matrices to risk networks in construction projects. *IEEE Trans. Eng. Manag.* **2019**, *68*, 1449–1460. [CrossRef]
9. Preiser, W.F. Post-occupancy evaluation: How to make buildings work better. *Facilities* **1995**, *13*, 19–28. [CrossRef]
10. Zimring, C.M.; Reizenstein, J.E. Post-occupancy evaluation: An overview. *Environ. Behav.* **1980**, *12*, 429–450. [CrossRef]
11. Eccleston, C.; Doub, J.P. *Preparing NEPA Environmental Assessments: A User's Guide to Best Professional Practices*; CRC Press: Boca Raton, FL, USA, 2012.
12. Mostofi, F.; Toğan, V.; Ayözen, Y.E.; Tokdemir, O.B. Construction safety risk model with construction accident network: A graph convolutional network approach. *Sustainability* **2022**, *14*, 15906. [CrossRef]
13. Cotoarbă, D.; Straub, D.; Smith, I.F. Probabilistic digital twins for geotechnical design and construction. *arXiv* **2024**, arXiv:2412.09432.
14. Reason, J. Understanding adverse events: Human factors. *BMJ Qual. Saf.* **1995**, *4*, 80–89. [CrossRef]
15. Larouzée, J.; Guarnieri, F. From theory to practice: Itinerary of Reasons' Swiss Cheese Model. In Proceedings of the Safety and Reliability of Complex Engineered Systems, ESREL, ETH, Zurich, Switzerland, 7–10 September 2015; pp. 817–824.
16. Qureshi, Z.H. A review of accident modelling approaches for complex critical sociotechnical systems. In Proceedings of the Twelfth Australian Workshop on Safety Critical Systems and Software and Safety-Related Programmable Systems, Adelaide, Australia, 30–31 August 2008.
17. Cassano-Piche, A.L.; Vicente, K.J.; Jamieson, G.A. A test of Rasmussen's risk management framework in the food safety domain: BSE in the UK. *Theor. Issues Ergon. Sci.* **2009**, *10*, 283–304. [CrossRef]
18. Rausand, M. *Risk Assessment: Theory, Methods, and Applications*; John Wiley & Sons: Hoboken, NJ, USA, 2013; Volume 115.
19. Fukuoka, K. Visualization of a hole and accident preventive measures based on the Swiss cheese model developed by risk management and process approach. *WMU J. Marit. Aff.* **2016**, *15*, 127–142. [CrossRef]
20. Reason, J. *Human Error*; Cambridge University Press: Cambridge, UK, 1990.
21. Reason, J. Human error: Models and management. *BMJ* **2000**, *320*, 768–770. [CrossRef] [PubMed]
22. Perneger, T.V. The Swiss cheese model of safety incidents: Are there holes in the metaphor? *BMC Health Serv. Res.* **2005**, *5*, 71. [CrossRef] [PubMed]
23. Salmon, P.M.; Stanton, N.A.; Lenné, M.; Jenkins, D.P.; Rafferty, L.; Walker, G.H. *Human Factors Methods and Accident Analysis: Practical Guidance and Case Study Applications*; CRC Press: Boca Raton, FL, USA, 2017.

24. Smith, S.; Wang, S. Reading and grammar learning through mobile phones. *Lang. Learn. Technol.* **2013**, *17*, 117–134.
25. Wiegmann, D.A.; Shappell, S.A. *A Human Error Approach to Aviation Accident Analysis: The Human Factors Analysis and Classification System*; Routledge: London, UK, 2017.
26. Shappell, S.A.; Wiegmann, D.A. *The Human Factors Analysis and Classification System—HFACS*; Routledge: London, UK, 2000.
27. Luxhøj, J.T.; Kauffeld, K. Evaluating the effect of technology insertion into the national airspace system. *Rutger Sch.* **2003**, *V*.
28. Leveson, N.G. Applying systems thinking to analyze and learn from events. *Saf. Sci.* **2011**, *49*, 55–64. [CrossRef]
29. Dekker, S. *The Field Guide to Human Error Investigations*; Routledge: London, UK, 2017.
30. Thomas, P.; Bratvold, R.B.; Bickel, E. The risk of using risk matrices. *SPE Econ. Manag.* **2014**, *6*, 56–66. [CrossRef]
31. Baybutt, P. Guidelines for designing risk matrices. *Process Saf. Prog.* **2018**, *37*, 49–55. [CrossRef]
32. Solís Toro, C. Constructing Operational Risk Matrices from Organizational Business Processes Using a Fuzzy Ahp Method. Bachelor's Thesis, Universidad EIA, Envigado, Colombia, 2019.
33. Hillson, D. *The Risk Management Handbook: A Practical Guide to Managing the Multiple Dimensions of Risk*; Kogan Page Publishers: London, UK, 2016.
34. Leveson, N. Improving the Standard Risk Matrix using STPA. *J. Syst. Saf.* **2019**, *55*, 13–22. [CrossRef]
35. Åkerberg, F. Risk Ranking Reversals and Classification Ranking Reversals in Risk Matrices. Master's Thesis, University of Gävle, Gavle, Sweden, 2021.
36. Clemen, R.T.; Reilly, T. *Making Hard Decisions with DecisionTools*; Cengage Learning: Belmont, CA, USA, 2013.
37. Anthony Cox, L., Jr. What's wrong with risk matrices? *Risk Anal. Int. J.* **2008**, *28*, 497–512. [CrossRef] [PubMed]
38. Baybutt, P. Designing risk matrices to avoid risk ranking reversal errors. *Process Saf. Prog.* **2016**, *35*, 41–46. [CrossRef]
39. Budescu, D.V.; Broomell, S.; Por, H.-H. Improving communication of uncertainty in the reports of the Intergovernmental Panel on Climate Change. *Psychol. Sci.* **2009**, *20*, 299–308. [CrossRef]
40. Markowski, A.S.; Mannan, M.S. Fuzzy risk matrix. *J. Hazard. Mater.* **2008**, *159*, 152–157. [CrossRef] [PubMed]
41. Scolese, C.J. *Improved Definition for Use of Risk Matrices in Project Development*; The George Washington University: Washington, DC, USA, 2016.
42. Pickering, A.; Cowley, S.P. Risk Matrices: Implied accuracy and false assumptions. *J. Health Saf. Res. Pract.* **2010**, *2*, 11–18.
43. Oliveira, M.D.; Bana e Costa, C.A.; Lopes, D.F. Designing and exploring risk matrices with MACBETH. *Int. J. Inf. Technol. Decis. Mak.* **2018**, *17*, 45–81. [CrossRef]
44. Olson, D.L.; Wu, D.D. *Enterprise Risk Management Models*; Springer: Berlin/Heidelberg, Germany, 2010.
45. Levine, E. Improving risk matrices: The advantages of logarithmically scaled axes. *J. Risk Res.* **2012**, *15*, 209–222. [CrossRef]
46. Pence, J.; Mohaghegh, Z. A discourse on the incorporation of organizational factors into probabilistic risk assessment: Key questions and categorical review. *Risk Anal.* **2020**, *40*, 1183–1211. [CrossRef]
47. Dreyer, M.; Bierwirth, M. The Impact of the Pandemic on German Academic IT Facilities in 2020 and beyond. Results of the Annual ZKI Survey on Trends at Academic IT Centres. *Proc. Eur. Univer* **2021**, *78*, 159–170.
48. Lu, Y.; Ding, N.; Wang, Y. Crime Prediction: An Empirical Study on the Impact of Housing Prices on the Regional Criminal Rate. *J. Phys. Conf. Ser.* **2021**, *1948*, 012045. [CrossRef]
49. Aznar-Crespo, P.; Aledo, A.; Melgarejo-Moreno, J.; Vallejos-Romero, A. Adapting social impact assessment to flood risk management. *Sustainability* **2021**, *13*, 3410. [CrossRef]
50. Weber, O. Measuring the Impact of Socially Responsible Investing. Available online: https://ssrn.com/abstract=2217784 (accessed on 14 February 2013).
51. Panychev, S.; Guzik, V.; Samoylenko, A.; Panychev, A. The prerequisites of forming a risk management system in the design of facilities space application. *MATEC Web Conf.* **2017**, *102*, 01030. [CrossRef]
52. Loh, B. Disaster risk management in Southeast Asia: A developmental approach. *ASEAN Econ. Bull.* **2005**, *22*, 229–239. [CrossRef]
53. Mensi, W.; Naeem, M.A.; Vo, X.V.; Kang, S.H. Dynamic and frequency spillovers between green bonds, oil and G7 stock markets: Implications for risk management. *Econ. Anal. Policy* **2022**, *73*, 331–344. [CrossRef]
54. Revilla, E.; Saenz, M.J. The impact of risk management on the frequency of supply chain disruptions: A configurational approach. *Int. J. Oper. Prod. Manag.* **2017**, *37*, 557–576. [CrossRef]
55. Diebold, F.X.; Hahn, J.; Tay, A.S. Multivariate density forecast evaluation and calibration in financial risk management: High-frequency returns on foreign exchange. *Rev. Econ. Stat.* **1999**, *81*, 661–673. [CrossRef]
56. Aven, T. On the new ISO guide on risk management terminology. *Reliab. Eng. Syst. Saf.* **2011**, *96*, 719–726. [CrossRef]
57. James, L.; James, S.; Hesketh, I. Evaluating the effectiveness of the fatigue and shift working risk management strategy for UK home office police forces: A pilot study. *Int. J. Emerg. Serv.* **2021**, *11*, 292–299. [CrossRef]
58. Crawford, M.; Stein, W. Risk management in UK local authorities: The effectiveness of current guidance and practice. *Int. J. Public Sect. Manag.* **2004**, *17*, 498–512. [CrossRef]
59. Williams, S.L.; To, Q.; Vandelanotte, C. What is the effectiveness of a personalised video story after an online diabetes risk assessment? A Randomised Controlled Trial. *PLoS ONE* **2022**, *17*, e0264749. [CrossRef] [PubMed]

60. Lundberg, R.; Willis, H.H. Examining the effectiveness of risk elicitations: Comparing a deliberative risk ranking to a nationally representative survey on homeland security risk. *J. Risk Res.* **2019**, *22*, 1546–1560. [CrossRef]
61. Jiang, Y.; Cukic, B.; Menzies, T. Cost curve evaluation of fault prediction models. In Proceedings of the 2008 19th International Symposium on Software Reliability Engineering (ISSRE), Seattle/Redmond, WA, USA, 10–14 November 2008; IEEE: New York, NY, USA, 2008.
62. Houck, O.A. Worst case and the Deepwater Horizon blowout: There ought to be a law. *Envtl. L. Rep. News Anal.* **2010**, *40*, 11033.
63. BP. *Deepwater Horizon Accident Investigation Report*; British Petroleum: London, UK, 2010.
64. Investigation Report Executive Summary, Drilling Rig Explosion and Fire at the Macando Well. 2016. Available online: https://www.dco.uscg.mil/Portals/9/OCSNCOE/Casualty-Information/DWH-Macondo/CSB/CSB-Macondo-Exec-Summary.pdf?ver=669ixQbnkL1l-OpgtcCpiA== (accessed on 25 April 2025).
65. National Commission on the BP Deepwater Horizon Oil Spill and Offshore Drilling. *Deep Water: The Gulf Oil Disaster and the Future of Offshore Drilling: Report to the President, January 2011: The Gulf Oil Disaster and the Future of Offshore Drilling*; Government Printing Office: Washington, DC, USA, 2011.
66. Deepwater Horizon Study Group. Final report on the investigation of the Macondo well blowout. *Cent. Catastrophic Risk Manag.* **2011**, *192*.
67. Crooks, E. *BP Case Told of 'Swiss Cheese' Theory*; Financial Times: London, UK, 2013.
68. Bly, M. *Deepwater Horizon Accident Investigation Report*; Diane Publishing: Darby, PA, USA, 2011.
69. Ahmed, F. Assessing the Impact of the Deepwater Horizon Oil Spill on Procedural Immunity and Crisis Response in the Industry: Lessons for Sustainable Offshore Management. *Int. J. Humanit. Soc. Sci. Manag.* **2023**, *3*, 604–609.
70. Xue, L.; Fan, J.; Rausand, M.; Zhang, L. A safety barrier-based accident model for offshore drilling blowouts. *J. Loss Prev. Process Ind.* **2013**, *26*, 164–171. [CrossRef]

buildings

MDPI

Article

Multiple Regression-Based Dynamic Amplification Factor Investigation of Monorail Tourism Transit Systems

Hong Zhang [1], Changxing Wu [2], Wenlong Liu [1,*], Shiqi Wei [2] and Yonggang Wang [2]

1 College of Mechanical and Intelligent Manufacturing, Central South University of Forestry and Technology, Changsha 410075, China; t20040415@csuft.edu.cn
2 School of Civil Engineering, Central South University, Changsha 410075, China; 224812290@csu.edu.cn (C.W.); 224811107@csu.edu.cn (S.W.); wyg2315592202@163.com (Y.W.)
* Correspondence: liuwnloong@csuft.edu.cn

Abstract: The monorail tourism transit system (MTTS) is a large-scale amusement facility. Currently, there is limited theoretical research on the vehicle–bridge coupling vibration and dynamic amplification factor (DAFs) of this system. The values specified in relevant standards are not entirely reasonable; for instance, the calculated value of the DAFs in the "Large-scale amusement device safety code (GB 8408-2018)" only takes speed into account and is set at 0.44 when the speed is between 20 and 40 km/h. This is overly simplistic and obviously too large. This paper aims to establish a reasonable expression of the DAFs for the MTTS and improve the design code of the industry. Firstly, using on-site trials of the project and the dynamics numerical simulation method, the dynamic response characteristics of the MTTS and the influencing factors of the DAFs were systematically analyzed. The rationality and accuracy of the model were verified. Secondly, combined with the joint simulation model, the dynamic influence mechanism of multifactor coupling on the DAFs was revealed. On this basis, the key regression parameters were selected by using the Pearson correlation coefficient method and the random forest algorithm, and the DAFs prediction model was constructed based on the least absolute shrinkage and selection operator (LASSO) regression theory. Finally, through cross-comparison of simulation data and specification verification, a recommended calculation expression of the DAFs for the MTTS was proposed. The research results show that the established prediction model can predict 94.50% of the variation information of the DAFs of the MTTS and pass the 95% confidence level and 0.05 significance test. The accuracy is high and relatively reasonable and can provide a reference for the design of the MTTS.

Keywords: monorail tourism transit system; vehicle–bridge coupling; dynamic amplification factor; multi-body dynamic; finite element method; LASSO regression

1. Introduction

With the rapid development of the tourism industry around the world and the increasing awareness of natural landscape protection, the monorail tourism transit system (MTTS), as an innovative solution for the low-carbon upgrading of scenic spots, has been widely applied all over the world with the advantages of lightweight construction and eco-friendly design concepts. Companies such as Intamin in Switzerland, Mack Rides in Germany, S&S in the United States, and CRRC in China have carried out a lot of work in this regard. As a key parameter in the design of MTTS, the DAFs is crucial to the safety design and comfort evaluation of MTTS. However, there is a significant gap between the theoretical research and engineering practice of the vehicle–bridge coupling and dynamic amplification factors

(DAFs) of MTTS. The industry design standards need to be improved, and there is a lack of research on key technical standards. The calculation and value selection of the DAFs are relatively rough. All these issues have restricted its development [1]. The design of this system is based on the "Large-scale amusement device safety code (GB 8408-2018) [2]". However, there are issues such as the DAFs being set less safely at low speeds and too conservatively at high speeds, with relatively single considerations. Adhering to the values stipulated in the code may lead to overly conservative designs and economic losses. MTTS has developed rapidly by taking advantage of the strengths of the traditional urban monorail transportation system, but there are also certain differences between the two. Specifically, the traditional urban monorail transportation system mainly uses prestressed concrete track beams and column structures, with a relatively large overall stiffness and a minimum curve radius of 100 m. In contrast, MTTS typically employs steel box girders and steel columns, and the modular construction can significantly reduce the construction period. Moreover, the minimum curve radius can be as small as 20 m, allowing it to flexibly adapt to the complex terrain of scenic areas. The diversity of structural parameters leads to significant differences in dynamic characteristics and impact effects between the two [3–5]. Therefore, the research on DAFs of MTTS is of positive significance for improving its design and evaluation system.

At present, the vibration characteristics of the transportation system are usually achieved through numerical simulation, such as solving the vehicle–bridge coupling vibration equation by programming with MATLAB [6–8]. However, during the derivation of the vibration equation, mechanical simplifications are made to the train and bridge, etc., which can easily cause a significant deviation between the numerical results and the actual running state of the train. The use of the finite element method (FEM) combined with multi-body dynamics (MBD) can effectively improve such problems [9]: Liu et al. established a simulation model of straddle-type monorail transportation using the multi-body dynamics software SIMPACK and analyzed the dynamic performance of monorail vehicles [10]. The results indicated that straddle-type monorail vehicles have good curve passing performance, and all static and dynamic performance evaluation indicators of the vehicles meet the requirements. Gao et al. developed a co-simulation model using the finite element method and MBD software to study the influence of vehicle speed and pier height on railway systems [11]. In addition, many studies have also considered the diversity and coupling effects of influencing factors in monorail transportation systems. For instance, Zhou et al. [12] improved the tire wear of monorail vehicles through dynamic parameter research and conducted a coupled dynamic analysis of monorail vehicles equipped with single-axle bogies [13]. Li et al. analyzed the dynamic response of long-span railway cable-stayed bridges and discussed the influence of factors such as train formation, track irregularity, driving direction, and train type on the impact coefficients of various components of cable-stayed bridges [14]. For the MTTS, Wang et al. [15] proposed a method for detecting the unevenness of the track beam in the MTTS, conducted multiple projects of unevenness measurement and spectral fitting, and compared the applicability of monorail spectral analysis with various spectral estimation and fitting methods. Finally, a joint model was established using ANSYS and UM software to analyze the dynamic response characteristics of the MTTS. Zhang et al. [16] established a general monorail vertical and horizontal unevenness spectrum for the MTTS based on measured data. GUO et al. established a comfort evaluation index for the monorail spectrum based on the vehicle–bridge coupling analysis model [17]. Zhang et al. established a wind–vehicle–bridge coupling model for the MTTS and analyzed the stiffness limit of the track beam under the influence of track unevenness and wind load [18]. The above studies mainly focus on the traditional transportation system and relatively few studies on MTTS; of course, these studies can also provide a methodology for MTTS.

Research on DAFs has been very extensive and in-depth in traditional transportation systems. Shervan et al. investigated the effect of train formation, train speed, span length, rise/span ratio, the first natural frequencies in vertical and lateral directions, and the combined elastic modulus of masonry and mortar on DAFs [19]. Babatope et al. investigated the deflection curve and the response characteristics of the DAFs of axially prestressed continuous beams under non-uniform speed loading [20]. Zhu et al. [21] established an 11-degrees-of-freedom three-dimensional vehicle model and a finite element model using MATLAB and ANSYS, respectively, and studied the DAFs of a three-span continuous arch-rigid frame bridge by using the vehicle–bridge coupling iteration method. They also proposed an improved damage identification method based on the DAFs [22], which enabled rapid detection and assessment of long-span suspension bridges. Zhou et al. took simply supported T-beam bridges, box girder bridges, and hollow slab bridges with standard spans as the research objects, derived the expression of the DAFs of simply supported beams under the action of three-axle vehicles based on the modal superposition method, and verified the variation law of the DAFs under different influencing factors through numerical simulation and field measurement [23]. Li et al. studied the DAFs and influencing factors of stay cables of cable-stayed bridges based on the vehicle–bridge coupling theory. The research shows that the pavement grade is the main influencing factor, and a reasonable value of the DAFs is proposed [24]. Deng et al. [25] compared the DAFs of strain and deflection of simply supported beam bridges and continuous beam bridges based on theoretical derivation and numerical simulation. The results showed that the DAFs of strain were basically less than those of deflection. In addition, they used the vehicle–bridge coupling vibration model to study and analyze the influence of parameters such as stiffness, mass, and damping of the vehicle model on the DAFs of the bridge [26] and verified the validity of the model through measured data. In summary, the existing research mainly focuses on traditional transportation systems and lacks studies on the DAFs of MTTS. Therefore, from the perspective of engineering application requirements and theoretical research value, it is of significant importance to conduct research on the DAFs of MTTS and establish a relatively reasonable calculation expression.

Based on a comprehensive analysis of existing research, the research approach of this paper is shown in Figure 1. Firstly, on-site tests of vehicle–bridge coupling were conducted based on actual projects, the dynamic response of the MTTS was analyzed, and a joint simulation model was established. Secondly, the joint model was used to analyze the influencing factors of the DAFs. Finally, due to the numerous influence factors of the DAFs and the interaction among them, the LASSO regression theory can effectively address such issues. Therefore, this paper employed the LASSO regression method to conduct a multiple regression analysis on the DAFs, establishing the calculation expression of the DAFs for the MTTS. Its rationality and accuracy were verified through comparison with simulation data and standards. Compared with the model established by the traditional research methods of monorail systems, the joint model established in this paper is more in line with the actual operation state, and the DAFs prediction model is more accurate and reasonable. This has a positive significance for improving the safety design and comfort evaluation of MTTS.

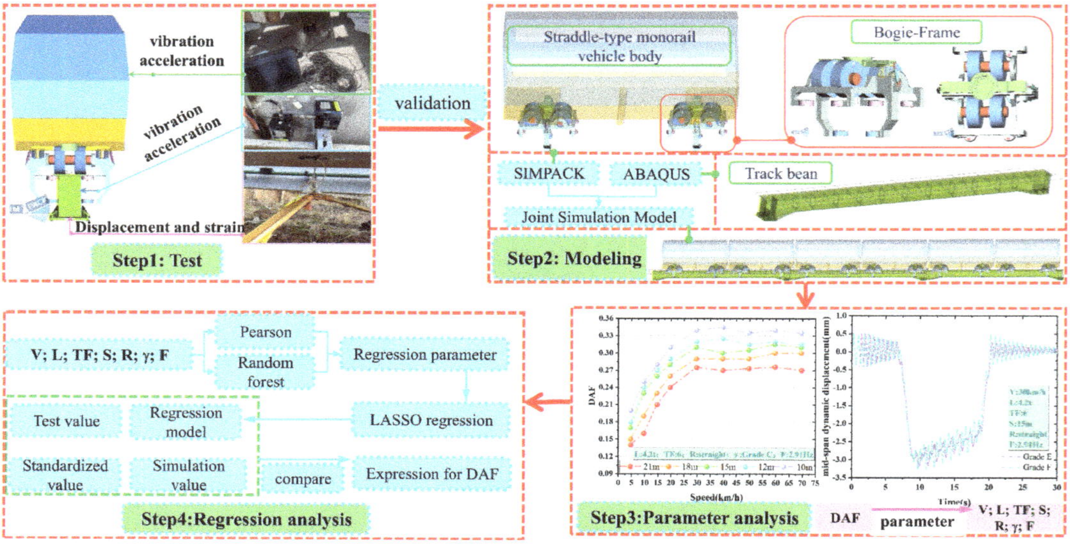

Figure 1. Flow diagram of the research.

2. Methodologies

This section first established a joint simulation model of the MTTS based on the actual project structure parameters. Then, on-site tests were carried out at the project site to analyze the influencing factors and variation patterns of the DAFs. Finally, through the comparative analysis of the test data and the simulation results, the rationality and accuracy of the joint simulation model were verified, which will provide a methodological basis for the subsequent expansion analysis of the DAFs influencing parameters.

2.1. Joint Simulation Model of MTTS

2.1.1. Modeling Monorail Trains

Based on the actual structural parameters of the CRRC Zhuzhou test line project, a 38-degree-of-freedom MTTS monorail vehicle dynamic model topology was established using the multi-body dynamics software SIMPACK (version 2021X) [27], as illustrated in Figure 2. In the SIMPACK modeling process, component appearances serve solely for visualization purposes, influencing only the graphical representation without materially affecting structural dynamics calculations or analyses. This study ensures strict consistency between core modeling parameters and the actual structure while appropriately simplifying the geometric characteristics of complex components, such as the monorail bogie. Specifically, the bogie and carriage are defined as independent subsystems, with the MTTS monorail vehicle structural dynamics analysis model constructed through primary and secondary suspension joints connecting the carriage and bogie subsystems.

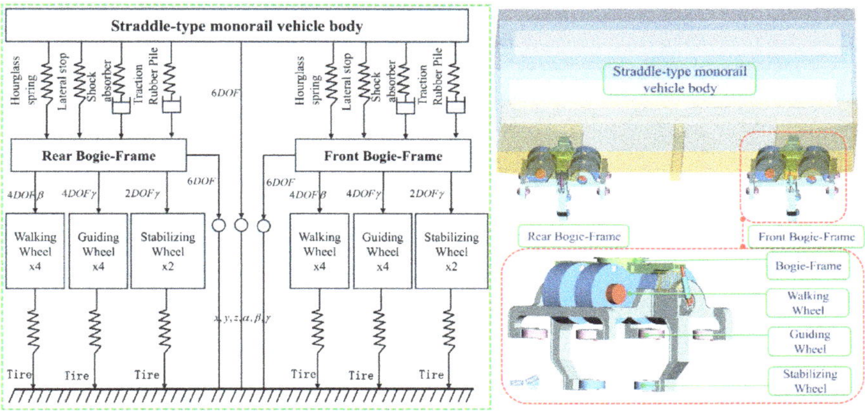

Figure 2. A 38-degree-of-freedom topological model of a monorail vehicle.

2.1.2. Modeling Track Beams

Based on the structural parameters of the CRRC Zhuzhou test line, a track beam model was established as shown in Figure 3 (using the 15 m span as an example). Figure 3a illustrates the cross-sectional parameters of the track beam, while Figure 3b presents the finite element model developed in ABAQUS [28]. The model employed solid elements, with simplifications applied to the stiffening ribs to balance computational efficiency and accuracy. Core parameters such as structural stiffness were calibrated to align with the actual structure. Subsequently, modal analysis and substructure generation were performed. Finally, through the ABAQUS-SIMPACK interface, the flexible track beam model was imported into SIMPACK, as depicted in Figure 3c.

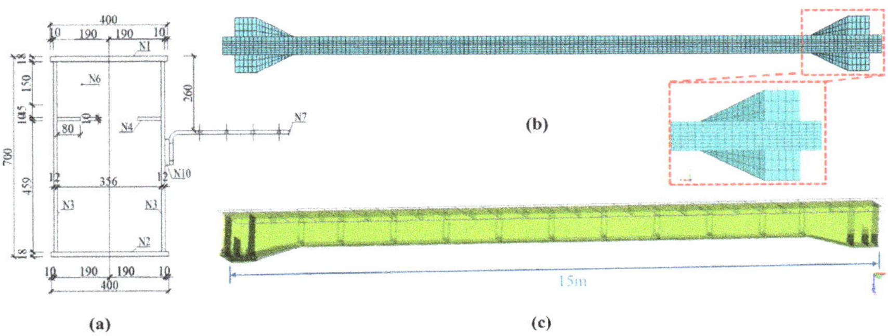

Figure 3. Track beam model for MTTS. (**a**) Track beam cross-section parameters; (**b**) finite element model; (**c**) flexible body model.

2.1.3. Joint Simulation Modeling

The main steps of the joint model establishment are shown in Figure 4 and can be described as follows: (1) ABAQUS was used to establish the rail beam model and generate inp and sim model data files that could be linked with SIMPACK; (2) SIMPACK was utilized to develop the car body, bogie, and tire models, respectively, and realized the coupling connection between the car body, bogie, and tires through the force element to establish the monorail car model; and (3) through the interface between SIMPACK and ABAQUS, the track beam data files were imported and Fbi model files were generated. Subsequently, the Ftr file was utilized to read the Fbi file information and write the track beam model data, establishing the track beam flexible body model. Finally, the wheel-rail contact relationship equations were employed to achieve the coupling of the monorail train with the track beam, completing the establishment of the ABAQUS-SIMPACK joint model. This approach enabled the simulation and analysis of the MTTS vehicle–bridge coupling system.

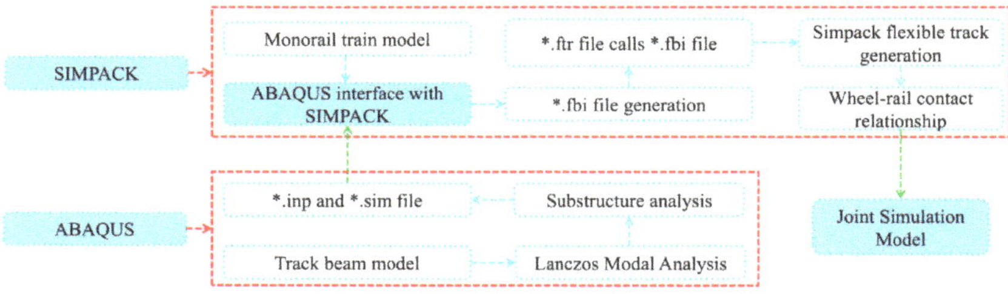

Figure 4. Joint simulation modeling process.

2.2. Vehicle–Bridge Coupling Field Trials of MTTS

To verify the rationality and accuracy of the joint simulation model, a typical beam span from the Zhuzhou MTTS test line project was employed for vehicle–bridge coupling dynamic response measurements. The monorail train speed was set across 2–15 km/h, with a total of five gears. The test group consisted of a 6-car formation fully loaded to 4.2 t; the train weighed approximately 24 t. The test spans were distributed as 12 m, 15 m, and 18 m. The MTTS vehicle–bridge coupling field test arrangement is shown in Figure 5 (taking the 15 m track beam as an example, the figure shows the specific distribution position of the 15 m span test section track in the line and the spans of the adjacent track beams on the left and right are 15 m and 18 m, respectively). Test monitoring equipment included a top bar displacement gauge, strain gauges, and vibration sensors (pickup). The data acquisition and analysis system utilized the YSV dynamic collector and DHDAS static collector. By varying parameters of the monorail train (speed, load) and track beam parameters (span), the dynamic and static responses of the track beam and vibration responses during the monorail train's travel were tested under different working conditions.

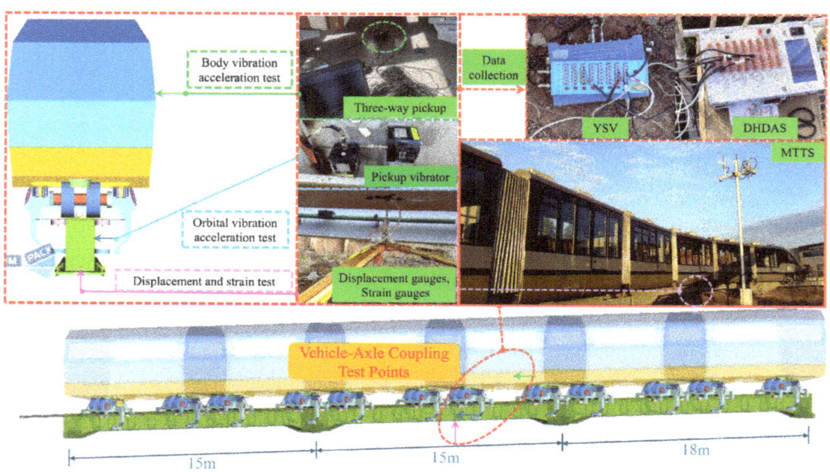

Figure 5. Vehicle–bridge coupling field test arrangement for MTTS.

The test results of the dynamic response and DAFs of the vehicle–bridge coupling of the MTTS are shown in Figure 6. The analysis reveals the following: (1) Within the speed range of 2 km/h to 10 km/h, both the displacement and strain DAFs show a positive correlation with the increase in speed and the growth rate is significant. After 10 km/h to 15 km/h, the increase in the strain DAFs gradually slows down and reaches its peak at 15 km/h, while the displacement DAFs still have room for increase, indicating that it has not yet reached the peak value. At the same time, it can be seen from the figures that the load has a certain influence on the DAFs, especially on the strain DAFs, which is more obvious. Compared with the light load, the maximum difference in the displacement and strain DAFs under heavy load is about 0.02. Within the current speed and load calculation range, the displacement DAFs are always greater than the strain DAFs, indicating that the displacement DAFs are more sensitive to the change in speed. (2) As the span increases, the DAFs show a significant negative correlation and change markedly. Under heavy load conditions, within the three calculated spans, the extreme differences between the maximum and minimum values of the DAFs are 0.08 (displacement DAFs) and 0.07 (strain DAFs), respectively. The increase in the displacement and strain DAFs under heavy load is more significant, indicating that the DAFs of the monorail train are more susceptible to span changes under heavy load, which is the result of their coupling effect. Moreover, under the same working conditions, the displacement DAFs is greater than the strain DAFs, and relevant research [25] also supports this conclusion.

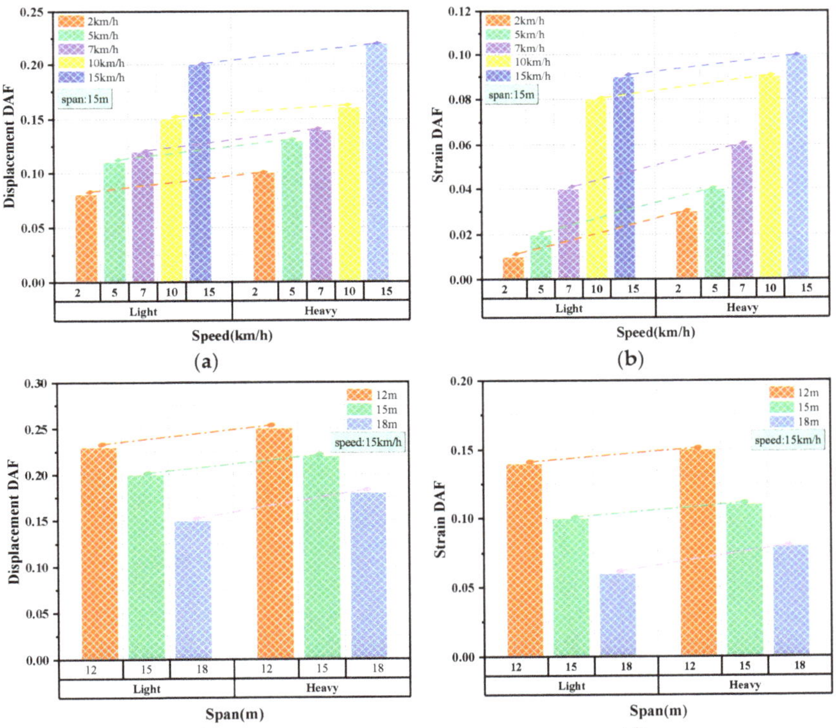

Figure 6. Field test results of MTTS. (**a**) Displacement DAFs for different speeds and loads; (**b**) Strain DAFs for different speeds and loads; (**c**) Displacement DAFs for different spans and loads; (**d**) Strain DAFs for different spans and loads.

2.3. Joint Simulation Model Validation

When a train with a speed of 15 km/h passed through a 15 m span track beam, the dynamic response at the mid-span was tested and compared with the simulation data, fully verifying the validity of the combined model. The results are shown in Figure 7. The analysis indicates that the extreme value differences of the vibration acceleration are 0.02 (in the lateral direction) and 0.21 (in the vertical direction), respectively; the extreme value difference of the mid-span dynamic displacement between the test and simulation is 0.05 mm. Overall, the simulation data are in good agreement with the test data, with the overall error extreme value being only 1.96%. This indicates that the combined simulation model has high rationality, reliability, and accuracy.

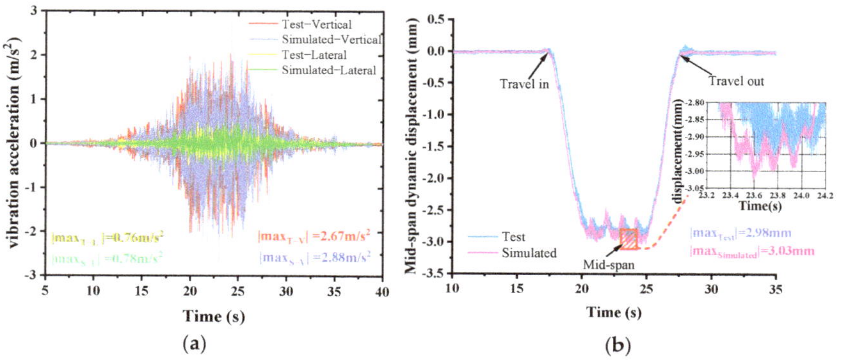

Figure 7. Comparison of tested and simulation results at the mid-span of a track beam. (**a**) Vibration acceleration time course curve; (**b**) time course curve of mid-span dynamic displacement.

3. Analysis of Factors Influencing the DAFs

This section conducted an analysis of the influencing factors of the DAFs of the MTTS based on the joint simulation model established in the previous section, exploring the influence mechanisms of train characteristic parameters (speed, load, and formation) and track beam characteristic parameters (span, curve radius, track unevenness spectrum classification, and fundamental frequency) on the DAFs. Based on the actual project operational conditions of MTTS and relevant specification requirements. The working conditions were set as follows: (1) speed (V): 10 km/h, 15 km/h, 20 km/h, 30 km/h, 40 km/h, 50 km/h, 60 km/h, and 70 km/h; (2) load (L): empty, 1/4 full load (1.05 t), half load (2.1 t), 3/4 full load (3.15 t), and full load (4.2 t); (3) train formation (TF): 1 to 6 train formations; (4) Span (S): 10 m, 12 m, 15 m, 18 m, and 21 m; (5) curve radius (R): 21 m span beams with radii of 50 m, 100 m, 150 m, 200 m, 300 m, 400 m; (6) track unevenness spectrum (γ): an irregularity spectrum based on measured data fitting was adopted, which was classified into seven grades from A to G in order of superiority [29], see Figure 8a; and (7) track beam fundamental frequency (F): 1.05 Hz (0.5 times stiffness), 2.91 Hz (1 times stiffness), 4.96 Hz (2 times stiffness), 7.03 Hz (3 times stiffness), 8.95 Hz (4 times stiffness), a total of five calculation conditions.

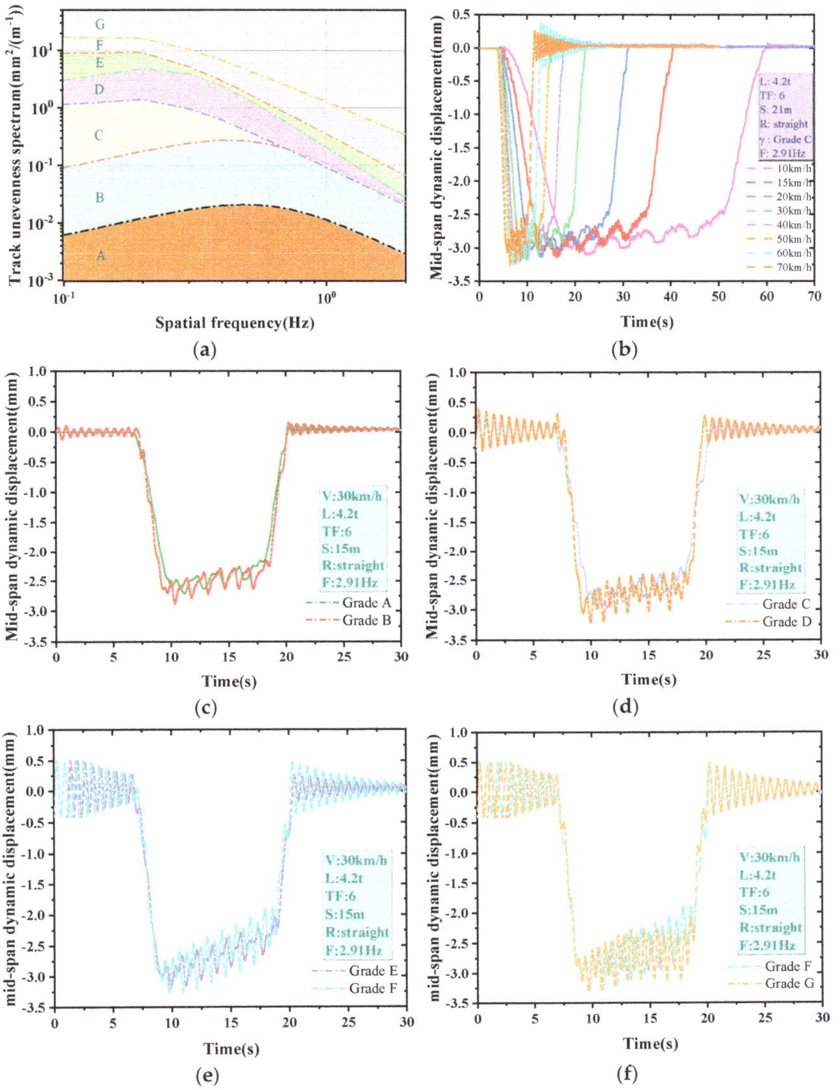

Figure 8. Track unevenness spectrum classification and dynamic response analysis results. (**a**) Track unevenness spectrum classification; (**b**) Time course of mid-span dynamic displacement at different velocities; (**c**–**f**) Effect of track unevenness on dynamic displacement in mid-span.

The mid-span dynamic displacement response result when the train speed was set at 0.1 km/h was taken as the static response. Based on the control variable method, the variation law of the DAFs under the influence of each control factor was explored. Due to the high consistency of the dynamic response curves of the track beam under the influence of various factors, this section only took the variation law of the mid-span dynamic response of the track beam under the influence of speed and irregularity as an example for a brief analysis. The results are as follows:

Based on the analysis of the dynamic displacement time curve shown in Figure 8b–f, with speed and track unevenness as control factors, it was observed that the dynamic displacement in the span exhibited a positive correlation with speed. As the speed increased, the extreme value of dynamic displacement demonstrated rapid growth below 30 km/h. When the speed exceeded 30 km/h, the growth rate of the dynamic displacement extreme value began to decelerate, reaching its peak at 70 km/h. Additionally, the dynamic displacement in the span demonstrated strong sensitivity to variations in track beam unevenness. With the deterioration of the track unevenness grading spectrum (from Grade A to Grade G), the dynamic displacement in the span is correspondingly accompanied by a marked increase in the overall volatility of the time curve. The density and amplitude of curve fluctuation sections exhibited a positive correlation and an approximately linear relationship with spectral grading changes.

The parameter analysis results for DAFs are shown in Figure 9. The conclusions are as follows: (1) Figure 9a shows that the train speed has a significant impact on the DAFs of the monorail track beam. Within the speed range of 0–30 km/h, the impact coefficient shows a logarithmic nonlinear growth trend overall with respect to the train speed and reaches its extreme value at 30 km/h. When the speed exceeds 30 km/h, the DAFs vary periodically around their extreme value with the train speed. (2) Figure 9b indicates that train loads constitute a minor proportion of the total loading effect and have a relatively insignificant impact on the structural impact effect. (3) Figure 9c demonstrates that the train formation significantly influences the dynamic load on the track beam DAFs. When the total train length is smaller than the track beam span, the DAFs exhibit strong sensitivity to train formation variations. However, when the total train length exceeds the track beam span, further increases in formation length no longer cause significant DAF growth. (4) Figure 9d reveals that the DAFs are significantly affected by the track beam span. Within the calculated span range, the DAF values exhibit an exponential decrease as the span increases. For small-span track beams, higher train loads intensify the dynamic response of key cross-sections, resulting in abrupt DAF changes. (5) Figure 9e illustrates that the DAFs of the track beam show an exponential nonlinear negative correlation with variations in curve radius. Compared with straight beams, smaller curve radii enhance the bending-twisting coupling effect, leading to significant fluctuations in the DAFs. (6) Figure 9f highlights that track unevenness substantially affects the vibration response of the monorail track beam. As the track irregularity grading spectrum deteriorates (from A to G), the sensitivity of the DAFs response exhibits gradual growth characterized by an S-shaped curve. (7) Figure 9g additionally indicates that fundamental frequency influenced the dynamic response of monorail track beams, with lower fundamental frequencies causing resonance phenomena and, consequently, larger DAF values.

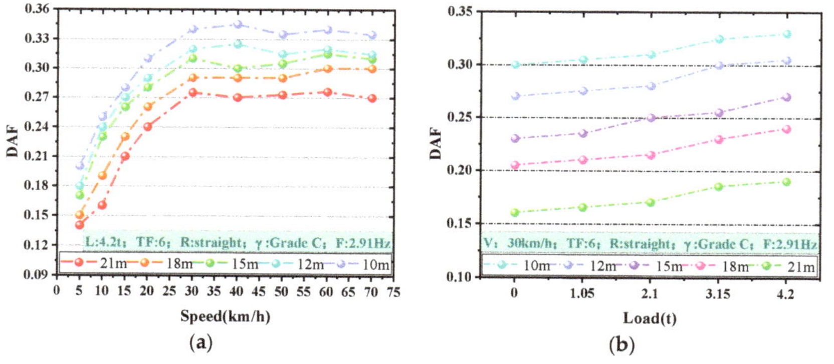

Figure 9. *Cont.*

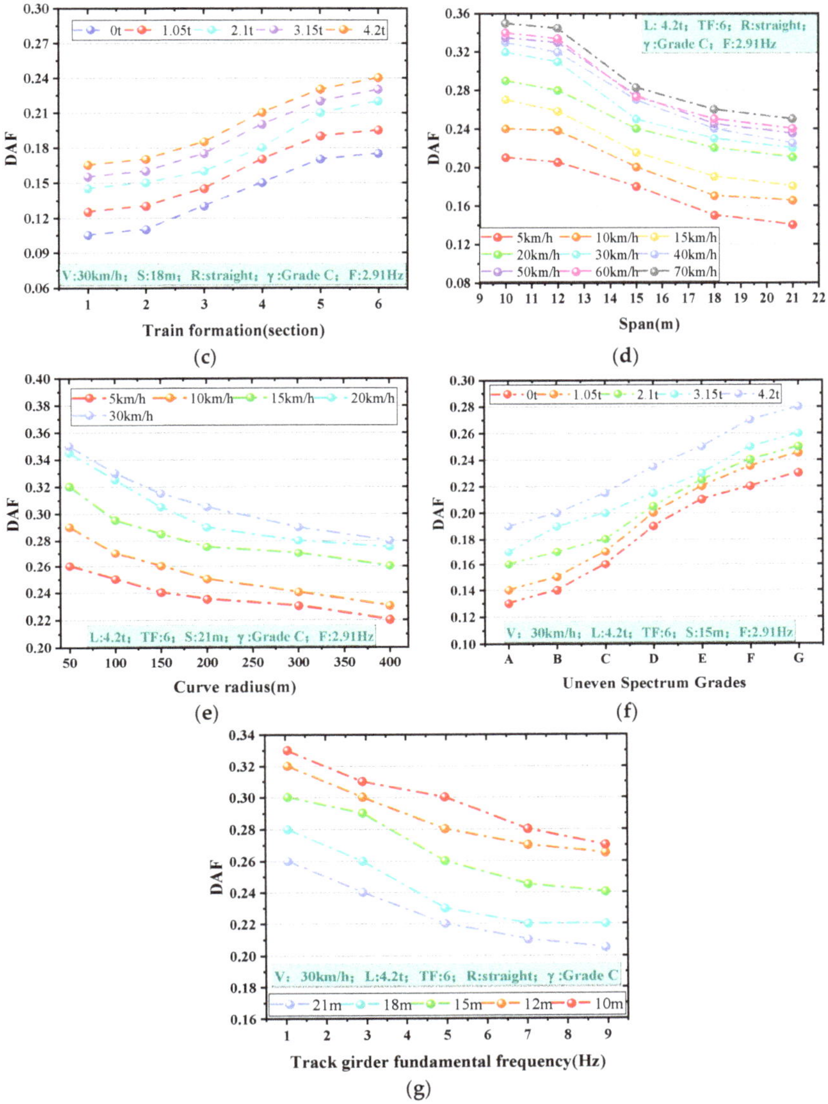

Figure 9. Results of the analysis of factors influencing the DAFs. (**a**) Effect of train speed on DAFs; (**b**) effect of load on DAFs; (**c**) Effect of train formation on impact coefficients; (**d**) effect of span on DAFs; (**e**) effect of curve radius on DAFs; (**f**) effect of track unevenness spectrum on DAFs; (**g**) effect of track beam fundamental frequency on DAFs.

4. LASSO-Based Regression Analysis of DAFs for Multifactor Coupling Effects

4.1. Regression Parameter Screening

The DAFs have many influencing factors, and there are mutual influences. This section using the Pearson correlation coefficient method and random forest algorithm to study the correlation between the influencing factors and the impact coefficients and multifactor coupling in the impact coefficients process of the degree of contribution to the screening of the regression parameters; however, there is a significant multiple covariance between the regression parameters, if the direct traditional regression analysis will increase the regression model overfitting and not. However, there is significant multicollinearity among the regression parameters, and direct traditional regression analysis will increase the risk of overfitting and instability of the regression model, while LASSO regression can better deal with the problem of multicollinearity, and at the same time, it can also further screen the regression variables of the model, which can increase the spatial dimension of the characteristics of the regression expression and simplicity at the same time of ensuring the model stability.

Based on the previous research results, this section used the Pearson correlation coefficient method [30] to explore the degree of correlation between each influencing factor and the DAFs and used the random forest algorithm [31] to explore the importance of each influencing factor in the coupling effect on the DAFs. The results were comprehensively analyzed to screen the regression parameters, reduce the dimension of the DAFs feature space, and improve the accuracy of the regression model.

The results of Pearson correlation analysis are shown in Figure 10a. The results indicate that the correlation coefficients between span, speed, track irregularity, fundamental frequency, and the DAFs are all greater than 0.8, suggesting a significant correlation. Moreover, the correlation coefficients among these four parameters are relatively large, indicating the existence of multicollinearity. The correlation coefficients of load, formation, and curve radius are relatively small, suggesting a slightly weaker correlation. The results of the feature importance analysis of the impact coefficient based on the random forest algorithm are shown in Figure 10b. It can be seen from the analysis that the importance proportion of the four influencing factors of span, speed, track irregularity, and fundamental frequency is 94.93%, indicating that these four factors cover most of the information in the process of multiple factors coupling to affect the DAFs. Therefore, these four influencing factors are selected as the regression parameters.

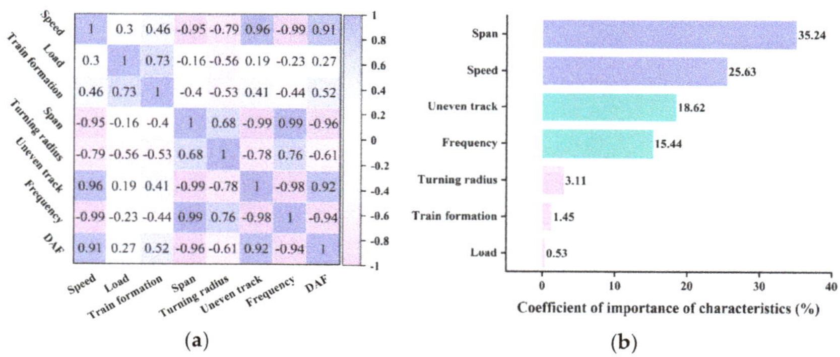

Figure 10. Regression parameter screening. (**a**) Pearson correlation analysis; (**b**) random forest feature significance analysis.

4.2. Multiple Regression Analysis of DAFs Based on LASSO Regression Theory

Based on the research results from the previous section, span, speed, unevenness, and base frequency were selected as regression parameters. However, there is multicollinearity among these characteristic factors. In this section, based on the LASSO regression theory [32], Python software (3.10.2) programming was used to conduct a multiple regression analysis of the DAFs. LASSO regression can effectively handle the problem of parameter multicollinearity and further screen the regression parameters by sacrificing some unbiasedness to reduce the dimension of the DAFs feature space, ensuring the stability of the regression model and increasing the simplicity and intuitiveness of the DAFs calculation expression. Since the relationship between a single control factor and the DAFs is approximately nonlinear, to facilitate subsequent calculations, the function relationship between a single factor and the DAFs was linearized first (Table 1).

Table 1. Regression parameters linearization correlation parameters.

Regression Parameters	Linearization Equation	Parameters	
		m	n
Speed (V)	$x^* = \frac{1}{1+(\frac{x}{m})^n}$	14.535	2.786
Span (S)		12.436	7.350
Frequency (F)		3.774	2.338
Unevenness (γ)		3.842	3.658

After linearizing the regression parameters, the Z-score method was used to standardize the original data to eliminate the influence of differences in the dimensions of various variables and ensure the accuracy of the regression model. As shown in Figure 11a,b, based on the LASSO regression theory, the regularization path diagram and MSE cross-validation diagram during the establishment of the multiple regression model of the DAFs are presented. Where Lambda denotes the regularization parameter, MSE represents the mean square error, and the standardization coefficient serves as an eigenvalue to quantify the influencing factors during the regularization calculation process. When the standardization coefficients of each regression parameter start to stabilize with the change of the regularization parameter Lambda, that is, the curve in the figure shows a relatively gentle trend, and the mean square error (MSE) of the established regression model reaches the minimum, the regularization parameter Lambda achieves the optimal value. At this point, the established regression model is the most stable and can effectively avoid the problem of multicollinearity. The analysis reveals that when Lambda is taken as 0.0056, the MSE reaches its minimum value. However, at this point, the regularization path diagram shows a relatively large degree of dispersion. Therefore, the Lambda1SE value obtained when Lambda is 0.0225 is selected as the optimal MSE.

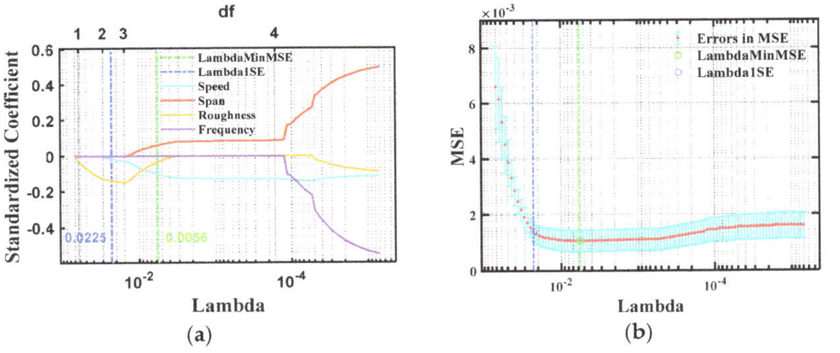

Figure 11. LASSO regression process. (**a**) Parameter regularization path; (**b**) MSE cross-validation.

Based on the optimal value, a LASSO regression model of the DAFs was established, and the regression expression is shown in Equation (1). The analysis reveals that the track beam fundamental frequency (F) and track unevenness (γ) are screened out of the LASSO model, while speed (V) and span (S) are retained as regression parameters. The calculation results from the program demonstrated that the coefficient of determination (R^2) of the LASSO regression model is 0.948, the adjusted R^2 is 0.945, and the root mean square error

(RMSE) is 0.0154. These results indicate that the regression model can predict 94.50% of the variation information of DAFs. Moreover, both the model and the variables have passed the 95% confidence interval test with a significance level of 0.05, indicating that the regression effect of the model is accurate and reliable.

$$\mu = 0.187 + \frac{0.185}{1 + \left(\frac{S}{14.535}\right)^{7.35}} - \frac{0.133}{1 + \left(\frac{V}{12.436}\right)^{2.786}} \tag{1}$$

Here, μ is the DAFs, S is the span (m), and V is the traveling speed (km/h).

4.3. Regression Model Validation

To verify the validity and accuracy of the LASSO regression model, this section calculated the intersection diagram of the DAFs for 200 groups of samples under the same operating conditions using both the regression model and the joint simulation model in Figure 12a. Considering the characteristics of the DAFs as a span function in most specifications, the predicted DAF values are calculated with span as the independent variable when the speed is set to 40 km/h. The comparison results are presented in Figure 12b.

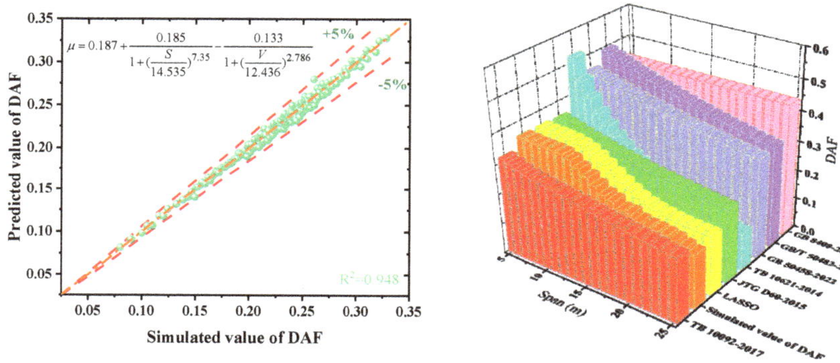

(**a**) (**b**)

Figure 12. Regression model validation. (**a**) Plot of the intersection of simulated and predicted values of the regression model; (**b**) comparison of regression models and specifications.

Analysis of Figure 12a shows that the predicted values from the regression model are in good agreement with the simulated values, and the DAFs values are basically in line with the characteristics of the normal distribution. Analysis of Figure 12b demonstrates that: (1) For the "Large-scale amusement device safety code (GB 8408-2018) [2]", "General Specifications for Design of Highway Bridges and Culverts (JTG D60-2015) [33]", "Standard for Design of Straddle Monorail Transit (GB 50458-2022) [34]", and "Code for Design of Urban Rail Transit Bridge (GB/T 51234-2017) [35]", the calculated DAFs values are consistently higher than the simulated DAFs values of the MTTS across different spans, with errors exceeding 10% in all cases. (2) For the "Code for Design on Railway Bridge and Culvert (TB 10092-2017) [36]", when the span is relatively small, the code-specified DAFs are lower than the simulated value. As the span increases, the difference between the two gradually decreases. When the span reaches 15 m, the difference becomes the smallest. However, as the span continues to increase, the code-specified value gradually exceeds the simulated value, and the difference shows a positive correlation with the span. Additionally, the DAFs values specified in GB 8408-2018 were overly simplified, and their direct adoption could potentially lead to significant errors. (3) The DAFs expression established based on the LASSO regression is generally consistent with the simulated values overall, with the error remaining within 4%, indicating that the DAFs expression can effectively represent the vast majority of the actual impact effect information.

The above research shows that the DAFs multiple regression model established based on LASSO has high reliability and accuracy and can be used as the recommended expression of DAFs for MTTS. This is of positive significance for improving the design and evaluation system of MTTSs.

4.4. Limitations and Future Work

MTTS and traditional monorail systems exhibit significant differences; therefore, the DAFs prediction model proposed in this paper is more suitable for large-scale amusement facilities that conform to the structural characteristics of MTTS. Furthermore, in the research of civil engineering, when the influencing factors of the research object are numerous, screening out the factors with significant correlation and high contribution and then conducting subsequent research (such as regression analysis, establishing prediction models, etc.) is important research content. When encountering such problems, the systematic research method proposed in this paper can be considered. This can not only avoid cumbersome analysis and derivation but also obtain conclusions that meet the requirements of analysis accuracy. In future research, the following work still needs to be carried out to make the research on the MTTS more complete.

i. A field of theoretical research requires extensive on-site trials of the project to provide support. Currently, the on-site trials for MTTS project research are relatively insufficient, especially in areas such as bearing force states, cross-sectional changes in steel box beams, train bogie structures, and tire forces and wear. Future studies could focus on conducting more detailed field tests on these aspects to enhance and refine the research system.
ii. There is still room for improvement in the research on the DAFs of MTTSs. Future studies could develop a probabilistic algorithmic prediction model for the DAFs and enhance prediction accuracy by comprehensively considering the uncertainty and complexity of influencing factors.

5. Conclusions

To address the limited knowledge of the DAFs for MTTS and general monorail systems, this study performed a comprehensive parametric analysis and predictive modeling based on a validated joint model. The main conclusions are summarized as follows:

i. Relying on the CRRC Zhuzhou test line project, a vehicle–bridge coupling field test of MTTS was conducted. The results show that the influence degrees of each test parameter on the DAFs are different, and the displacement DAFs are greater than the strain DAFs.
ii. The joint simulation model established based on the vehicle–bridge coupling test parameters of the MTTS and the actual project parameters was demonstrated to be reasonable and reliable, and the simulation results of the structural dynamic response model were in good agreement with the test results.
iii. Based on the joint simulation model for analyzing DAF impact parameters, the results showed that each impact parameter exhibited varying degrees of influence on the variability of the DAFs. Notably, the span and speed were identified as having the most significant impact.
iv. Based on the Pearson correlation coefficient method and the random forest algorithm, the regression parameters of the DAFs were screened. The results show that there is a significant correlation between speed, span, unevenness, fundamental frequency and the DAFs, and the total contribution degree of these factors in the process of the

DAFs under the coupling effect of multiple factors is 94.93%. Therefore, speed, span, unevenness, and fundamental frequency were selected as the regression parameters of the DAFs.

v. Utilizing LASSO regression theory to establish the DAFs multivariate regression model, the calculation results of the regression model and the simulation model were compared with the specifications. The results indicated that the regression model was reasonable and effective, passing the 95% confidence level and 0.05 significance test while demonstrating high precision. Therefore, it is recommended as a calculation expression for the DAFs of MTTSs.

Author Contributions: Methodology, H.Z. and C.W.; software and formal analysis, C.W., S.W. and Y.W.; data curation, C.W., S.W. and Y.W.; writing—original draft preparation, H.Z. and C.W.; writing—review and editing, C.W. and W.L.; supervision and project administration, W.L. All authors have read and agreed to the published version of the manuscript.

Funding: This research received no external funding.

Data Availability Statement: The data presented in this study may be available upon reasonable request.

Conflicts of Interest: The research was conducted in the absence of any commercial or financial relationships that could be construed as potential conflicts of interest.

References

1. Guo, F.; Chen, K.; Gu, F.; Wang, H.; Wen, T. Reviews on current situation and development of straddle-type monorail tour transit system in China. *J. Cent. South Univ. Sci. Technol.* **2021**, *52*, 4540–4551.
2. *GB 8408-2018*; Large-Scale Amusement Device Safety Code. State Administration of Work Supervision: Beijing, China, 2018.
3. Lee, J.; Kim, H.; Lee, K.; Kang, Y.J. Effect of load combinations on distortional behaviors of simple-span steel box girder bridges. *Metals* **2021**, *11*, 1238. [CrossRef]
4. Cheng, J.; Yao, H. Simplified method for predicting the deflections of composite box girders. *Eng. Struct.* **2016**, *128*, 256–264. [CrossRef]
5. Zhou, M.; Liu, Z.; Zhang, J.; An, L.; He, Z. Equivalent computational models and deflection calculation methods of box girders with corrugated steel webs. *Eng. Struct.* **2016**, *127*, 615–634. [CrossRef]
6. Li, H.; Xia, H.; Soliman, M.; Frangopol, D.M. Bridge Stress Calculation Based on the Dynamic Response of Coupled Train–Bridge System. *Eng. Struct.* **2015**, *99*, 334–345. [CrossRef]
7. Wang, H.; Zhu, E. Dynamic response analysis of monorail steel-concrete composite beam-train interaction system considering slip effect. *Eng. Struct.* **2018**, *160*, 257–269. [CrossRef]
8. Li, X.; Ge, Y.; Jin, Z. Calculation and analysis of the straddle-type monorail vehicle-track beam coupling vibration. *J. Railw. Eng. Soc.* **2018**, *035*, 78–83.
9. Zheng, K. Simulation Analysis and Optimization on Dynamics Performance of Straddle-Type Monorail Vehicle. Master's Thesis, Chongqing Jiaotong University, Chongqing, China, 2010.
10. Liu, G. Study on the influence of track-beam stiffness on straddle monorail transit ride comfort. *J. Railw. Investig. Surv.* **2016**, *42*, 85–88. [CrossRef]
11. Gao, Q.; Cui, K.; Li, Z.; Li, Y. Numerical investigation of the dynamic performance and riding comfort of a straddle-type monorail subjected to moving trains. *Appl. Sci* **2020**, *10*, 5258. [CrossRef]
12. Zhou, J.; Du, Z.; Yang, Z.; Xu, Z. Dynamic parameters optimization of straddle-type monorail vehicles based multi-objective collaborative optimization algorithm. *J. Veh. Syst. Dyn.* **2020**, *58*, 357–376. [CrossRef]
13. Zhou, J.; Du, Z.; Yang, Z. Dynamic Response of the full-scale Straddle-type Monorail Vehicles with Single-axle Bogies. *J. Mech.* **2019**, *25*, 17–24. [CrossRef]
14. Li, Y.; Bao, Y.; Dong, S.; Zeng, Y.; Xiang, H. Influencing factors of impact coefficient for long-span railway cable-stayed bridges. *J. Vib. Shock.* **2015**, *34*, 138–143. [CrossRef]
15. Wang, P.; Guo, F.; Zhang, H.; Jin, J.; Liao, Q.; Yan, Y. Acquiring the High-Precision Spectrum of Track Irregularity by Integrating Inclination in Chord Methods: Mathematics, Simulation, and a Case Study. *Mathematics* **2024**, *12*, 2197. [CrossRef]
16. Zhang, H.; Wang, P.; Chen, C.; Ji, Y.; Wei, S.; Guan, B.; Liu, Z.; Zhang, J. Numerical Study on the Sensitive Wavelength and Amplitude Management of Track Irregularities in Monorail Systems. *Buildings* **2024**, *14*, 4041. [CrossRef]

17. Guo, F.; Ji, Y.; Liao, Q.; Liu, B.; Li, C.; Wei, S.; Xiang, P. The Limit of the Lateral Fundamental Frequency and Comfort Analysis of a Straddle-Type Monorail Tour Transit System. *Appl. Sci* **2022**, *12*, 10434. [CrossRef]
18. Zhang, H.; Wang, P.; Li, Q.; Jin, J.; Wei, S.; Guo, F.; Feng, C.; Deng, Q. An Experimental and Numerical Study on the Lateral Stiffness Limits of Straddle-Type Monorail Tour-Transit Systems. *Buildings* **2024**, *14*, 3111. [CrossRef]
19. Shervan, A.; Amin, M. Investigating dynamic amplification factor of railway masonry arch bridges through dynamic load tests. *J. Constr. Build. Mater.* **2018**, *183*, 693–705. [CrossRef]
20. Babatope, O.; Emmanuel, O.A. Dynamic amplification factor and interactions of a beam under compressive axial force and load travelling at varying velocity. *J. Forces Mech.* **2023**, *13*, 100241. [CrossRef]
21. Zhu, J.; Xu, Y. Research on the impact factor of the three-span continuous bean-arch combined bridge based on the vehicle-bridge coupled vibration. *J. Railw. Eng. Soc.* **2019**, *16*, 959–967. [CrossRef]
22. Zhu, J.; Chen, X. Damage Detection Method of Suspension Bridge Based on Index of Strain Impact Factor. *J. Vib. Meas. Diagn.* **2023**, *43*, 450–458+617. [CrossRef]
23. Zhou, Y.; Xue, Y.; Gao, X.; Li, R.; Wang, Y.; Zhao, Y. Research on dynamic amplification factor of highway simply supported girder bridge based on modal superposition method. *J. Traffic Transp. Eng.* **2023**, *23*, 146–155. [CrossRef]
24. Li, Z.; Li, J.; Feng, D. Analysis of dynamic impact coefficients and affecting factors of stay cables in cable-stayed bridges based on vehicle-bridge interaction theory. *J. Vib. Shock.* **2024**, *43*, 67–74. [CrossRef]
25. Deng, L.; Duan, L.; Zhou, Q. Comparison of dynamic amplification factors calculated from bridge strain and deflection. *J. Eng. Mech.* **2018**, *35*, 126–135. [CrossRef]
26. Deng, L.; Duan, L.; He, W.; Ji, W. Study on Vehicle Model for Vehicle-bridge Coupling Vibration of Highway Bridge in China. *J. China J. Highw. Transp.* **2018**, *31*, 92–100.
27. SIMPACK. Version 2021x. Dassault Systemes: Velizy-Villacoublay, France, 2021. Available online: https://blog.3ds.com/brands/simulia/simpack-r2021x-multibody-system-simulation-updates/ (accessed on 20 September 2024).
28. ABAQUS. *ABAQUS User's Manual*, Version 2016; Dassault Systemes: Velizy-Villacoublay, France, 2016. Available online: http://62.108.178.35:2080/v2016/index.html (accessed on 20 September 2024).
29. Ji, Y. Study on the Unevenness of the Top Plate of Track Beams of Monorail Tour Transit System. Master's Thesis, Central South University, Changsha, China, 2023.
30. Zhang, Z.; Zhou, A.; Shi, Z.; Zhang, H.; He, X.; Wang, Y.; Bai, Z.; Xi, D. Explaining Relationships between Chemical Structure and Tar-Rich Coal Pyrolysis Products Yield Based on Pearson Correlation Coefficient. *Fuel* **2025**, *395*, 135029. [CrossRef]
31. Liu, Y.; Ren, Q. Online Incremental Learning Approach of Heat Pump and Chiller Models Based on the Dynamic Random Forests in Queue Structure. *Energy* **2025**, *323*, 135840. [CrossRef]
32. Lasheen, A.; Sindi, H.F.; Zeineldin, H.H.; Morgan, M.Y. Online Stability Assessment for Isolated Microgrid via LASSO Based Neural Network Algorithm. Energy Convers. *Manag. X* **2025**, *25*, 100849. [CrossRef]
33. *JTG D60-2015*; General Specifications for Design of Highway Bridges and Culverts. Ministry of Transport of the People's Republic of China: Beijing, China, 2015.
34. *GB 50458-2022*; Standard for Design of Straddle Monorail Transit. Ministry of Housing and Urban-Rural Development of the People's Republic of China: Beijing, China, 2022.
35. *GB/T 51234-2017*; Code for Design of Urban Rail Transit Bridge. Ministry of Housing and Urban-Rural Development of the People's Republic of China: Beijing, China, 2017.
36. *TB10092-2017*; Code for Design on Railway Bridge and Culvert. China National Railroad Administration (CNRA): Beijing, China, 2017.

buildings

MDPI

Article

Seismic Fragility and Loss Assessment of a Multi-Story Steel Frame with Viscous Damper in a Corrosion Environment

Wenwen Qiu [1,2], Haibo Wen [3,*], Chenhui Gong [4], Zhenkai Zhang [1], Wenjing Li [1] and Shuo Li [1]

1 School of Economics and Management, Gandong University, Fuzhou 344000, China; wenwenqiu@gdc.edu.cn (W.Q.); zhenkaizhang@gdc.edu.cn (Z.Z.); wenjingli@gdc.edu.cn (W.L.); shuoli@gdc.edu.cn (S.L.)
2 School of Business Management, University Utara Malaysia, Sintok 06010, Kedah, Malaysia
3 School of Civil Engineering, Central South University, Changsha 410075, China
4 School of Applied Engineering, Gandong University, Fuzhou 344000, China; chenhuigong@gdc.edu.cn
* Correspondence: haibowen_2000@163.com

Abstract: Corrosion can accelerate the deterioration of the mechanical properties of steel structures. However, few studies have systematically evaluated its impact on seismic performance, particularly with respect to seismic economic losses. In this paper, the seismic fragility and loss assessment of a multi-story steel frame with viscous dampers (SFVD) building are investigated through experimental and numerical analysis. Based on corrosion and tensile test results, OpenSees software 3.3.0 was used to model the SFVD, and the effect of corrosion on the seismic fragility was evaluated via incremental dynamic analysis (IDA). Then, the economic losses of the SFVD during different seismic intensities were assessed at various corrosion times based on fragility analysis. The results show that as the corrosion time increases, the mass and cross-section loss rate of steel increase, causing a decrease in mechanical property indices, and theprobability of exceedance of the SFVD in the limit state increases gradually with increasing corrosion time, with an especially significant impact on the collapse prevention (CP) state. Furthermore, the economic loss assessment based on fragility curves indicates that the economic loss increases with corrosion time. Thus, the aim of this paper is to provide guidance for the seismic design and risk management of steel frame buildings in coastal regions throughout their life cycle.

Keywords: steel frame; corrosion; seismic fragility; loss assessment

1. Introduction

Steel structures have been extensively employed in modern construction due to their high strength, excellent ductility, and easy construction [1,2]. However, traditional steel frames exhibit relatively low lateral stiffness and insufficient ductility capacity, particularly in high-rise buildings, leading to challenges in meeting practical seismic performance requirements. The equipment of structural dampers [3,4] serves as an effective seismic retrofitting measure, significantly enhancing the seismic performance of steel frames and reducing their seismic-induced responses.

Steel buildings are prone to corrosion in humid or other corrosive environments, which degrades their mechanical properties, thereby affecting the overall safety and seismic performance of the structure. Salt spray corrosion is one of the main types of corrosion that steel buildings face in coastal areas, and chloride ions can accelerate the corrosion process of steel [5]. Corrosion has no significant effect on the mechanical or chemical properties of steel, and the degradation of mechanical properties is attributed mainly to the reduction in

https://doi.org/10.3390/buildings15142515

steel section thickness caused by corrosion [6]. Guo et al. [7] investigated the mechanical behavior of Q345 steel after long-term corrosion in a salt spray environment and observed a decrease in yield strength, ultimate strength, and fracture strength with increasing corrosion degree (mass loss rate η), especially when $\eta \geq 10\%$. Corrosion affects different steel types in various ways. Kong et al. [8] analyzed the corrosion products and microstructures of Q235 and Q355 steels after corrosion using XRD and SEM, and the results showed that although the corrosion products of both steels were similar, their microstructural morphologies differed significantly. Compared to Q235 steel, corrosion has a smaller effect on the strength and ductility of Q355 steel. Furthermore, corrosion weakens the fatigue strength of steel, with this effect being more pronounced at high temperatures [9,10].

In recent years, numerous studies have evaluated the effect of corrosion on the seismic performance of steel structures from the perspective of steel structures and steel members, and the results showed that corrosion can reduce the load-carrying capacity, deformation capacity, and energy-dissipation capacity of steel frames [11–14]. Zhang et al. [12] investigated the seismic performance of steel beam-to-column connections after corrosion, and the results showed that corrosion lowers the load-bearing capacity, initial stiffness and energy-dissipation capacity of steel beam-column connections and accelerates local buckling. Zhang et al. [14] investigated the effect of corrosion on the seismic behavior of steel frames using experiments and finite element analysis, quantified the corrosion level with the mass loss rate and developed a deterioration model for the mechanical properties of steel as corrosion increases. Therefore, it is worthwhile to investigate the seismic performance of corrosion on steel frames.

Seismic fragility, defined as the probability that a structure will reach or exceed a specific damage index, is crucial for assessing a structure's seismic performance. Lad et al. [15] used IDA to assess the seismic fragility of corroded steel frames and local components and reported that corrosion reduces the load-bearing and energy-dissipation capacity of steel frames, increasing their damage probability during earthquakes. Notably, corrosion most significantly affects the fragility of columns. Xu et al. [16] performed accelerated corrosion tests on a 5-story steel frame and used IDA to obtain seismic fragility curves under various corrosion times, and the results showed that the structure's probability of exceedance gradually increases with increasing corrosion time, with the most significant impact observed on the CP state. Lekeufack et al. [17] analyzed the fragility curves of three kinds of steel frames (NCSF, CCSF, and CWSF) under various seismic intensities, and the results showed that corrosion significantly increases structural fragility. Moreover, CWSF shows lower fragility than CCSF, particularly in the CP state. Additionally, assessing seismic-induced economic losses via fragility or fragility functions is effective. For example, Jiang et al. [18] considered the uncertainty between failure probability and damage cost, proposed a cost-based seismic fragility analysis method, and calculated the seismic life cycle cost of steel frames with steel plate walls. Significant endeavors have been witnessed to quantify the effects of corrosion on the seismic performance and seismic fragility of steel frames. However, few studies are related to the influence of corrosion on seismic loss assessment. In addition, although many studies have evaluated the seismic life cycle cost of steel frames, the effect of natural corrosion has not been considered.

To this end, this paper attempts to investigate the effect of corrosion on the fragility and economic loss of a multi-story SFVD for potential use in the seismic design and risk management of steel frame buildings in coastal regions throughout their life cycle. The difference between the present investigation and other studies is that it takes into account the impact of corrosion on the fragility of steel frame structures and establishes a link between corrosion and economic losses via fragility analysis.

2. Experimental Study on Steel Corrosion

2.1. Materials

The material used for the corrosion test in this work was Q355 construction steel. Specimens were designed according to Chinese standard GB/T228.1-2010 [19] and GB/3075-2008 [20]. The specimen thickness was 18 mm, as shown in Figure 1.

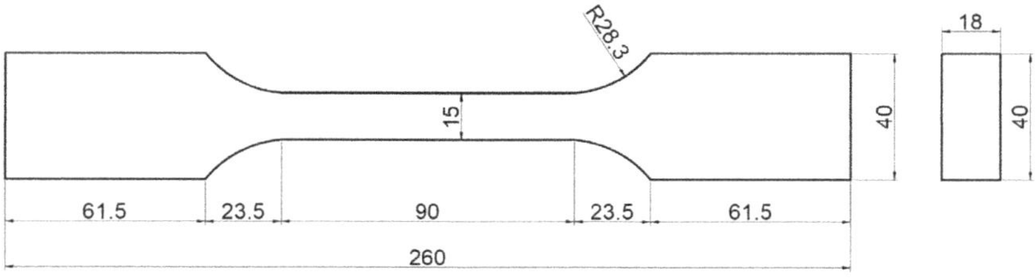

Figure 1. Specimen dimensions (unit: mm).

2.2. Corrosion Test

In this paper, an artificial spray accelerated corrosion test was conducted on Q355 construction steel. According to References [7,16], the corrosion tests were carried out at a room temperature of 25 °C and a relative humidity of 44%, and the corrosive solution used was a 50 g/L NaCl solution with a pH of 6.5–7.5. To ensure uniform corrosion on both sides of the specimens, they were fixed at a 45° angle to the horizontal plane. The spray method used was an intermittent spray with a 1-day cycle, and all specimens were flipped every eight days. Prior to the experiment, a 10% oxalic acid solution was prepared, and the processed specimens were ultrasonically cleaned to remove surface oils and residues, followed by drying. After the experiment, rust was removed from the specimens via a 20% oxalic acid solution. The specimens were subsequently neutralized with a 20% sodium carbonate solution, dried, weighed, and stored in plastic bags.

After corrosion, the specimens were numbered and grouped according to the corrosion time (32, 64, 92, 128, 164, and 192 days) into six groups, with three specimens per group to reduce randomness. They were labeled in the form of Si (i = 0, 1, 2. . .18). Among them, S0 represents the uncorroded specimen, and S1, S2, and S3 denote the three groups of specimens corroded for 32 days, and so on.

2.3. Tensile Test

To investigate the mechanical properties of steel at various corrosion times, tensile tests were conducted on both corroded and uncorroded specimens via a T-shaped working test bench. A total of 19 groups of tests were carried out, as shown in Figure 2. In accordance with the Chinese standard GB/T 228.1 [19], tensile tests were performed on corroded specimens via displacement control. The loading rates for the elastic, yield, and strain-hardening stages were 1.3 mm/min, 1.3 mm/min, and 10.8 mm/min, respectively.

Prior to the experiment, to measure the elongation of the corroded steels, a gauge length of 90 mm was used to mark the tensile section of each specimen. After fracture, the final length within the gauge region was measured via a digital Vernier caliper. In addition, during the test, the average thickness and width of each specimen were input to obtain the mechanical properties of the steel.

Figure 2. T-shaped working test bench.

2.4. Results

Here, the mass loss η and cross-section loss rate η_s of specimens at various corrosion times were analyzed, which can be expressed as follows:

$$\eta = \frac{W_0 - W_1}{W_0} \times 100\% \quad (1)$$

$$\eta_s = \frac{A_0 - A_1}{A_0} \times 100\% \quad (2)$$

where W_0 and W_1 are the mass of the specimen before and after corrosion, respectively; A_0 and A_1 are the initial area and residual area of the specimen before and after corrosion, respectively.

Figure 3 shows the relationship between the mass loss rate and corrosion time of the specimens, where the scatter points represent the experimental results and the solid line represents the fitting result. The results show that the mass loss rate gradually increases with increasing corrosion time. Compared with the uncorroded specimens, the mass loss rate of the specimens after 192 days of corrosion decreases by a maximum of 11.00%. It is worth noting that the mass loss rate shows a significant power-function [7] relationship with corrosion time, with a correlation coefficient exceeding 99%. In addition, comparing the mass loss rates of specimens at the same age reveals small numerical differences, which suggests uniform corrosion.

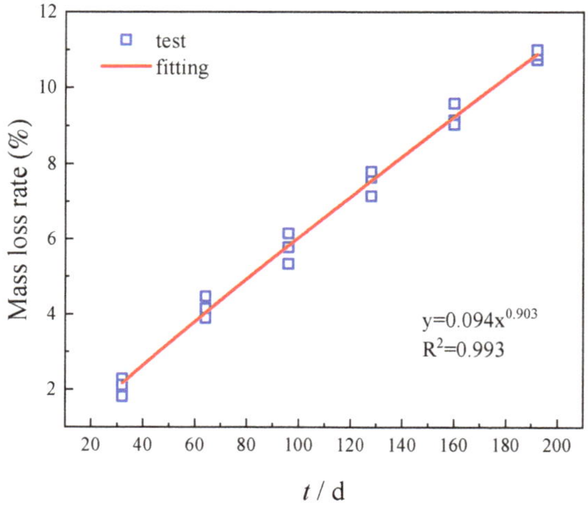

Figure 3. Mass loss rate for various corrosion times.

Figure 4 shows the cross-section loss rate at various corrosion times. The results show that the cross-section loss rate increases with increasing corrosion time, and the cross-section loss rate has a strong linear relationship with corrosion time.

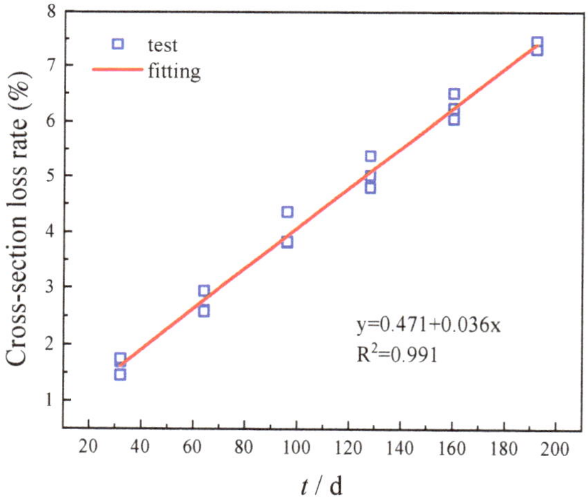

Figure 4. Cross-section loss rate for various corrosion times.

Common parameters used to characterize the degree of steel corrosion include mass loss rate, cross-section loss rate, and corrosion time. However, due to variations in testing methods and specimen dimensions among different corrosion tests, these parameters cannot be directly compared. Here, a relationship between mass loss rate, cross-section loss rate, and corrosion time was established, which can predict corrosion degree parameters at different corrosion times and provide a reference for conversion.

Figure 5 shows the nominal stress–strain curves of the corroded specimens after the tensile test. The results show that with increasing corrosion, the yield strength, ultimate tensile strength, and elongation of the steel all exhibit a downward trend, which is mainly attributed to the formation of rust pits on the steel surface due to corrosion, thereby reducing the effective cross-sectional area of the steel [6,21]. In the elastic stage, the stress–strain curves of all the corroded specimens are highly similar, indicating that corrosion has little effect on the elastic modulus of steel. Additionally, the corroded specimens show no distinct yield plateau, a phenomenon that may be due to the excessive carbon content in the steel during processing which reduces its ductility.

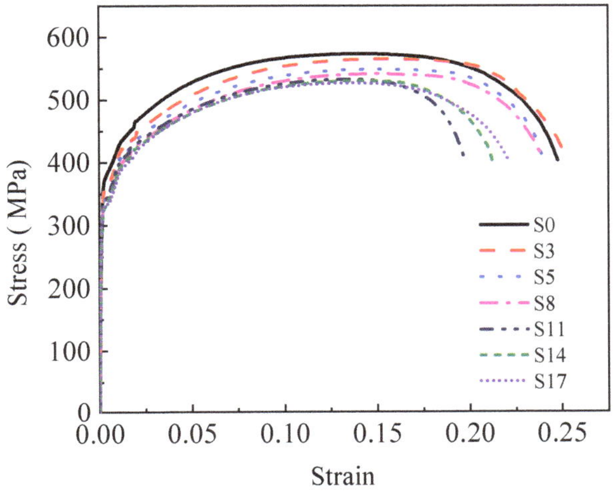

Figure 5. Stress–strain curve of the corrosion specimens.

It is well known that the rough surface of corroded steel can affect the accuracy of measuring the cross-sectional area of the parallel section. Hence, this work uses the nominal stress as the evaluation index. Table 1 shows the main parameters from the tensile test, including elastic modulus (E_s), yield strength (f_y), ultimate tensile strength (f_u), ultimate load (F_u), yield-strength ratio (f_y/f_u), and elongation (δ). In addition, owing to the lack of an obvious yield plateau in the corroded specimens, the conditional yield strength is used as the yield strength of the specimens here.

Table 1. Specimen mechanical property indices.

t/d	No.	E_s/MPa	f_y/tMPa	f_u/MPa	F_u/MPa	f_y/f_u	δ
0	S0	198.38	375.86	573.22	155.88	0.656	24.77%
	S4	188.63	347.26	551.43	146.06	0.630	23.20%
64	S5	187.66	342.39	548.45	144.76	0.624	24.01%
	S6	184.43	338.61	543.93	144.10	0.623	23.97%
	S10	183.33	339.59	538.61	139.44	0.630	19.82%
128	S11	183.87	339.62	532.68	137.06	0.638	19.74%
	S12	184.61	345.95	521.68	134.74	0.663	18.54%
	S16	——	——	——	——	——	——
192	S17	177.56	329.32	526.54	132.50	0.625	22.10%
	S18	179.45	332.30	532.11	134.14	0.625	20.16%

As shown in Table 1, all the indices of the corroded specimens decrease gradually with increasing corrosion time. Compared to the uncorroded specimen S0, the nominal elastic modulus, nominal yield strength, nominal tensile strength, and ultimate load of the S-series corroded specimens decrease by a maximum of 10.50%, 12.38%, 8.14%, and 15.00%, respectively. Furthermore, the elongation of steel decreases by up to 25.15% under the most severe corrosion, indicating that corrosion has a significant impact on the ductility of steel. Although the yield-strength ratio of steel decreases, the trend is not remarkable. The ratio fluctuates around 0.63 for the S-series corroded specimens, which meets the Chinese standard GB/T 19879-2005 [22] with a value of less than 0.85. This suggests that the mechanical property degradation of corroded steel does not significantly affect its safe reserve capacity.

3. Fragility Analysis of the SFVD

In this paper, the OpenSees software was employed to model a 5-story, 2-span SFVD presented in reference [23], as shown in Figure 6. The X-direction and Y-direction spans of the steel frame are 4000 mm and 3500 mm, respectively, and each floor is 3000 mm high. The section size of the beams is H200 × 600 × 20 × 20, and the section size of the columns is H600 × 600 × 20 × 20. In addition, the viscous damper was installed on Axis 2 of the SFVD, as shown in Figure 6a, with a stiffness of 100 kN/mm, a damping coefficient of 3 kN(s/m), and a damping coefficient of 1.

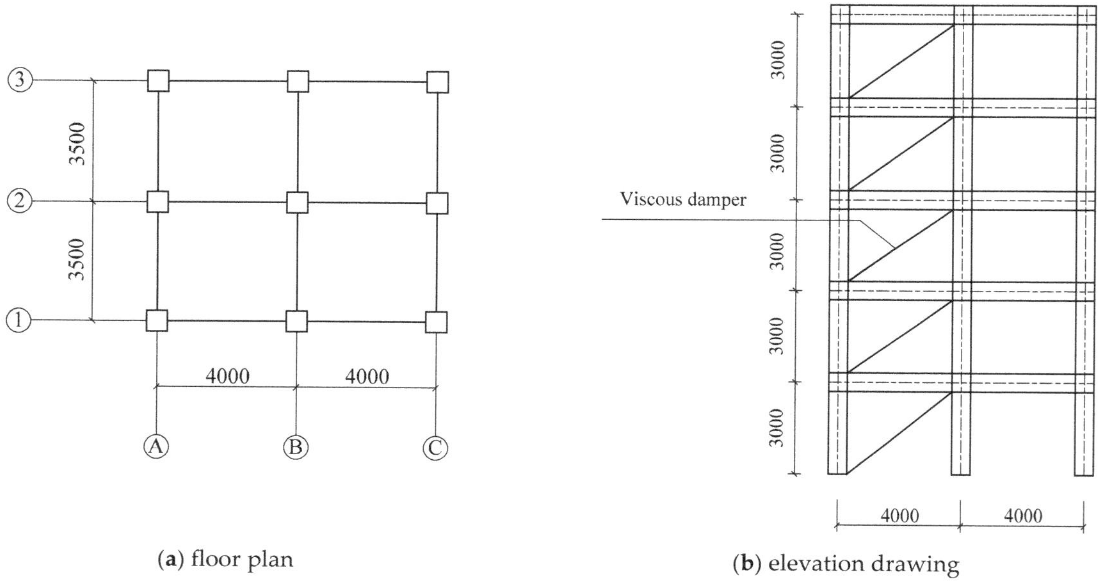

Figure 6. Schematic diagram of the SFVD (unit: mm).

In the modeling process, the mechanical parameters of the S-series steel with corrosion times of 0, 64, 128, and 192 days from Section 2 are used to predict the mechanical properties of the 5-story steel frame after corrosion. Frame columns are rigidly connected to the ground and frame beams, while viscous dampers adopt hinged connections; the simulation of viscous dampers employs Maxwell elements. Given that the SFVD model has been validated in reference [23], more details are not reiterated here.

3.1. Seismic Wave Selection

To determine the uncertainty of seismic hazards, it is generally necessary to select representative real-world seismic waves from a region for measurement. In this paper, 20 representative seismic ground motion records suitable for seismic analysis of regions such as Los Angeles, Boston, and Seattle in the United States were selected for IDA, which was based on a SAC project [24] funded by FEMA for steel structure design, as shown in

Figure 7. Here, the spectral acceleration $S_a(T_1)$ was adopted as the intensity measure (*IM*) in the IDA.

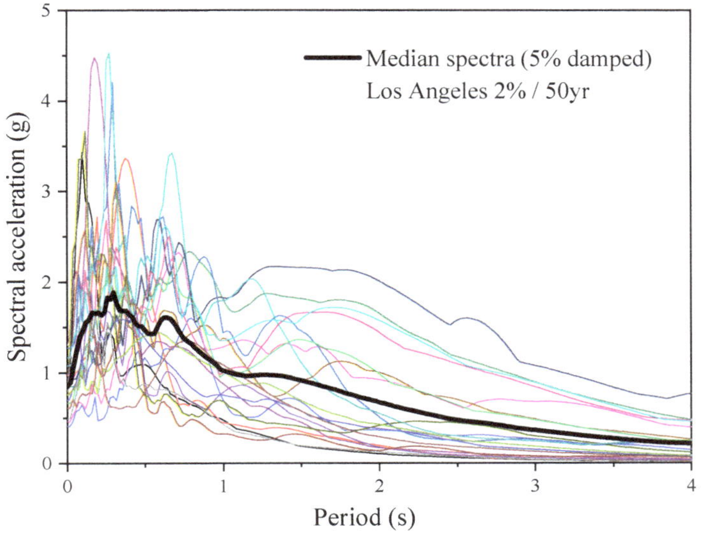

Figure 7. Response spectra of the 20 selected ground motions.

3.2. Determination of the Damage Measure

In this paper, the IDA method is employed to assess the seismic performance of the structure, which has the advantages [17,25] of capturing the complete range of the structural responses, accounting for uncertainties, and evaluating multiple limit states. The selection of the structural damage measure (*DM*) is critical prior to performing IDA analyses. Ye and Jiang et al. [26,27]. selected the maximum inter-story drift ratio as the *DM* and confirmed its reliability. In addition, the Chinese standard GB50010-2010 [28] specifies the maximum inter-story drift ratio as the threshold value for seismic deformation checking and proves that this parameter can be commonly used as a significant index for seismic design and performance evaluation. Therefore, in this paper, the maximum inter-story drift ratio is selected as the *DM* for the IDA of the SFVD. A brief description of the IDA process is as follows: IDA of the SFVD structure was performed using 20 selected ground motions from Figure 7, with $S_a(T_1)$ as the IM and scaled in increments of 0.25 g. The analysis was terminated when the maximum inter-story drift ratio reached 4%. The relationship between $S_a(T_1)$ and the maximum inter-story drift ratio was then obtained.

3.3. Determination of the Damage States

The Federal Emergency Management Agency (FEMA) defined structural damage into three levels [29]: immediate occupancy (IO), structural damage (SD) and collapse prevention (CP), as shown in Figure 8. These states divide structural performance into four levels: immediate occupancy, remaining operation, life-safety maintenance and near collapse. The current Chinese seismic design code defined structural damage into four levels: no damage, slight damage, moderate damage, and severe damage. The two classification approaches are highly consistent. Hence, in this paper, the maximum inter-story drift ratios for the IO, SD, and CP states are defined as 1%, 2%, and 4%, respectively.

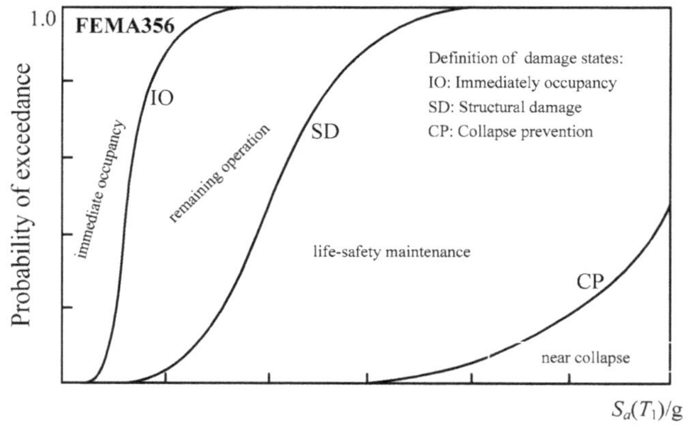

Figure 8. FEMA 356 definition of damage states.

3.4. Fragility Curve Analysis

The *IM* and *DM* follow a lognormal distribution, and their relationship is expressed as follows:

$$DM = \alpha(IM)^{\beta} \tag{3}$$

where α and β are coefficients determined by regression.

In this section, $S_a(T_1)$ and the maximum inter-story drift are selected as the *IM* and *DM*, respectively. The cloud method is employed to develop a linear relationship between θ_{max} and $S_a(T_1)$ on a log–log coordinate system, as shown in the following equation.

$$\ln \theta_{max} = a + \beta \ln[S_a(T_1)] \tag{4}$$

where $a = \ln \alpha$ is a constant.

According to the three damage states (IO, SD, and CP) of the 5-story SFVD building, the spectral acceleration of each state can be retrieved. Then, the logarithmic standard deviations of different random variables are calculated. The calculation formula is as follows:

$$\beta = \sqrt{\frac{\sum\limits_{i=1}^{N} (\ln Sa_i - \ln m)^2}{N-2}} \tag{5}$$

where m and β are the median value and the logarithmic standard deviation of the seismic intensity in the fragility function, respectively. And N represents the number of input ground motion records.

According to Equation (5), the logarithmic standard deviations for each damage state at various corrosion times were calculated, as shown in Table 2. Here, the seismic fragility function based on seismic intensity measures is used to reflect the seismic fragility. This function follows a standard lognormal distribution, which can be expressed as follows:

$$F(x) = \varphi\left[\frac{(\ln x - \ln m)}{\beta}\right] \tag{6}$$

where $\varphi(*)$ represents the integral of the standard normal probability [30,31].

Table 2. Fragility parameters for various corrosion times.

Corrosion Time (d)		IO	SD	CP
		$\theta_{max} = 1\%$	$\theta_{max} = 2\%$	$\theta_{max} = 4\%$
0	m	1.1434035	2.286815	4.573625
	β	0.4282113	0.428214	0.428215
64	m	1.0984700	2.1969300	4.3938750
	β	0.4423005	0.4422966	0.4427271
128	m	1.0741750	2.1483550	4.2967100
	β	0.4431142	0.4431152	0.4431155
192	m	1.0574000	2.1148017	4.2296018
	β	0.4746999	0.4797504	0.4797512

According to Equation (6), the seismic fragility curves of the SFVD in the IO, SD, and CP performance limit states at various corrosion times can be obtained, as shown in Figure 9. The results show that the trend of the fragility curves remains essentially consistent across all corrosion times. Within the test seismic motion range, the probability of exceedance increases gradually with increasing $S_a(T_1)$, indicating an increase in the fragility of the SFVD.

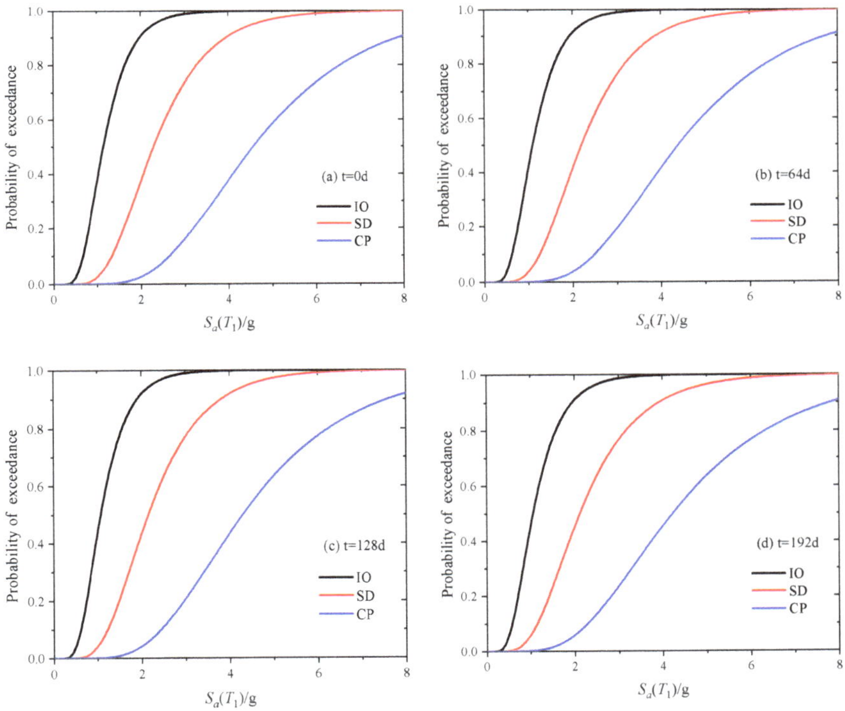

Figure 9. Fragility curves for various corrosion times: (**a**) 0 days; (**b**) 64 days; (**c**) 128 days; and (**d**) 192 days.

To gain a clearer understanding of the impact of corrosion time on fragility, all the results from Figure 8 are plotted on the same coordinate system, as shown in Figure 10. It is evident that at the same $S_a(T_1)$, the probability of exceedance gradually increases with increasing corrosion time, which is due to the fact that corrosion reduces the cross-sectional area of structural components, thereby decreasing the seismic resistance of the SFVD. In addition, comparing the fragility curves in the three limit states, it is observed that the fragility curves in the IO and SD stated do not significantly change, whereas the curve for the CP state exhibits a more pronounced variation. For example, when the probability of exceedance is 50%, the corresponding differences in $S_a(T_1)$ between the SFVDs with 0 days and 192 days of corrosion are 0.08 g and 0.16 g for the IO and SD states, respectively, while this difference for the CP state reaches 0.35 g. From another perspective, when $S_a(T_1)$ is 3 g, compared to the uncorroded SFVD, the SFVD with 192 days of corrosion shows an increase of 7.46% in probability of exceedance in the CP state, and an increase of 3.00% in the SD state, while their probabilities in the IO state are nearly the same. This indicates that corrosion has an especially significant impact on the probability of exceedance of the SFVD in the CP state.

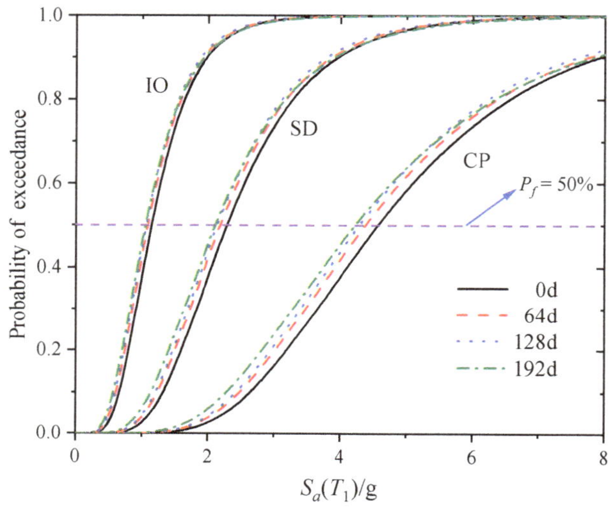

Figure 10. Comparison of fragility curves for various corrosion times.

4. Loss Assessment of the SFVD

4.1. Economic Losses

In the previous sections, the occurrence probabilities at various damage states (IO, SD, and CP) during different seismic intensities were obtained from performance-based fragility functions. Considering the impact of corrosion on seismic economic losses, this paper proposes a fragility analysis-based methodology for quantifying seismic-induced economic losses. By establishing a relationship between the probability of exceedance at different seismic intensities and corresponding economic losses, this method calculates the economic losses of buildings under various seismic intensity levels, offering a scientific basis for earthquake risk management and disaster prevention planning.

Seismic economic losses are generally comprised of the following three components: (1) direct economic losses encompassing structural damage to buildings and property losses; (2) indirect economic losses resulting from income disruption due to structural functional downtime; and (3) human casualty costs associated with injuries or fatalities. Silva et al. [32] and Nettis et al. [33] conducted a detailed assessment of the direct economic losses of buildings under earthquake action, establishing a relationship between economic losses and initial costs through damage factors. However, they neglected the contributions of building downtime and casualties. Here, the economic loss of steel frames is regarded as the sum of the maintenance costs, the recovery costs of damaged equipment and accessories inside the building, compensation expenses for casualties, and costs related to revenue loss from other shutdowns. Moreover, considering that the earthquake occurrence probability varies with different damage states, $C_{seismic}$ can be expressed by the following equation:

$$C_{seismic} = \sum_{j=1}^{K} C_j P_j \tag{7}$$

where C_j and P_j represent the loss cost and occurrence probability of the jth damage state, C_j and P_j can be written as

$$C_j = C_j^{damage} + C_j^{content} + C_j^{relocation} + C_j^{economic} + C_j^{injury} + C_j^{fatality} \tag{8}$$

$$P_j = \begin{cases} 1 - P_1 \ (j = 1) \\ P_{j-1} - P_j \ (j = 2, 3) \\ P_4 \ \ (j = 4) \end{cases} \tag{9}$$

where C_j^{damage}, $C_j^{content}$, $C_j^{relocation}$, $C_j^{economic}$, C_j^{injury} and $C_j^{fatality}$ represent the damage repair cost, building appurtenances loss cost, relocation cost, rental income loss, medical expenses for minor and major casualties, and fatality cost, respectively, and P_j can be obtained from the performance-based fragility curves (Figure 9).

4.2. Assessment of Economic Losses

Through calculation, the total floor area of the SFVD was determined to be 280 m^2, and according to Reference [18], the initial cost of the SFVD was estimated to be about 0.812 million. According to the population density definition in ATC-13 [34], the occupancy rate is 4 persons per 1000 square feet (approximately 92.9 m^2). The total floor area yields an estimated occupancy of approximately 13 individuals. Regional variations introduce significant uncertainty in assessing seismic economic losses. Given the building's location in the Los Angeles area, the basic cost data from the Los Angeles region were referenced for this evaluation, as shown in Table 3.

Table 3. Value of basic loss and calculation formula in the damaged state.

Categories	Cost Value	Calculation Formula [35]
Demolition	USD 404.9/m^2	USD 404.9/m^2 × Total area × Damage factor
Contents costs	USD 875.4/m^2	USD 875.4/m^2 × Total area × Damage factor
Relocation	USD 49.0/month/m^2	USD 49.0/month/m^2 × Total area × Loss of function
Rental loss	USD 17.94/month/m^2	USD 17.94/month/m^2 × Total area × Loss of function
Income loss	USD 252.6/month/m^2	USD 252.6/month/m^2 × Total area × Loss of function
Minor injury	USD 2600/person	USD 2600/person × Occupancy × Minor injury
Serious injury	USD 26,000/person	USD 26,000/person × Occupancy × Serious injury
Human death	USD 4,758,000/life	USD 4,758,000/life × Occupancy × Fatality

It should be noted that this cost data corresponds to 2019 values, and their current value needs to be adjusted using an annual growth rate. According to the International Construction Market Survey 2018 [36], which reported an average annual growth rate of 5% over the past three years in Los Angeles, the following conversion formula was proposed:

$$C_{2025} = C_{2019}(1 + 0.05)^6 \tag{10}$$

To calculate the economic losses of the SFVD at various damage states, the loss assessment methodology from Reference [18] was adopted. This methodology, which utilizes damage probability matrices, was determined by ATC-13 [34] and FEMA-227 [37] for seismic damage and loss assessment of multiple types of buildings. The details are shown in Table 4, which defines the corresponding inter-story drift ratio for various damage states and gives the associated damage factors, loss of function days, minor injury rates, serious injury rates, and fatality rates.

Table 4. Definitions of damage states and loss indices.

Performance Level	Story Drift Ratio (%)	Damage Factor	Loss of Function (d)	Minor Injury (%)	Serious Injury (%)	Fatality (%)
I	$\theta < 0.2$	0.0	0	0	0	0
II	$0.2 < \theta < 0.4$	0.005	3.4	0.003	0.0004	0.0001
III	$0.4 < \theta < 0.7$	0.05	12.08	0.03	0.004	0.001
IV	$0.7 < \theta < 1.5$	0.2	44.72	0.3	0.04	0.01
V	$1.5 < \theta < 2.5$	0.45	125.66	3	0.4	0.1
VI	$2.5 < \theta < 5.0$	0.8	234.76	30	4	1
VII	$5.0 < \theta$	1.0	346.93	40	40	2

According to the cost data and loss indices in Tables 3 and 4, as well as Equations (7)–(10), the economic losses for various seismic intensities can be calculated, as shown in Figure 11. The results show that economic losses increase gradually with increasing corrosion time. However, the influence of corrosion time on the losses is not significant. The loss difference between the SFVDs with 0 days and 192 days of corrosion reaches a maximum value of USD 103,582 when the $S_a(T_1)$ is 3.26 g, which is about 9.52% of the initial cost. Notably, under relatively strong or weak seismic intensities, the economic loss curves of SFVDs at different corrosion times exhibit significant overlap. In addition, early-stage corrosion has a more pronounced impact on economic losses. Hence, it should be feasible for builders to transform economic losses from corrosion into prevention investments through systematic risk conversion.

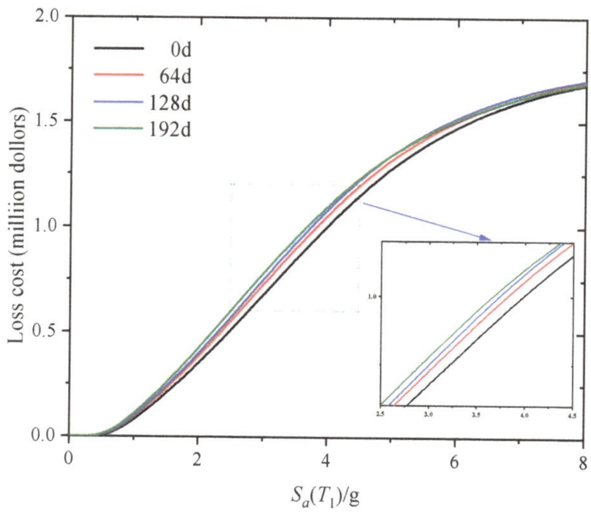

Figure 11. Economic losses for various corrosion times.

5. Conclusions

In this paper, a salt spray accelerated corrosion test was performed on Q355 steel, followed by tensile tests to investigate the effects of corrosion on its mechanical properties. DIA was conducted to perform seismic fragility analysis on a 5-story SFVD at various corrosion times, and the economic losses were assessed. The main conclusions obtained are as follows:

1. With the increase of corrosion time, the mass loss rate and cross-section loss rate of steel gradually increase. Corrosion causes rust pits on the surface of steel, leading to a reduction in its effective cross-sectional area and thereby significantly lowering the mechanical property indices of steel, such as yield strength, ultimate tensile strength, and elongation. Although corrosion degrades the mechanical properties of steel, it does not affect its safety performance.
2. The probability of exceedance of the SFVD in the limit state increases gradually with increasing corrosion time. This effect is relatively minor in the IO and SD states, while it becomes significant in the CP state.
3. The economic losses of the SFVD increase with increasing corrosion time, with the maximum loss difference within the test seismic motion range being USD 103,582, which is about 9.52% of the initial cost. In addition, early-stage corrosion has a significantly pronounced impact on economic losses.

This study reveals the effects of corrosion on the seismic fragility and economic losses of steel frames and provides some preliminary results. Future research on the long-term behavior of steel frames should account for corrosion. In addition, the quantified results of the economic loss impact of corrosion indicate that transforming corrosion-induced economic losses into preventive investments, such as anti-corrosion coatings, should be feasible.

Author Contributions: Conceptualization, W.Q. and H.W.; methodology, H.W.; software, Z.Z. and W.L.; validation, C.G.; formal analysis, W.Q.; investigation, W.Q.; resources, W.Q.; data curation, W.Q.; writing—original draft preparation, W.Q.; writing—review and editing, W.Q. and H.W.; visualization, S.L.; supervision, W.Q.; project administration, W.Q.; funding acquisition, W.Q. All authors have read and agreed to the published version of the manuscript.

Funding: This research received no external funding.

Data Availability Statement: The raw data supporting the conclusions of this article will be made available by the authors on request.

Conflicts of Interest: The authors declare no conflicts of interest.

References

1. Gatheeshgar, P.; Poologanathan, K.; Gunalan, S.; Shyha, I.; Sherlock, P.; Rajanayagam, H.; Nagaratnam, B. Development of affordable steel-framed modular buildings for emergency situations (COVID-19). *Structures* **2021**, *31*, 862–875. [CrossRef]
2. Wasse, A.D.; Dai, K.; Wang, J.; Sharbati, R. State-of-the-art review: Seismic design and performance assessment of special concentrically braced frames developed for complex industrial building structures. *Int. Steel. Struct.* **2024**, *24*, 280–295. [CrossRef]
3. Pourzangbar, A.; Vaezi, M.; Mousavi, S.M.; Saber, A. Effects of brace-viscous damper system on the dynamic response of steel frames. *Int. J. Eng. Trans. B Appl.* **2020**, *33*, 720–731.
4. Zhou, C.; Wen, H.; Zhao, J.; Jiang, L.; Xu, X.; Zheng, H.; Tian, Y.; Liang, M.; Wang, X.; Zhang, H. Investigation of the Hysteresis Performance of Multi-Story Y-Shaped Eccentrically Bare Braced Steel Frame with Block Slit Damper (BSD). *Buildings* **2025**, *15*, 451. [CrossRef]
5. Zhao, X.; Liao, X.; Hu, Z.; Li, X.; Nie, Y.; Liu, J.; Xu, Y. Seismic fragility analysis of steel pipe pile wharves with random pitting corrosion. *Buildings* **2023**, *13*, 2619. [CrossRef]
6. Rajput, A.; Paik, J.K. Effects of naturally-progressed corrosion on the chemical and mechanical properties of structural steels. *Structures* **2021**, *29*, 2120–2138. [CrossRef]

7. Guo, Q.; Wu, Z.; Xing, Y.; Lu, Y.; Zhang, F.; Li, W. Corrosion evolution and mechanical property deterioration of Q355NH weathering steel in long-term neutral salt spray environment. *Constr. Build. Mater.* **2024**, *411*, 134193. [CrossRef]
8. Kong, Z.; Jin, Y.; Hossen, G.M.S.; Hong, S.; Wang, Y.; Vu, Q.; Truong, V.; Tao, Q.; Kim, S. Experimental and theoretical study on mechanical properties of mild steel after corrosion. *Ocean. Eng.* **2022**, *246*, 110652. [CrossRef]
9. Gkatzogiannis, S.; Weinert, J.; Engelhardt, I.; Knoedel, P.; Ummenhofer, T. Correlation of laboratory and real marine corrosion for the investigation of corrosion fatigue behaviour of steel components. *Int. J. Fatigue.* **2019**, *126*, 90–102. [CrossRef]
10. Wu, Q.; Chen, X.; Fan, Z.; Nie, D.; Wei, R. Corrosion fatigue behavior of FV520B steel in water and salt-spray environments. *Eng. Fail. Anal.* **2017**, *79*, 422–430. [CrossRef]
11. Wang, H.; Liu, X.; Xu, S.; Hu, S. Experimental investigation on seismic performance of CHS T-joints in salt spray environment. *J. Constr. Steel Res.* **2024**, *221*, 108910. [CrossRef]
12. Zhang, Y.; Fang, C.; Wang, W.; Wu, L.; Cao, P. Influence of corrosion on seismic performance of steel beam-to-column connections. *Eng. Struct.* **2024**, *312*, 118284. [CrossRef]
13. Zhang, X.; Zheng, S.; Zhao, X. Experimental and numerical study on seismic performance of corroded steel frames in chloride environment. *J. Constr. Steel Res.* **2020**, *171*, 106164. [CrossRef]
14. Zhang, X.; Zheng, S.; Zhao, X. Experimental and numerical investigations into seismic behavior of corroded steel frame beams and columns in offshore atmospheric environment. *J. Constr. Steel Res.* **2023**, *201*, 107757. [CrossRef]
15. Lad, D.; Freddi, F.; Ghosh, J.; Rossetto, T. Impact of corrosion deterioration on the seismic performance of steel frame structures. In Proceedings of the SECED 2023 Conference, Cambridge, UK, 14 September 2023.
16. Xu, S.; Zhang, Z.; Li, R.; Wei, L. A method for the seismic vulnerability assessment of corroded steel structures. *J. Eng. Mech.* **2018**, *35*, 107–115. (In Chinese)
17. Lekeufack, B.C.; Fu, B.; Samuel, M.A.; Alam, S. Seismic resilience assessment of coastal carbon and weathering steel moment frame considering uniform atmospheric corrosion effects. *Structures* **2024**, *69*, 107274. [CrossRef]
18. Jiang, L.; Jiang, L.; Hu, Y.; Ye, J.; Zheng, H. Seismic life-cycle cost assessment of steel frames equipped with steel panel walls. *Eng. Struct.* **2020**, *211*, 110399. [CrossRef]
19. *GB/T 228.1-2010*; Metallic Materials Tensile Testing Part 1: Method of Test Method at Room Temperature. China Standards Press: Beijing, China, 2010. (In Chinese)
20. *GB/T 3075-2008*; Metal Materials-Fatigue Testing Axial Force-Controlled Method. China Standards Press: Beijing, China, 2008. (In Chinese)
21. Aditya, P.A.; Navvar, K.M.; Bahubali, J.K.; Pandikkadavath, M.S.; Mangalathu, S. Near-fault seismic vulnerability assessment of corrosion inflicted steel moment resisting frames. *J. Constr Steel. Res.* **2025**, *229*, 109527. [CrossRef]
22. *GB/T 19879-2005*; Steel Plates for Building Structures. China Standards Press: Beijing, China, 2005. (In Chinese)
23. Chen, X.; Lin, Z. *Structural Nonlinear Analysis Program Opensees Theory and Tutorial;* China Architecture & Building Press: Beijing, China, 2014. (In Chinese)
24. Cornell, C.A.; Jalayer, F.; Hamburger, R.O.; Da, F. The probabilistic basis for the 2000 SAC/FEMA steel moment frame guidelines. Submitted to. *J. Struct. Eng.* **2002**, *128*, 526–533. [CrossRef]
25. Cheng, X.; Jiang, L.; Yang, J.; Zhang, X.; Hu, Y. Seismic performance and risk mitigation of CFS shear walls with CFRP-reinforced screw connections and special-shape L-section end studs. *J. Build. Eng.* **2024**, *88*, 108824. [CrossRef]
26. Ye, J.; Jiang, L. Seismic risk assessment of mid-rise cold-formed steel structures in China considering various uncertainties. *China Civ. Eng. J.* **2021**, *54*, 74–83. (In Chinese)
27. Jiang, L.; Ye, J. Quantifying the effects of various uncertainties on seismic risk assessment of CFS structures. *Bull. Earthq. Eng.* **2020**, *18*, 241–272. [CrossRef]
28. *GB 50011-2010*; Code for Seismic Design of Buildings. Architecture and Building Press: Beijing, China, 2016. (In Chinese)
29. FEMA356. *NEHRP Guidelines for the Seismic Rehabilitation of Buildings;* Federal Emergency Management Agency: Washington, DC, USA, 2002.
30. Jiang, L.; Ye, J. Risk-based overall seismic robustness assessment of cold-formed steel structures considering vulnerability and fragility. *Bull. Earthq. Eng.* **2022**, *20*, 7161–7184. [CrossRef]
31. Liu, Y.; Hu, Y.; Zhao, J.; Jiang, L.; Sun, H.; Wang, D. Seismic risk mitigation of concrete-filled double-skin steel tube frames with beam-only-connected shear walls. *Structures* **2025**, *71*, 107949. [CrossRef]
32. Silva, A.; Macedo, L.; Monteiro, R.; Castro, J.M. Earthquake-induced loss assessment of steel buildings designed to Eurocode 8. *Eng. Struct.* **2020**, *208*, 110244. [CrossRef]
33. Nettis, A.; Di Mucci, V.M.; Ruggieri, S.; Uva, G. Seismic fragility and risk assessment of isolated bridges subjected to pre-existing ground displacements. *Soil. Dyn. Earthq. Eng.* **2025**, *194*, 109335. [CrossRef]
34. ATC-13. *Earthquake Damage Evaluation Data for California;* Applied Technology Council: Redwood City, CA, USA, 1985.
35. Zhu, J.; Zhao, J.; Tan, P.; Zhou, F. Seismic life-cycle loss estimation of single-story factory buildings. *Earthq. Eng. Eng. Dyn.* **2018**, *38*, 51–64. (In Chinese)

36. Turner & Townsend and Federal Emergency Management Agency (FEMA). *International Construction Cost Survey 2018–2019*; Turner & Townsend: Mumbai, India, 2018.
37. FEMA 227. *A Benefit-Cost Model for the Seismic Rehabilitation of Building*; Federal Emergency Management Agency: Washington, DC, USA, 1992; Volume 1.

MDPI AG
Grosspeteranlage 5
4052 Basel
Switzerland
Tel.: +41 61 683 77 34

Buildings Editorial Office
E-mail: buildings@mdpi.com
www.mdpi.com/journal/buildings

www.ingramcontent.com/pod-product-compliance
Lightning Source LLC
Chambersburg PA
CBHW061455210926
51850CB00008B/1729
* 9 7 8 3 7 2 5 8 4 9 9 5 6 *